Radioecology: Nuclear Energy and the Environment

Volume I

Authors

F. Ward Whicker, Ph.D.
Professor
Department of Radiology and
Radiation Biology
Colorado State University
Fort Collins, Colorado

Vincent Schultz, Ph.D.
Professor
Department of Zoology
Washington State University
Pullman, Washington

CRC Press, Inc.
Boca Raton, Florida

Library of Congress Cataloging in Publication Data

Whicker, F. Ward.
Radioecology: nuclear energy and the environment.

Includes bibliographies and index.
1. Radioecology. I. Schultz, Vincent, joint author. II. Title.
QH543.5.W48 574.5'222 80-26184
ISBN 0-8493-5353-X (v.1)
ISBN 0-8493-5354-8 (v.2)

Direct all inquiries to CRC Press, Inc., 2000 Corporate Blvd., N.W., Boca Raton, Florida, 33431.

International Standard Book Number 0-8493-5353-X (Volume I)
International Standard Book Number 0-8493-5354-8 (Volume II)

Library of Congress Card Number 80-26184
Printed in the United States

PREFACE

These volumes attempt to address many of the complex questions of nuclear energy and the environment. It is intended as a broad survey of the field of radiation ecology, with in-depth coverage of selected topics. The rationale, scope, and social relevance of the field, as well as a brief historical review, are treated in the introductory chapter. Volume I, Chapters 2 and 3, covering ecological and radiological principles as they relate to radiation ecology, are included for readers who may need such a background for enhanced understanding of subsequent chapters. Environmental radioactivity, both natural and man generated, is covered in some depth from the standpoint of sources and environmental distribution in Volume I, Chapter 4. Volume I, Chapter 5 on radionuclide movement in ecosystems is offered with respect to individual radionuclides and observations in selected land and water environments. A reasonably detailed treatment of mathematical models and their use in predicting rates of movement and degree of concentration in ecosystem components is presented in Volume II, Chapter 1 with emphasis on linear, first-order kinetics of single and multicompartment systems. Comprehension of the material presented in this chapter is not a requirement for understanding other chapters in this book. Volume II, Chapter 2 is a survey of knowledge on the effects of ionizing radiation on individual organisms, natural populations, and communities of terrestrial and aquatic environments and is presented in order that in Volume II, Chapter 3 the implications of radioactivity in the environment from nuclear activities of man can be more fully explored.

Within the appendix (Volume II) we have included a detailed citation of the major books and bibliographies on radiation ecology, a table of radionuclides and their physical characteristics, and a list of review articles on specific radionuclides as related to the environment.

In order that the reader may more completely explore the literature we have followed each chapter with suggested additional readings. Many of these publications and others cited within the text are governmental releases and rather difficult to locate. To facilitate their procurement we have included references to abstract sources, for example *Nuclear Science Abstracts,* when we thought it would be helpful. Publications of the U.S. Department of Energy (DOE) and its predecessors, Atomic Energy Commission (AEC) and Energy Research and Development Administration (ERDA) are available in many depository libraries throughout the U.S. and in some foreign countries. Within these depositories one can also find selected foreign publications.

Although scientists are likely to be the major readers, we believe that administrators, managers, and laymen can find some relevant reading herein. These volumes represent the impressions gained by the authors over two decades of research and teaching of the subject on radiation ecology.

Radiation ecology deals with natural and man-produced radioactivity, the movement of such materials in the environment, and the effects of ionizing radiation on populations and biotic communities. It is not a science unto itself; rather it is a hybrid that contains elements of several basic physical and biological disciplines. However, it has been found to be extremely relevant in the practical sense to a society that wants both more energy and a quality environment and to some of the more esoteric aspects of basic science as well. From a rather practical beginning which was concerned with the environmental implications of the first nuclear detonations, radiation ecology has enlarged its scope to the myriad questions raised in regard to nuclear power reactors and the associated fuel cycle and other potential means of gaining energy from nuclear technology. In addition, techniques utilizing radionuclides have contributed to many disciplines within the environmental sciences.

Some 10 years ago, the authors began writing a textbook on radiation ecology and attempted to write a complete, authoritative and accurate book, incorporating most of the pertinent literature. That was our mistake...we never finished the book. It became a full-time task merely to catalogue the stream of new papers, let alone to read and assimilate them. Therefore, our resolve in these volumes is to tell it as we see and understand it, utilizing only a small proportion of the reference material available, and realizing that our knowledge is incomplete and subject to error. In this effort we have primarily utilized those papers with which we have had personal contact. Knowingly, we have not included many excellent papers, particularly from the foreign literature. However, we have attempted to rectify this shortcoming by including many of these papers in the additional readings following each chapter.

We are indebted to a very long list of colleagues, students, friends, and family, each of whom contributed in some important way to the preparation of this volume. We are particularly indebted to our wives and students who provided much encouragement and inspiration to finish these volumes. We specifically acknowledge Susan L. White and Patricia L. Schultz for accomplishing the tremendous task of typing several manuscript drafts and April D. Whicker for artwork and proofreading. Finally, we are grateful to Dr. Allyn H. Seymour, friend and colleague, who contributed useful suggestions on the final manuscript, and to Janelle Sparks of CRC Press for her patience and guidance during the publication process.

F. Ward Whicker
Fort Collins, Colo.

Vincent Schultz
Pullman, Wash.

THE AUTHORS

F. Ward Whicker, Ph.D., is a Professor in the Department of Radiology and Radiation Biology at Colorado State University, Fort Collins. He teaches and conducts research in radioecology and allied subjects.

Dr. Whicker received a B.S. degree in applied chemistry in 1962 and a Ph.D. degree in radiation biology in 1965 from Colorado State University. In 1965, he received training in the use of radioisotopes in ecological research at the University of Hannover in West Germany. His academic and research history includes visiting professorships at the Institute of Arctic Biology, University of Alaska, Fairbanks, and the Department of Botany, University of North Carolina, Chapel Hill.

Dr. Whicker is a Fellow of the American Association for the Advancement of Science and currently holds membership in the Ecological Society of America, the Wildlife Society, Health Physics Society, and the Society of Sigma Xi. He also belongs to several academic honor societies including Phi Kappa Phi, Kappa Mu Epsilon, and Beta Beta Beta.

Dr. Whicker has published approximately 60 journal articles dealing with various topics in radioecology and basic biology. He has authored several review articles and book chapters on radioecology and has prepared over 40 technical reports on various research contracts and consultantships.

The research activities of Dr. Whicker have included studies on the behavior of radionuclides in aquatic and terrestrial ecosystems, the effects of radiation on natural populations, environmental impacts of energy technologies, radionuclide transport modeling and dose assessment, and description and analysis of natural ecosystems. He has served as an advisor to governmental agencies and industrial firms on problems of nuclear waste disposal, uranium production, nuclear gas stimulation, and assessment of environmental radioactivity.

Vincent Schultz, Ph.D., is Professor of Zoology and Wildlife Biology, Washington State University, Pullman, and a member of the Graduate Faculty of the Program in Environmental Science. Dr. Schultz graduated in 1946 from Ohio State University, Columbus with a B.Sc. degree in agriculture and obtained a M.Sc. degree in zoology in 1948 and a Ph.D. degree in zoology from the same university. In 1954, he was granted an M.Sc. in statistics by Virginia Polytechnic Institute, Blacksburg. In addition, he attended the University of Connecticut, Storrs, Yale University, New Haven, Conn., and the University of Pennsylvania, Philadephia under the Army Specialized Training Program, completing the premedical program in 1945. Following his graduation from Virginia Polytechnic Institute he was for 2 years a U.S. Public Health Service Fellow in biostatistics at The Johns Hopkins University, Baltimore, Md. During the period 1959 to 1966, he was employed as an ecologist by the U.S. Atomic Energy Commission, Washington, D.C., in which he was involved in developing a program in radiation ecology under the supervision of the late Dr. John N. Wolfe. Since that time he has been associated with Washington State University teaching courses in mathematical, statistical, and radiation ecology. Dr. Schultz has served as a consultant to the U.S. Department of Energy and its predecessors, the U.S. Nuclear Regulatory Commission, Battelle Pacific Northwest Laboratories, and Sandia Laboratories. He has presented invited lectures on aspects of radiation ecology and has published over 100 manuscripts on a variety of subjects.

Dedicated
to
April D. Whicker
Patricia L. Schultz
and
our children and students

RADIOECOLOGY:
NUCLEAR ENERGY AND THE ENVIRONMENT

F. Ward Whicker and Vincent Schultz

Volume I

Chapter 1
Introduction and Historical Perspective

Chapter 2
Ecological Principles Applied to Radioecology

Chapter 3
Radiological Principles

Chapter 4
Environmental Radioactivity

Chapter 5
Radionuclide Behavior in Ecosystems

Volume II

Chapter 1
Quantitative Aspects of Radionuclide Transport

Chapter 2
Effects of Ionizing Radiation on Species, Populations, Communities, and Ecosystems

Chapter 3
Consequences of Radioactivity in the Environment

Appendix A
Sources of Information

Appendix B
Literature Reviews of Specific Radionuclides in the Environment

Appendix C
Table of Radionuclides: Physical Half-Lives and Major Radiations

TABLE OF CONTENTS

Volume I

Chapter 1
Introduction and Historical Perspective 1
References 7
Additional Readings 8

Chapter 2
Ecological Principles Applied to Radioecology 11
I. Introduction 11
II. The Ecosystem Concept 12
III. Structure and Organization in Ecosystems 15
A. Populations 16
B. Communities 19
C. Ecosystems 23
IV. Functional Relationships within Ecosystems 25
A. Energy Flow in Ecosystems 25
B. Mineral Cycles of Ecosystems 27
C. Interactions of Organisms 31
References 34
Additional Readings 35

Chapter 3
Radiological Principles 37
I. Introduction 37
II. The Radiations and Radioactivity 37
A. Types of Ionizing Radiations 37
1. Alpha Particles 37
2. Beta Particles 38
3. Neutrons 40
4. X- and Gamma Rays 41
B. Radioactivity 43
1. Radioactivity and Half-Life 43
2. Specific Activity 44
3. The Curie 46
III. Interaction of Radiation with Matter 46
A. Alpha Particle Absorption 46
B. Beta Particle Absorption 47
C. X- and Gamma Ray Interactions 48
1. Total Attenuation 48
2. Rayleigh Scattering 49
3. Compton Effect 50
4. Photoelectric Absorption 50
5. Pair Production 51
6. Relative Importance of Photon Attenuation Processes 51
IV. Radiation Measurement 52
A. Instruments and Apparatus 52
1. Ionization Chambers 52
2. Proportional and Geiger-Mueller Detectors 53
3. Scintillation Detectors 55

4. Semiconductor Detectors 55
5. Photographic Emulsions 56
6. Thermoluminescent Detectors 57
B. Radiation Counting 57
1. Sample Collection and Preparation 58
2. Counting Yield 59
3. Counting Statistics 61
C. Radiation Dosimetry 63
1. Concept of Radiation Dose and Exposure 63
2. Dosimeters 65
3. Calculation of Dose 66
4. Field Dosimetry 68
V. Fission and Fusion 69
References 70
Additional Readings 72

Chapter 4
Environmental Radioactivity 73
I. Introduction 73
II. Natural Sources of Radiation 73
A. Cosmic Radiation 73
B. Cosmogenic Radionuclides 76
C. Primordial Radionuclides 77
D. Summary of Dose Rates from Natural Radiation 82
III. Man-Generated Radioactivity 84
A. The Nuclear Fuel Cycle 84
1. Mining 86
2. Milling 88
3. Purification and Conversion 90
4. Isotopic Enrichment 90
5. Fuel Element Fabrication 91
6. Radionuclides in a Reactor 91
7. Fuel Reprocessing 95
8. Waste Disposal 95
9. Radiation Dose from the Nuclear Fuel Cycle 101
B. Nuclear Explosions 103
1. Historical Perspective 103
2. Formation of Fallout 105
3. Properties of Fallout 109
4. Transport of Fallout 112
5. Peaceful Applications of Nuclear Explosives 117
C. Other Sources of Environmental Radioactivity 121
References 123
Additional Readings 127

Chapter 5
Radionuclide Behavior in Ecosystems 129
I. Introduction 129
II. General Concepts 130
A. General Transport Processes 130
B. Factors Affecting Radionuclide Transport and Accumulation 133

1. Properties of Radionuclides ... 135
2. Characteristics and Behavior of Organisms ... 137
3. Ecosystem Attributes ... 141

III. Behavior of Specific Groups of Radionuclides ... 147
A. Nonmetals (H, C, P, I) ... 147
B. Light Metals ... 149
1. Group IA (K, Rb, Cs) ... 149
2. Group IIA (Ca, Sr, Ba, Ra) ... 151
C. Noble Gases (A, Kr, Xe, Rn) ... 152
D. Heavy Metals (Cr, Mn, Fe, Co, Zn, Zr, Tc, Ru, Pb, Po) ... 153
E. Rare Earths (Y, La, Ce, Pr, Pm) ... 157
F. Actinides (Th, U, Pu) ... 158
G. Summary: General Properties of Selected Radionuclides ... 162

IV. Behavior of Radionuclides in Specific Environments ... 168
A. Terrestrial Ecosystems ... 168
1. Temperate Mountains ... 168
2. Arctic Regions ... 173
3. Lower Coastal Plain of the Southeastern U.S. ... 175
B. Aquatic Ecosystems ... 176
1. Freshwater Ecosystems ... 177
2. Marine Ecosystems ... 181

References ... 185
Additional Readings ... 192

Index ... 197

Chapter 1

INTRODUCTION AND HISTORICAL PERSPECTIVE

Life has evolved in a natural ionizing radiation environment of incessant bombardment by cosmic rays and radiations from primordial radionuclides in the crust of the earth. However, the omnipresent radiation background has risen perceptibly during the last 30-plus years as a result of man's activities, and concern has been expressed for the welfare of life exposed to this additional radiation. The growth of knowledge and the population growth of mankind have both led to fundamental needs that are increasing in most nations. Three of these need categories are energy, defense, and other technologies. Fulfillment of these needs increasingly involves technological innovations in areas that employ nuclear energy and radioactive materials. Specific examples are diagrammed in Figure 1. It is difficult, and in some cases, impossible, to employ nuclear energy and radioactive materials without some release of radioactivity and chemicals to the environment. These facts underscore the social relevance of ecology, radioecology, radiobiology, health physics, and related fields.

Human population growth and conflict may ultimately result in sufficient strife among nations to lead to the use of nuclear weapons, which have the potential for enormous physical destruction of man and his environment. Possibility of a nuclear war has raised the specter of destruction of our environment by means of blast, fire, and ionizing and ultraviolet radiation. Of primary concern is whether or not the various ecosystems throughout the world are capable of sustaining themselves, or, if modified, what are the time spans and successional patterns involved for self-repair? It is rather obvious in the event of an all-out nuclear warfare that survival of human societies as they are now known is intimately related to these matters. Radioecology provides significant input into the effort of trying to grasp the full implications of such a catastrophic event.

The peaceful use of nuclear energy for electrical power and its potential uses for things such as earth excavation, natural gas stimulation, etc., although not having the dramatic effects of a nuclear war, have the potential for insidious long-term effects unless the ecological consequences of these uses are understood and given serious consideration. These considerations are not necessarily unique to nuclear energy, but are an integral part of any modification of the environment by man. Unique to the production and use of nuclear energy are the problems of disposal of tremendous quantities of radioactive waste.

The impact of the Nuclear Age and increased environmental pollution from all technologies have been a matter of public interest for many years; yet, it has been mainly within the last 2 decades that people have expressed specific concern as to effects on the environment rather than just upon the health and safety of man. The question of the direct effects of ionizing radiation on man has been exhaustively addressed through research and through the process of determining and enforcing radiation protection standards. While direct exposure of man is of principal concern to society, it is increasingly recognized that man's welfare is closely related to the quality of the environment. Some species of plants and animals are as sensitive to ionizing radiation as man, and, in many situations, nonhuman populations receive substantially greater exposures than people. These volumes are principally directed toward relationships between ionizing radiation and native plant and animal populations and their environments.

Research into the matter of environmental radioactivity provides information useful not only to persons concerned with radiation hazards, but also to those concerned with other forms of pollution and to those studying normal functioning of ecological sys-

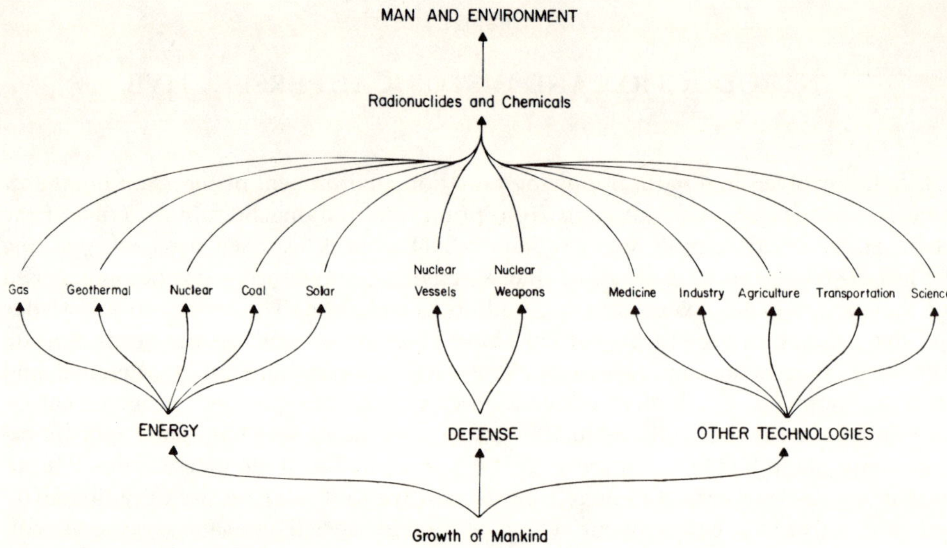

FIGURE 1. The growth of mankind has led to three major societal need categories, each of which embodies activities that result in releases of radioactive substances to the environment.

tems. The relative ease of measurement of minute amounts of radioactive material has helped make radionuclides a ''model pollutant'' for study. Ecological processes that govern radionuclide behavior also effect behavior of smog, pesticides, and other chemical substances that may threaten the environment. Since the ecosystem itself largely regulates the movement and concentration of radionuclides, observations of radionuclide behavior can, in conjunction with suitable models, yield basic information about functional processes of ecological systems.

Because of the tremendous complexity involved in the behavior and effects of radionuclides in the environment and the comparative youthfulness of radiation ecology, there is still much to be learned. Despite the fact that knowledge in the field has made impressive gains in the last 3 decades, pressing questions still exist. It takes time, often years or decades, to understand ecosystems in their normal and stressed conditions, and quick, simple answers cannot be expected. Nonetheless, considerable literature addressing the fate and effects of radionuclides in the environment does exist. Unfortunately, it is rather widely scattered. It is our purpose in preparing this book to bring together in one place a brief history of radioecology, particularly in the U. S., some pertinent scientific principles, and ways of evaluating the problems of radiation and the environment. This book is not intended as an ultimate reference source for radioecology, but rather a synthesis of information and a general guide to the original literature.

Radioecology or ''radiation ecology'' can be considered a field of endeavor which encompasses the relationships between ionizing radiation or radioactive substances and the environment, or subunits within the environment. These subunits may be populations, communities, ecosystems, biomes, or even the biosphere. Radiobiology, in contrast, deals with radiation and levels of biological organization from the molecule to the organism. The primary subdivisions of radiation ecology include:

1. Radionuclide movement within ecological systems and accumulation within specific ecosystem components such as soil, air, water, and biota

2. Ionizing radiation effects on individual species, populations, communities, and ecosystems
3. Use of radionuclides and ionizing radiation in studies of structure and function of ecosystems and their component subsystems

Questions commonly asked regarding movement, accumulation, and effects of radionuclides in ecological systems include:

1. How rapidly do radionuclides move from one part of the system to another and what are the mechanisms and pathways of such transfers?
2. What are the relative concentrations and radiation doses from radionuclides in important biological components relative to the total amount of environmental contamination?
3. What is the long-term behavior and ecological effect of radionuclides in the environment?
4. At what environmental concentrations are effects, if any, seen in populations?
5. Environmental studies using radioisotopic tracers may answer such questions as: Where do animals move? What do they eat? How much and which elements and nutrients are used? What are the dynamics of upper-lower atmospheric mixing? What are the circulation patterns of lakes and oceans?

Behavior of radionuclides in the environment is governed by a highly complex set of natural processes and circumstances. Nearly every component of an ecological system will have some influence upon the fate of radioactive materials introduced into the system. Of major significance, however, is the fact that natural processes may concentrate radionuclides in biological tissues to levels that are orders of magnitude higher than in the ambient media of air and water. Further, there is no doubt that radionuclides, in sufficient concentrations, are harmful to all living systems.

Despite substantial research, our ability to specify the levels of environmental contamination with radionuclides that would just begin to cause demonstrable ecological change is poor. Part of the problem is that the radiation dose to biological tissues from environmentally dispersed contaminants is highly variable, owing to the physical and chemical behavior of the material within a complex system. Another problem is that of quantifying subtle biological effects in complex systems at comparatively low radiation exposures. Large, fixed radiation sources have been used to irradiate ecosystems *in situ* at levels sufficient to produce observable effects; however, this research is only partially relevant to lower radiation levels from dispersed contaminants. The authors are still confronted with the task of determining the quantities of specific radionuclides that can be allowed in specific ecological systems with the assurance that no undesirable effects will result. Further, the task is compounded with disagreement as to what are "undesirable effects".

The authors believe that radioecology is an important, complex, and diverse science, dictating an interdisciplinary approach, which is accomplished in practice by the varied and unique academic training of radioecologists. The field draws heavily on information from the basic physical and biological sciences, and current curricula of the radioecologist may encompass physics, chemistry, mathematics, geology, plant and animal ecology and physiology, taxonomy, radiation biology, dosimetry, and more. Because radioecology integrates these fields, to some extent, it frequently leads to interdisciplinary contributions which can aid the extension of each discipline.

Since the early 1940s, growth of radiation ecology has paralleled the development of nuclear technology in the world, with its greatest advancements being in the U.S.

and the U.S.S.R. Very significant contributions have also been made by investigators in Finland, France, Italy, Japan, and the U.K., as well as by individuals associated with the International Atomic Energy Agency (IAEA), Austria. Its development is woven within a web of expectations, frustrations, and accomplishments. To the historian, it is associated with places such as Sevastopol (U.S.S.R.), Helsinki (Finland), Cadarache (France), Eniwetak (Atoll), Anaktuvuk Pass and Amchitka Island (Alaska), Monaco, and Oak Ridge (Tennessee); to the researcher of literature in the field, with persons such as Aleksakhin, Auerbach, Baranov, Davis, Egami, Grauby, Hanson, Miettinen, Nakatani, Odum, Peredel'skiy, Polikarpov, Platt, Ravera, Seymour, Wolfe, and Woodwell, to mention a few.

Although some individuals have objected to calling this endeavor "radiation ecology" or "radioecology", these names are established in the literature and will be used throughout this book. To the best of our knowledge, the expression arose independently and simultaneously in the U.S. and the U.S.S.R. in 1956.[1,2] However, the discipline had a beginning earlier than 1956. In Russia, Tarkhanov[3] reported in 1896 on a study of the influence of X-rays on the eggs of a lamprey; Gajewskaya[4] collected *Artemia salina* near Sevastopol and reported in 1923 the effects of X-rays in an ecological context; and for many years Vernadsky[5] and Vinogradov[6] investigated the natural radioactivity of organisms. Investigators such as these laid the foundation for an aquatic radiation ecology program in the U.S.S.R. second to none in the world. Similarly, the authors can list early investigators and investigations in the U.S. and other countries that were concerned with the biological effects of X-rays and the accumulation of natural radioactivity by organisms. However, the effects of ionizing radiation on the natural environment and its components were of little general concern prior to man's release of fission and activation products into the environment.

Perhaps the Atomic Age was born at 3:25 p.m., December 2, 1942, beneath the west stands of the University of Chicago's Stagg Field, Illinois, with the first sustained nuclear chain reaction, the release of enormous amounts of energy, and the production of other elements from uranium. The history of nuclear energy to and following this date is summarized by Asimov.[7-9] At Trinity, N.M. on July 16, 1945, the first successful test of an atomic weapon was conducted. Almost immediately following the test and for a period of years thereafter, field surveys were conducted at Trinity to discover the extent and degree of contamination by radionuclides, including radiological assay of animal tissues, and external evidence of radiation damage.[10,11] Permanent study transects were established at Trinity and utilized some 20 years later in a study of revegetation of the site.[12]

Radiation ecology as a discipline probably began in 1943 in the U.S. with establishment of The Applied Fisheries Laboratory, University of Washington, Seattle, under the supervision of Donaldson, with the support of the Manhattan Engineer District, forerunner of the U.S. Atomic Energy Commission (USAEC). The function of the Laboratory was to study the impact of the Hanford reactors on biota of the Columbia River and its plume. It was at Hanford, Wash. in late 1944 that the first significant releases of radionuclides to the environment occurred. In 1946, the Laboratory's activities were extended to the U.S. Nuclear Testing Program in the Pacific.[13] Although the term "radiation ecology" was not used at the time, the work of the Laboratory was oriented in that direction as evidenced by the 1946 doctoral dissertation of Welander[14] on the effects of ionizing radiation on chinook salmon (*Onchorhynchus tschawytscha*) and that of Foster[15] on effects of ionizing radiation on rainbow trout (*Salmo gairdneri*). If this was not the beginning, surely the pre- and post-nuclear detonation surveys of the environment conducted by the Laboratory in 1946 and 1947 in the Pacific were. The Laboratory is now called the Laboratory of Radiation Ecology. Concurrently with these activities, other endeavors were developing at Hanford under the jurisdiction of the General Electric Co., contractor with the USAEC. Initially these activities were plant-operation oriented, but later developed into a full-fledged ecological program with Davis as its first director. Currently, environmental research is conducted at Hanford by the Ecosystems Department, Battelle Pacific Northwest Laboratories, with principal support from the U.S. Department of Energy (DOE), Washington, D.C., formerly the U.S. Energy Research and Development Administration (ERDA), successor in 1974 to the USAEC.

The development of a radiation ecology program at Oak Ridge National Laboratory (ORNL), operated for the USAEC (now DOE) by Union Carbide Corp. at Oak Ridge, Tenn. is a story of frustrations and progress culminating in the development of an organization known for its staff and excellence in ecological research. Further, it is a story of Auerbach and his supervisors' attempts to enlighten a bureaucracy as to the needs for ecological research in the Atomic Age. This program began with the establishment of a cooperative program in 1950 between the USAEC and the Tennessee Valley Authority (TVA) for the purpose of investigating the physical and biological effects of radionuclides resulting from plant operations on the environment of the general area. The research activity under this agreement was centered around the ecology of White Oak Lake, Roane County, Tenn., an artificial impoundment containing radioactive wastes. Investigators included Krumholz, fisheries biologist; Nease, botanist, and Ross and Helm, limnologists. In the background were administrators Shoup, McCrady, and Wiebe, who provided moral support and the necessary link between field personnel and management. By February 1954, the research and reports[16-19] were completed, and all field personnel left for what must have seemed to be more fertile fields. The real potential for a broad ecological program of acknowledged excellence at ORNL was apparently not yet recognized.

In 1941, Struxness, who was later to become associated with the Health Physics Division of ORNL, took an ecology course at Northwestern University, Evanston, Ill., leading to a series of events that resulted in the establishment of an ecology program at ORNL. Struxness persuaded the instructor of the course, Park, to become a consultant on ecology at ORNL. Subsequently, Park visited Oak Ridge, Hanford, and the Nevada Test Site, Mercury, and prepared reports on his observations and recommendations. Frustrations now began to develop as many individuals did not share Park's enthusiasm for ecology, nor did many even understand what he was trying to say. To a lesser degree even today ecologists are having similar difficulties in explaining ecological concepts and the need for an ecological approach to solving environmental problems.

Struxness, who became leader of Waste Disposal Research and Engineering of the Health Physics Division at ORNL, and Morgan, head of the Division, pushed for the development of an ecology program at ORNL which resulted in the hiring in 1955 of Auerbach, a student of Park, with an initial fiscal year (FY) 1955 budget of $56,000 involving 1.7 man-years. This program has developed over the years into a multidisciplinary effort with annual budgets in excess of $1 million.

Development of this program was not easy. Some of the negative comments made would appear irresponsible in light of today's environmental concern. In a memorandum dated September 16, 1954, one federal administrator commented in regard to establishing an ecology program at ORNL, " . . . the argument that we do not know the consequences of radiation damage to the environment is not a valid argument for support of such a program." Such a comment was not atypical as the co-author of this book heard many similar expressions while associated with the USAEC during the period 1959 to 1966. The concept that man was an integral part of his environment and that concern for the environment was not independent of man was only beginning to be generally accepted.

Despite opposition, a program was established at ORNL, later to become the Environmental Sciences Division. Ecology was rapidly becoming an important component of the USAEC programs. The history of the development of radiation ecology programs at other sites, such as Argonne National Laboratory (Illinois), Idaho National Engineering Laboratory (Idaho Falls), Los Alamos Scientific Laboratory (New Mexico), Lawrence Livermore Laboratory (California), Nevada Test Site, and the Savannah River Ecology Laboratory (Aiken, S.C.), is as interesting as that at ORNL. In addition, professional scientific groups began in the 1950s to recognize the need for radioecological research. For instance, The Ecological Society of America established a Committee on the Effects of Radioactivity on Natural Populations,[20] and the National Academy of Science, Washington, D.C., formed the Committee on Effects of Atomic Radiation on Oceanography and Fisheries.[21]

Much of the development of radioecology in the U.S. was due to the farsighted vision and encouragement of Dunham, who became chief of the Division of Biology and Medicine (DBM) of the USAEC in 1955. In 1958 the Environmental Sciences Branch of DBM was established under the leadership of Wolfe, a plant ecologist. While on leave of absence from Ohio State University, Columbus, in 1955 and 1956, Wolfe was instrumental in developing an ecology program in DBM. After returning to the university in 1956, he was recalled to DBM in 1958 to head the new Environmental Sciences Branch. The period from 1955 to 1958 was a time of ecological research activity and growth within DBM. Meetings were called to formulate plans for a substantial ecology program within the USAEC, and the list of early planners reads like a "Who's Who in Ecology". One such meeting called by DBM in 1955, May 19th to the 21st to discuss the ecology program was attended by Billings, Buell, Curtis, Ketchum, Krumholz, Odum, Park, Patrick, Pitelka, Sears, and Wolfe, who listened to presentations by Larson, Lindberg, Lowman, Held, Dunning, Dudley, Odum, Patrick, Morgan, Auerbach, Pearce, Foster, Krumholz, Hungate, and Shoup.

The recognition of the role of ecology in the Nuclear Age resulted in increased funding and research accomplishments. The total DBM budget in FY 1957 was $30.5 million; this grew to $90.8 million in FY 1970. Environmental radiation studies within DBM consisted of three categories: terrestrial and freshwater ecology, marine sciences, and atmospheric radioactivity and fallout, with responsibility for the ecology and marine science categories assigned to the Environmental Sciences Branch. The FY 1957 DBM budget for environmental radiation studies was 2.3 million dollars. In FY 1970 the budget had increased approximately to the following:

	Millions ($)
Terrestrial and freshwater ecology	6
Marine sciences	4
Atmospheric radioactivity and fallout	8.9
Total	18.9

These funds were utilized to support research on USAEC-controlled areas as well as under contracts with academic and other organizations.

Today the DOE, replacement for ERDA, supports one of the largest ecological research programs in the U.S. Emphasis has changed to some degree as a result of modified responsibilities of the agency, passage of the National Environmental Protection Act, legal decisions concerning the agency's environmental responsibilities, and the growth of the science of ecology. Until the early 1960s most funded radioecological research involved fission and activation products produced in connection with the development and testing of nuclear weapons. Since the early 1960s emphasis in radiological areas has shifted more toward the problems of nuclear energy development, with increasing concern about heavy elements such as uranium and plutonium, and to materials difficult to confine such as tritium and krypton. Within the past several years, the DOE responsibilities have broadened to include environmental problems associated with all forms of energy development, including radiological and nonradiological impacts.

A similar picture of growth and accomplishment apparently occurred in the U.S.S.R. as evidenced by the tabulation of Yegorov and Kurilova[22] in which they graphically portray the increase in the number of publications on the subject. Polikarpov[23,24] presents an excellent discussion of the scope of radioecology and the endeavors in marine radioecology in the U.S.S.R. If one peruses the listings in *Nuclear Science Abstracts* and *Energy Research Abstracts,* it is evident that investigators in the U.K., Finland, Sweden, Japan, France, and Italy have also been increasingly active in the field until the last few years when a decrease in activity became apparent.

The authors hope the following chapters will help make the reader aware of the magnitude of research efforts, knowledge accumulated in the discipline, and the enthusiasm of investigators during this trying and exciting developmental period of radioecology. A measure of the maturity of radiation ecology as a discipline is reflected in the quality and quantity of publications and the organization of professional meetings. The literature is rather voluminous and continues to grow. The major sources of information are listed in Volume II, Appendix A.

REFERENCES

1. **Kuzin, A. M. and Peredel'skiy, A. A.,** Nature conservancy and some aspects of radioactive-ecological relations (in Russian), *Okhr. Prir. Zap. SSSR,* 1, 65, 1956.
2. **Odum, E. P.,** Consideration of the total environment in power reactor waste disposal, *Proc. Int. Conf. Peaceful Uses of Atomic Energy,* 13, 350, 1956.
3. **Tarkhanov, I. R.,** Experiments on the action of Roentgen's X-rays on the animal organism (in Russian), *Izv. St. Peterburg. Biol. Lab.,* 1(3), 47, 1896.
4. **Gajewskaya, N. W.,** Der Einfluss der Roentgenstrahlen auf *Artemia salina, Assoc. Theor. Appl. Limnol.,* 1, 359, 1923.
5. **Vernadsky, V. I.,** Concentration of radium by living organisms (in Russian), *Dokl. Akad. Nauk SSSR, Ser. A,* 2, 33, 1929.
6. **Vinogradov, A. P.,** *The Elemental Composition of Marine Organisms* (English transl.), Memoir 2, Efron, J. and Setlow, J. K., Eds., Sears Foundation for Marine Research, Yale University, New Haven, Conn., 1953.
7. **Asimov, I.,** Worlds Within Worlds: The Story of Nuclear Energy, Vol. 1, Office of Information Services, U.S. Atomic Energy Commission, Washington, D.C., 1972.
8. **Asimov, I.,** Worlds Within Worlds: The Story of Nuclear Energy, Vol. 2, Office of Information Services, U.S. Atomic Energy Commission, Washington, D.C., 1972.
9. **Asimov, I.,** Worlds Within Worlds: The Story of Nuclear Energy, Vol. 3, Office of Information Services, U.S. Atomic Energy Commission, Washington, D.C., 1972.
10. **Bellamy, A. W., Leitch, J. L., Larson, K. H., and Dunn, D. B.,** The 1948 Radiological and Biological Survey of Areas in New Mexico Affected by the First Atomic Bomb Detonation, U.S. AEC Rep. UCLA-32, University of California at Los Angeles, 1949.
11. **Leitch, J. L.,** Summary of the Radiological Findings in Animals from the Biological Surveys of 1947, 1948, 1949, and 1950, U.S. AEC Rep. UCLA-111, University of California at Los Angeles, 1951.
12. **Dunn, D. B. and Spellman, D. L.,** The Revegetation at the New Mexico Atomic Bomb Crater, U.S. AEC Rep. COO-1296-1, University of Missouri, Columbia, 1965.
13. **Hines, N. O.,** *Proving Ground: An Account of the Radiobiological Studies in the Pacific, 1946-1961,* University of Washington Press, Seattle, 1962.

14. **Welander, A. D.**, Studies of the Effects of Roentgen Rays on the Growth and Development of the Embryos and Larvae of the Chinook Salmon (*Onchorhynchus tschawytscha*), Ph.D. thesis, University of Washington, Seattle, 1946.
15. **Foster, R. F.**, Some Effects on Embryo and Young Rainbow Trout (*Salmo gairdneri* Richardson) from Exposing the Parent to X-rays, Ph.D. thesis, University of Washington, Seattle, 1948.
16. **Krumholz, L. A.**, A Summary of Findings of the Ecological Survey of White Oak Creek, Roane County, Tennessee, 1950-1953, U.S. AEC Rep. ORO-132, Oak Ridge Operations Office, Oak Ridge National Laboratory, Oak Ridge, Tenn., 1954.
17. **Krumholz, L. A.**, An Ecological Survey of White Oak Creek, 1950-1953, Vol. 1, U.S. AEC Rep. ORO-587, Oak Ridge Operations Office, Oak Ridge National Laboratory, Oak Ridge, Tenn., 1954.
18. **Krumholz, L. A.**, An Ecological Survey of the Limnology of White Oak Creek and Lake, Vol. 2, U.S. AEC Rep. ORO-587, Oak Ridge Operations Office, Oak Ridge National Laboratory, Oak Ridge, Tenn., 1954.
19. **Krumholz, L. A.**, An Ecological Survey of the Vertebrate Fauna of White Oak Lake and Environs, Vol. 3, U.S. AEC Rep. ORO-587, Oak Ridge Operations Office, Oak Ridge National Laboratory, Oak Ridge, Tenn., 1954.
20. **Billings, W. E., Ketchum, B. H., Odum, E. P., and Park, O.**, Report of the committee on the effects of radioactivity on natural populations, *Bull. Ecol. Soc. Am.*, 37(1), 21, 1956.
21. National Academy of Sciences-National Research Council, Oceanography, fisheries and atomic radiation. Summary report of Committee on the Effects of Atomic Radiation on Oceanography and Fisheries, *Science*, 124(3210), 13, 1956.
22. **Yegorov, V. N. and Kurilova, N. V.**, Informational aspects of the development of marine radioecology (in Russian), in *Bibliographical Index of Publications of the Institute of Biology of South Seas A. S. Ukr SSR on the Problem "Radiation and Chemical Biology" (1957-1972)*, Polikarpov, G. G., Ed., Naukova Dumka, Kiev, 1973, 53.
23. **Polikarpov, G. G.**, Author's preface to revised edition, in *Radioecology of Aquatic Organisms*, Schultz, V. and Klement, A. W., Jr., Eds., North-Holland, Amsterdam, 1966, xx; (English transl.).
24. **Polikarpov, G. G.**, The subject matter and scope of marine radioecology, in *Radioecology of Aquatic Organisms*, Schultz, V. and Klement, A. W., Jr., Eds., North-Holland, Amsterdam, 1966, 3; (English transl.).

ADDITIONAL READINGS

Aleksakhin, R. M., Achievements and some actual tasks of water radioecology, in *Radioecology of Water Organisms. II. Distribution and Migration of Radionuclides in Freshwater and Seawater Biocenoses*, Andrushaitis, G. P., Ed., Zinatne, Riga, 1973, 7; AEC-tr-7607, U.S. Atomic Energy Commission, Washington, D.C., 1975; (English transl.).

Baturin, G. N., Ilyina, T. D., and Popov, N. I., Soviet investigations of natural radionuclides in marine and ocean sediments, *Proc. R. Soc. Edinburgh Sect. B*, 72(16), 181, 1971/72.

Buchsbaum, R., Species response to radiation; radioecology, in *Radiation Biology and Medicine*, Claus, W. D., Ed., Addison-Wesley, Reading, Mass., 1958, 124.

Davis, J. J., The Atomic Energy Commission's interest in bioenvironmental research, *BioScience*, 21(12), 600, 1971.

Foster, R. F., The history of Hanford and its contribution to radionuclides to the Columbia River, in *The Columbia River Estuary and Adjacent Ocean Waters: Bioenvironmental Studies*, Pruter, A. T. and Alverson, D. L., Eds., University of Washington Press, Seattle, 1972, 3.

Gromadska, M., Radioekologia — nowa galaz ekologii, *Ekol. Pol. Ser. B*, 7(3), 175, 1961.

Grosch, D. S., Ecology, pure and applied, in *Biological Effects of Radiations, Part 4*, Blaisdell, New York, 1965, 237.

Ichikawa, T., Eto, K., and Egami, N., Radiation, ichthyology, and fisheries (in Japanese), in *Radioactivity and Fishes: Contamination, Injuries, and Utilization*, Egami, N., Ed., Koseisha Kosei Kaku, Tokyo, 1973, 30.

Klechkovskiy, V. M., Polikarpov, G. G., and Aleksakhin, R. M., Introduction, in *Radioecology*, Vol. 2, Klechkovskiy, V. M., Polikarpov, G. G., and Aleksakhin, R. M., Eds., Atomizdat, Moscow, 1971, 5; *Radioecology*, Greenberg, D., Ed., John Wiley & Sons, New York, 1973; (English transl.).

Odum, E. P., Ecology and the atomic age, *Assoc. S. E. Biol. Bull.*, 4(2), 27, 1957.

Odum, E. P., Ed., Environmental aspects of nuclear radiation, *Bull. Ga. Acad. Sci.*, 15, 127, 1957.

Odum, E. P., Radiation ecology, in *Fundamentals of Ecology*, 3rd ed., W.B. Saunders, Philadelphia, 1971, 451.

Martin, W. E., Radioecology and the study of environmental radiation, *Bull. Torrey Bot. Club,* 91(4), 283, 1964.

Peredel'skiy, A. A., The principles and problems of radioecology (in Russian), *Zh. Obshch. Biol.*, 18(1), 17, 1957; Rep. CEA-tr-1369, French Comissariat a l'Energie Atomique, Paris; (French transl.).

Peredel'skiy, A. A., Questions of radioecology (in Russian), *Priroda,* 8, 27, 1958.

Peredel'skiy, A. A., Radioecology, *Vopr. Ekol.*, 4, 66, 1962; JPRS 19,496; (English transl.).

Polikarpov, G. G., Development of marine radiation hydrobiology, *Gidrobiol. Zh.*, 13(5), 57, 1977.

Schultz, V. and Whicker, F. W., Jr., Radiation ecology, *CRC Crit. Rev. Environ. Control,* 4(4), 423, 1974.

Seymour, A. H., Introduction, in *Radioactivity in the Marine Environment,* National Research Council, National Academy of Sciences, Washington, D.C., 1971, 1.

Verkhovskaya, I. N., Tasks, methods and perspectives of radioecological investigations in different biogeocoenoses (in Russian), in *Methods of Radioecological Investigations,* Verkhovskaya, I. N., Ed., Atomizdat, Moscow, 1971, 3.

Witherspoon, J. P., Jr., Radioecology, in *Medical Radiation Biology,* Dalrymple, G. V., Gaulden, M. E., Kollmorgen, G. M., and Vogel, H. H., Jr., Eds., W.B. Saunders, Philadelphia, 1973, 250.

Wolfe, J. N., Impact of atomic energy on the environment and environmental science, in *Radioecology,* Schultz, V. and Klement, A. W., Jr., Eds., Reinhold, New York, 1963, 1.

Wolfe, J. N., Radioecology: retrospection and future, in Symposium on Radioecology, U.S. AEC Rep. CONF-670503, Nelson, D. J. and Evans, F. C., Eds., U.S. Atomic Energy Commission, Washington, D.C., 1969, xi.

Wolfe, J. N., Foreword, in A Tropical Rain Forest: A Study of Irradiation and Ecology at El Verde, Puerto Rico, U.S. AEC Rep. TID-24270, Odum, H. T. and Pigeon, R. F., Eds., U.S. Atomic Energy Commission, Washington, D.C., 1970, iii.

Chapter 2

ECOLOGICAL PRINCIPLES APPLIED TO RADIOECOLOGY

I. INTRODUCTION

The purpose of this chapter is to provide a brief introduction to the science of ecology. A basic understanding of concepts and terminology of ecology is assumed throughout much of this volume and it is realized that a portion of the audience to whom this book is addressed will not have undertaken formal study of the subject. Another reason for including this chapter is to show the importance of ecological processes in controlling transport and accumulation of radioactive materials and in affecting biological responses to radiation stress. The authors also want to show how radionuclides may be used to answer important questions in basic ecological research. Finally, they would like to point out that radiation is a normal and necessary feature of our environment. Since there are many excellent texts on basic ecology,[1-5] the authors are not attempting a comprehensive treatment of it. Instead, they stress those ecological principles which are most relevant to the broad question of ionizing radiation in the environment.

The importance of fundamental relationships between ionizing radiation and the total biotic environment in which we and other organisms live is poorly appreciated. The frequently irrational fear of radiation, and the nuclear reactions and transformations which produce such radiations, might not be so irrational if it were realized that life on earth is totally dependent upon nuclear processes. Electromagnetic radiations from the sun, which are generated by the almost countless numbers of thermonuclear reactions occurring each second, are absolutely essential to the photosynthetic process, which in turn supports all life on earth. In addition, the surface of the earth might be too cold to support life, at least as one would know it, were it not for the heat generated from the radioactive decay of naturally occurring radionuclides in the crust and core of the earth. So, while man ponders the question of nuclear vs. other forms of energy, the biospheric machine cranks persistently onward under nuclear power! Yet, another way in which natural radiation impinges upon life processes is in helping to produce genetic variability which is one of the key driving forces in the evolutionary process.[6] The point then, is that the structure and function of life evolves toward a state compatible with all features of the environment. Radiation is one of these features.

As the authors have stated, radiation is an inherent feature of the environment and the environment in turn plays a dominant role in determining the behavior and effects of radionuclides which find their way into it. However, in addition to these natural relationships between radiation and the environment, there are legal and social motivations for interest in ecology and radioecology. The National Environmental Policy Act (NEPA) of 1969 generally requires environmental reports and environmental impact statements on any major projects or developments which may have measurable ecological consequences. Environmental impact statements are subject to review by the public and by governmental agencies, such as the Council on Environmental Quality (CEQ) and the Environmental Protection Agency (EPA), Washington, D.C., before permits to proceed with the specific project are issued. In cases of projects involving radiation or radioactive materials, the Nuclear Regulatory Commission (NRC), EPA and the individual states may also become involved in the environmental impact review process. Since radioactive materials in sufficient concentrations are potentially harmful to organisms, and hence, the environment, impact statements must include evaluations of radionuclides whenever there is potential for their release to the environment. Such evaluations require knowledge and understanding of radioecological concepts, which in turn are based upon physical and ecological principles.

While ecology involves the scientific study of ecosystems, there is much misconception of what ecology really is. Professionals involved with ecology and natural resource management view ecology as a science, a complex science involving many others. The field of ecology has its college courses, professional societies, journals, meetings, and consulting firms. However, the public sometimes views ecology as a place, a thing, a political movement, or something else. They may view an "ecologist" as a barefoot person with shaggy hair living in a van, or a physics professor who gives impassioned speeches on overpopulation, pollution, and dwindling natural resources. The difference between an ecologist and environmentalist is frequently obscure to the lay person.

The fact is that most environmentalists are not ecologists. Almost anyone who is actively concerned about the environment may be called an "environmentalist". On the other hand, a person who deserves the title "ecologist", is probably one who is capable of objective analysis and scientific investigation of ecosystem structure and function. Such a person most likely has had substantial formal training and experience in ecology and related sciences. It is interesting, though, that most ecologists are environmentalists, at least to the degree of sharing in the concern for preservation of natural resources and improvement of environmental quality.

A basic premise of ecology is that the various living and nonliving entities of the environment are interdependent and fit together to form a functional system, the value of which is greater than the sum of its individual parts. This "holistic" view is not new. Throughout recorded history the dependence of man on his surroundings has been recognized. However, the recognition of ecology as a bona fide science developed more slowly than many other scientific endeavors. The tools of the early ecologist were not sophisticated: perhaps fieldglasses, notebook, and hiking shoes were the major necessities. Modern ecology, however, involves much in the way of scientific hardware and quantitative theory. It is now realized by many that ecosystem functions are probably every bit as complex as the biochemical reactions which occur in a living cell. It is also increasingly realized that ecology is extremely relevant to the problems faced by a growing society in a finite environment.

Let us not get involved with a strict definition of ecology, because the concept of it is inherently complex. Nevertheless, one can ponder some of the definitions that have been proposed such as: (1) ecology is the study of interrelationships among living systems and their environment; (2) ecology is the investigation of the total relation of organisms to their organic and inorganic environment; (3) ecology is the study of systems at a level in which individuals or whole organisms may be considered elements of interaction either among themselves or with a loosely organized environmental matrix; and (4) ecology is the biology of ecosystems. The most important concept to grasp at this point is that of viewing the environment as a functional system composed of parts that are interdependent.

II. THE ECOSYSTEM CONCEPT

Ecosystems are assemblages of specific biota and abiotic entities within a given space or area. Each entity provides certain functions which are of critical importance to the whole system. The spatial boundaries of the system may be physically obvious, or they may be indistinct. In some cases, the boundaries of an ecosystem are defined rather arbitrarily for the purpose of observation and study.

A lake or pond is comparatively easy to recognize as an ecosystem because its boundary, or shoreline, is usually delineated clearly. A lake or pond contains plants, animals, dissolved minerals, and suspended particles in a medium of water, all of which is underlaid by bottom detritus and sediments. These components function in a mutually dependent fashion, more or less independently of the surrounding area. There are inputs to and losses from the lake ecosystem, however. Inputs include solar radiation, water, minerals, and organic material. Losses include heat, water, minerals, and organic material. Since some of these inputs and losses are necessary for the maintenance of structure and function of the system, it is not entirely independent of its surroundings.

Terrestrial ecosystems are more difficult to visualize because their boundaries are usually less distinct. For instance, coniferous forests may extend over hundreds of miles without interruption. They may blend into shrublands or grasslands, but the boundaries of each system may not be sharply defined. However, like lakes, terrestrial ecosystems contain specific structures such as plants, animals, gases, particulates, organic detritus, and soil, each of which provides certain functions of importance to the whole system. Inputs to and losses from terrestrial ecosystems also occur and some are of critical importance, such as sunlight and precipitation.

The ocean is far too large and heterogeneous to be considered a single ecosystem. Rather, an ocean is an assemblage of marine ecosystems, each of which occupies a certain geographical portion of the biosphere and has certain distinctive characteristics. An estuary, a rocky coastal shelf, a coral reef, and a portion of the deep ocean floor are but a few examples of recognizable marine ecosystems. As with terrestrial ecosystems, the boundaries of marine ecosystems may be distinct, obscure, or even arbitrary.

Other examples of "ecosystems" include those which are artificial, such as aquariums, terrariums, and spaceships. Each can contain all the physical materials necessary to support and maintain life. Each, however, must depend upon a continual energy source, or at least a store of energy in some form that will maintain life until that store is consumed. One can fill a jar with pond water, set it in a window where sunlight can reach it, and presto, an ecosystem appears! It will be self-sustaining, except for the sunlight continually needed for plant growth and replacement of water lost by evaporation. A spaceship is a life support system consisting of substances borrowed from the biosphere of the earth. Such artificial "ecosystems" may be regarded as small volumes of the biosphere of the earth encapsulated and projected temporarily into space. A spaceship is a self-sustaining ecosystem, but it can remain so only for a short period of time. The authors are more concerned in this book with natural or cultivated ecosystems which make up the biosphere of the earth and upon which man and other species must depend.

Ecosystems contain specific groups of organisms which perform certain functions and they also contain nonliving components. Figure 1 illustrates the major components which are characteristic of virtually all types of ecosystems. The key elements of the ecosystem are the producers, the consumers, the decomposers, and a nutrient pool. Nutrients cycle through the system in a continuous fashion and, theoretically, they can be used over and over again, providing they are not lost from the system. Energy, on the other hand, does not cycle. It is eventually lost by respiration in the form of heat. This is shown by the arrows which lead from the biotic entities to the environment. Because of this constant but necessary energy loss, a continual energy source, the sun, is required to drive the system.

A fraction of the radiant solar energy which falls on the earth at an average rate of about 2 cal/cm^2/min, is converted to chemical energy in the form of carbon-containing compounds. This conversion is accomplished by chlorophyll-bearing plants through the process of photosynthesis. In photosynthesis, gaseous carbon dioxide from the atmosphere is combined with water to form organic compounds which can later be oxidized with the release of stored energy. Green plants are termed "producers" because only this group of organisms is capable of converting solar energy to forms that

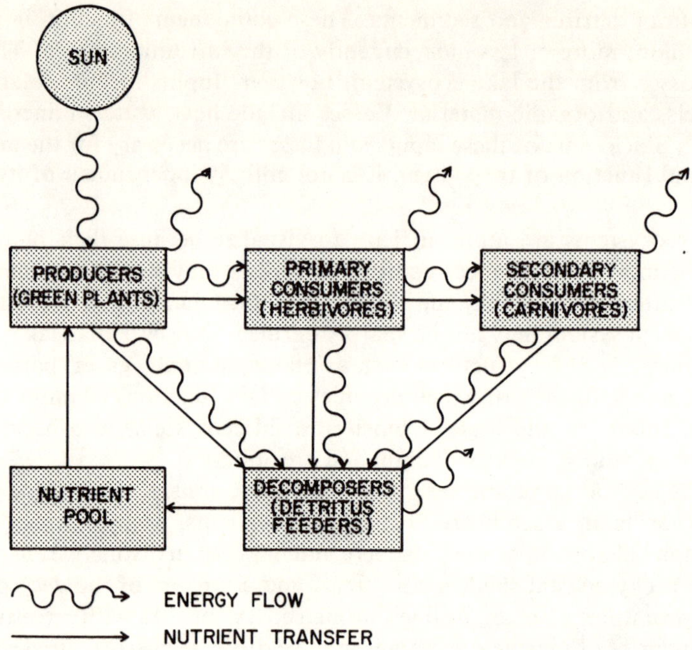

FIGURE 1. Simplified diagram of energy flow and nutrient transfer in ecosystems.

can be utilized by consumer organisms. Since producers are self-feeding, that is, they provide their own food for growth and metabolism, they are described as being autotrophic. Consumer organisms must eat energy containing food for their growth and metabolism and they are, therefore, described as heterotrophic.

The producers ultimately provide the food base for all aquatic and terrestrial ecosystems. They require solar energy and nutrients from the environment in order to function. The nutrient pool refers to essential elements and chemical compounds found in the atmosphere, water, and soils or sediments. The essential nutrients include carbon dioxide, water, and various inorganic minerals. The producers must undergo respiration in order to maintain themselves and reproduce, and thus, part of the energy they store is lost to the environment as irretrievable heat. Part of the chemical energy stored in tissues of producers is transferred to consumer organisms and part is transferred directly to decomposer organisms which feed upon dead plant material or detritus. Consumption of living or dead plant material also results in the transfer of minerals to consumers or decomposers.

Consumer organisms include herbivores, which feed directly on plant material, carnivores, which eat herbivores or other carnivores, and omnivorous animals which consume both plant and animal tissue. Consumers generally receive all their energy intake in the form of organic food and they give up energy to growth, maintenance, reproduction, and predatory animals. Minerals received through ingestion and sometimes by assimilation are deposited in tissues or excreted. Minerals and part of the energy in the tissues of consumers ultimately wind up in the decomposer compartment through excretion and death. A hierarchy of producers and consumers is sometimes referred to as a food chain, or a food web in cases where individual species have alternative food choices.

Decomposers include organisms which feed on and break down dead organic material. Detritus-feeding arthropods and earthworms, for instance, function to assist the breakdown of organic matter by passing it through their bodies and extracting a portion of the available energy and minerals. However, the microorganisms, including fungi, actinomycetes and bacteria are the chief decomposers. The action of these organisms results in the virtually complete decomposition of organic detritus, with the release of carbon dioxide from respiration and minerals. The minerals are released in physical and chemical forms which can be readily assimilated once again by plants. In some ecosystems, a large fraction of the total nutrient inventory is contained in the biotic components. In such systems, the decomposition of detritus is essential to the continued growth of plants.

It should be clear to the thoughtful reader at this point that energy and nutrient flows serve as threads which bind components of ecosystems together, setting up a web of interdependencies. However, energy and nutrients are not the only crucial links within ecosystems. Interrelationships have evolved around many other factors. These factors include physical features such as temperature, moisture, and light intensity, and biological phenomena such as mutualism, competition, and a host of other behavioral attributes. Natural ecosystems tend to evolve toward a state of stability which is compatible with the abiotic features of a particular place. Biological species composition of ecosystems tends toward some array or hierarchy in which mutually compatible relationships win out over those which have deleterious effects on the system. Much like a successful business in an industrial society, a species which endures in a natural ecosystem is likely successful because it can find what it needs and it can form compatible interrelationships with other species and with the abiotic environment. In many ways, ecosystems are analogs to modern cities. Both have their driving forces, producers and consumers, currency, resource pools, and organization. Both are highly interwoven networks of physical and biological entities which have holistic characteristics. A pinyon-juniper (*Pinus edulis — Juniperus* spp.) woodland ecosystem has unique characteristics, as does Chicago.

What has all this to do with radioecology? Plenty. For instance, the mineral nutrient cycles and the individual transfers which comprise these cycles provide the first clue as to how many radionuclides might behave in the environment. Strontium-90 and ^{226}Ra, for instance, will follow fairly close to the geochemical pathways of calcium. Tritium (^{3}H) behavior can be predicted quite accurately from the hydrologic cycle. Energy flows determine the rates at which radionuclides are passed through food chains, and ecosystem structure determines how radionuclides will eventually be partitioned in the environment. The overall sensitivity of ecosystems to radiation is closely related to the ability of indigenous biologic communities to withstand natural stresses and environmental extremes. Finally, many of the intricate relationships within ecological systems can be sorted out using radionuclide tracers. In essence, ecology is fundamental to radioecology and radioecology is one facet of ecology, the study of ecosystems.

III. STRUCTURE AND ORGANIZATION IN ECOSYSTEMS

Ecosystems are not a haphazard array of soils, plants, and animals. They have recognizable, predictable structure, and organization. In fact, the whole realm of biology has obvious structure and organization. Biological organization may be viewed from the standpoint of taxonomic groups, as well as from that of hierarchial levels of organization.[1] The hierarchial level of organization, for instance, refers to cells vs. organisms, vs. ecosystems. Studies in biology might focus on molecules of biological importance (molecular biology), or they might focus at the cellular level (cellular biology).

Other points of focus in the biological realm are tissues, organs, and discrete organisms. Each of these levels, while interrelated, can receive specific attention using techniques unique to each.

Ecology is concerned primarily with "higher" levels of organization, namely populations, communities, and ecosystems. Populations are made up of numerous organisms capable of interbreeding, while communities are composed of numerous populations or species which coexist in a given area. Ecosystems are composed of a biotic community and the supporting abiotic environment. Ecology is also concerned with biomes, or larger units which are sufficiently distinct from one another that each is clearly recognizable. Examples of biomes are deserts, coniferous forests, and grasslands. Ecology might even venture one step farther and consider the whole biosphere, or that envelope around the earth which contains living organisms. Let us now look at some attributes of populations, communities, and ecosystems in more detail.

A. Populations

A group of organisms having sufficiently similar biological characteristics and which occupy an area such that interbreeding is possible, may be termed a "population". By virtue of interbreeding, or some other form of reproduction, a population shares a common pool of genetic information. Each member of a population belongs to the same species. However, a species may have distinct populations that have become isolated from one another by some form of barrier. Over long periods of time, through evolutionary processes, genetically isolated populations of a common species may change sufficiently that interbreeding would no longer be possible even if the barriers of isolation were removed. Through this mechanism, new species may evolve.

Populations have numerous recognizable and sometimes measurable attributes. Such attributes are not characteristic of individuals which make up the population, but rather they are determined by the collective characteristics of the individuals. Such collective characteristics supercede the sum of individual characteristics. For instance, a population has a given total number of individuals which occupy a certain geographical area. The total number may be partitioned by sex and age classes. A certain fraction of the population will die and a certain number of individuals will be born within the population in some length of time. Individuals within various sex and age classes have a mean-life expectancy and a certain probability of dying from disease, predation, starvation, or other cause of mortality. These sorts of statistical parameters are characteristic of populations, and when quantified, can form the basis for the subject of population dynamics. An understanding of the basic concepts of population dynamics is essential to the investigation of how environmental modifications, whether positive or negative, affect populations.

Let us very briefly consider in an oversimplified way, a population parameter of fundamental importance, the total number of individuals (N). If N_o individuals are introduced at time, t_o, into an area (or habitat) suitable for growth of the population, N will likely increase with time (t) (Figure 2). As long as the environment or the population itself is not limiting the growth of the population, it will increase exponentially (Curve A). In this unlimiting situation, the population can grow at its "intrinsic rate of natural increase", termed "r". Expressed mathematically,

$$\frac{dN}{dt} = rN \qquad (1)$$

i.e., the rate of change of the population with respect to time is proportional to population size. Integration of this expression and evaluation of the constant of integration

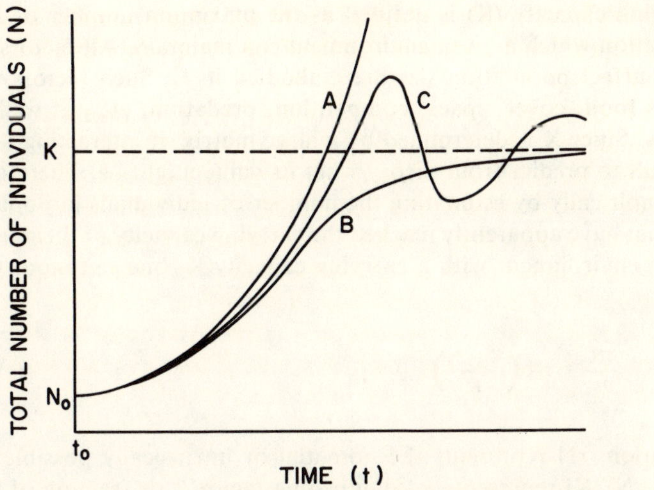

FIGURE 2. Forms of population growth curves. Curve A represents growth in an unlimited environment, while B represents growth in a limited environment with a population carrying capacity, K. Curve C represents populations which temporarily overshoot K and subsequently oscillate above and below K.

for the stipulation that $N = N_o$ at $t_o = 0$ leads to:

$$N = N_0 e^{rt} \tag{2}$$

Now r, the intrinsic rate of natural increase expresses the net fractional growth of the population per unit time. However, this net growth rate must account for natality (birth rate) and mortality (death rate). This can be accomplished by defining r as the difference between the potential birth rate (b) and the "normal" death rate (d):

$$r = b - d \tag{3}$$

where each term is expressed as a fraction of the population per unit time. So long as the actual birth rate exceeds the actual death rate, r is positive and the population will increase according to the model (Equation 2). If, on the other hand, conditions are such that mortality exceeds natality, r becomes negative and the population declines. When $b = d$, $r = o$ then the population is said to be "stable".

Of fundamental importance in population dynamics and in ecology as a whole is the fact that populations cannot expand indefinitely. Populations need space and other resources and the biosphere and all habitats within it are of limited size and therefore have limited capacity to provide the needs of any population. Therefore, as populations get larger, there is increased competition for the resources needed to sustain population growth. At some point in time, per capita resources decline to the point where population growth begins to decline. The actual value of r becomes smaller than its potential value in an unlimited environment. Since time-variable rate constants such as r are mathematically difficult to work with, another concept and term, "carrying capacity", is commonly employed to act as feedback which limits population growth to that which the environment can support.

The carrying capacity (K) is defined as the maximum number of individuals of a given population which a given environment can maintain. All factors of the environment which affect population size are embodied in K. Such factors include obvious ones such as food, cover, space, competition, predation, etc., as well as more subtle relationships. Since K is determined by a large matrix of interacting factors, it is normally difficult to predict from theory what its value might be. Therefore, K is usually estimated empirically by estimating the number of individuals in populations that are stable and that have apparently reached the carrying capacity of their environment.

In a given environment with a carrying capacity K, one can modify Equation 1 as follows:

$$\frac{dN}{dt} = rN\left(\frac{K - N}{K}\right) \tag{4}$$

In this equation, rN represents the potential or intrinsically possible increase, while the term [(K−N)/K] represents a "dampening factor", or fraction of the carrying capacity utilized by N individuals. Note that when N is small compared to K, the dampening factor is nearly unity and the population can grow at a rate close to its theoretical maximum. However, as N approaches K, the dampening factor severely limits the growth rate, dN/dt. When N = K, dN/dt = 0. Equation 4 may be integrated to yield the expression:

$$N = \frac{K}{1 + e^{[\ln(K - N_o)/N_o] - rt}} \tag{5}$$

which defines N in terms of K, N_o, r, and t. This equation, which yields N as a function of time, as is shown by Curve B in Figure 2, is termed the logistic equation.[1] Note that the logistic equation yields a sigmoid growth curve which has an upper asymptotic limit, K. The rate at which N approaches K is determined by the value of r, as well as the value of N at some time, t.

Some populations may exceed or overshoot K as shown by Curve C in Figure 2. This can occur from the momentum of rapid population growth in fecund species. However, a drastic drop in population numbers can be expected soon after the peak is reached because the environment can no longer meet the needs of the population and mortality will exceed natality. Following such a population crash, growth may again resume when K exceeds N. These phenomena can produce cyclic oscillations in N. In some species, the oscillations will gradually dampen and population numbers stabilize. Other species may continue to undergo cyclic oscillations indefinitely. Rather obviously population growth is more complex than modeled above, but this over simplified presentation is presented to familiarize the reader with simple concepts of population growth.

Stressing agents, such as radiation, can affect population numbers by affecting natality (b), mortality (d), and carrying capacity (K). If the effects of stress directly affect b or d, then the effects are considered *direct.* It is well known, for example, that ionizing radiation can kill reproductive cells, therefore, damaging reproductive capability or fecundity, which lowers b. Radiation can also cause direct mortality, increasing d. On the other hand, radiation may *indirectly* affect a population by altering the carrying capacity of its environment. An example of this is an experiment conducted by Cadwell and Whicker in which radiation-resistant ant populations declined in response to radiation-induced mortality of plant populations upon which the ants were dependent.[7]

Another, more common example of indirect effects on populations is habitat destruction caused by road building and other surface disturbances. Bulldozers and other heavy equipment may not cause significant direct mortality to alert, mobile animals such as sage grouse (*Centrocercus urophasianus*) or pronghorn antelope (*Antilocapra americana*) but they do destroy a portion of the habitat needed by such species, and in doing so, K is reduced. Animals displaced from destroyed habitat cannot simply move over into surrounding areas that are already filled to capacity. The population must pay the price through adjustments in numbers to the lowered value of K.

B. Communities

Biotic communities are assemblages of populations which coexist in common environments. Populations, in addition to having *intra*specific relationships, have *inter*specific relationships with other populations. Interrelationships among populations, which are extremely varied and range from mutual dependencies to competition and predation, lead to structure and organization at the community level.

Communities have an autotrophic component (plants) and a heterotrophic component (consumers). Each component has an assemblage of populations which exhibit mutual interactions. Communities have attributes not shared by the individual populations which comprise them. They can be characterized by the kinds of populations present, as well as by the relative importance of each population to the entire community. Communities can respond to stress or altered environmental factors, and, at the same time, communities themselves alter their immediate environment. Mineral and energy flows can be quantified at the community level, or such flows could be classified by populations.

The study of communities is usually termed "synecology", which contrasts with the study of individual populations, "autecology". In order to properly manage a mule deer (*Odocoileus hemionus*) population, an autecological study focused upon that population and factors affecting it may be needed. On the other hand, proper management of a national park or forest should rely heavily upon synecological investigations. In the latter case, concern is with many species and among multiple uses of the land such as water, timber, grazing, recreation, and aesthetics.

The study of form and physical structure in biotic communities is termed "physiognomy", which is parallel to morphology at the organism level.[3] Physiognomy is determined largely by the existing growth forms of plants because of their importance to the general appearance of the community. Examples of major growth forms on land include trees of several types; shrubs, which are woody like trees but of smaller stature; epiphytes, which grow on other plants; herbs, which include ferns, grasses, sedges, and forbs; and thallophytes, which include lichens, mosses, and liverworts. Physiognomy is a convenient way to describe communities and the presence of certain growth forms in certain geographical areas implies the presence of certain plant and animal species.

Climate and soil factors largely determine the physiognomy of a given region. For example, in the temperate latitudinal zone, dry climates give rise to deserts, semiarid climates produce grasslands, and moist climates are characterized by forests. Extreme environments subject to intense cold, wind, or drought are tolerated best by plants of low stature such as thallophytes, herbs, and small shrubs. Trees are virtually absent in such areas and instead occur in environments in which climatic extremes have become ameliorated by topography or other factors.

Community composition is a more refined description of communities than is physiognomy. Composition refers to the specific populations which exist within the community. The collective properties of these populations and their interactions determine the nature of the community. The composition of two separate communities can be compared by a parameter known as the "coefficient of community" (CC).[8] Although CC can be calculated in several ways, the following equation is commonly used:

$$CC = \frac{2c}{a + b} \qquad (6)$$

where a is the number of species in one community or study plot, b is the number in another, and c is the number common to both. The CC is relatively easy to ascertain since all that is needed is a species list for the two communities being compared. A shortcoming of the CC is that it does not account for quantitative differences in species abundance or other measures of importance.

An index which compares species composition of two communities and also accounts for differences in abundance or importance is "percentage similarity" (PS), which is defined as:

$$PS = 1 - 0.5 \sum_{i=1}^{n} \left| a_i - b_i \right| \qquad (7)$$

where a_i and b_i are measures of importance for the ith species of n in communities A and B, respectively.[3]

Both indices of comparative community composition, CC and PS, are useful in determining the degree of difference between separate communities and changes in composition of one community over time. Changes in community composition in response to environmental stress can be measured in this manner. For example, Fraley and Whicker[9] have shown that high levels of chronic radiation stress elicit dramatic changes in the coefficient of community calculated on the basis of species present in control and irradiated grassland plots in Colorado.

An extremely important attribute of the biotic community is diversity. Diversity is a measure of the structural complexity of communities and it is commonly calculated on the basis of the number of species in the community and the apportionment of individual organisms among the species. In general, the more species there are in a community, the more diverse it is considered to be. Also, the more evenly the numbers of individual organisms are apportioned among the species present, the more diverse the community (Figure 3). The greater the number of species and the more equitably the individuals are apportioned among the species, the greater is the opportunity for "functional diversity". Functional diversity implies a complex network of interspecific interactions and the possibility of alternative resources to supply needs such as food and shelter.

Diversity is considered by many ecologists to be related in some way to stability. It can be argued that the greater the species diversity, the greater is the potential for alternative functional relationships, the greater is the "protective structure" of the ecosystem, and the more readily the system can adjust to short-term stresses and perturbations. This is not unlike the concept of corporations which attempt to achieve financial stability by diversifying their investments. This idea is supported in geographic areas where mature, stable communities are more diverse than nearby communities which exist only temporarily following landscape disturbance. It is also supported if one compares the stability of single-crop agricultural systems (which are nondiverse, unstable, and expensive to maintain) with natural ecosystems (which are usually diverse, stable, and self-maintained). However, there are natural communities which do not necessarily follow this pattern. For instance, some mature (climax) communities are less diverse than preceding, less stable communities. Also, some stable communities, especially in harsh environments, are not particularly diverse.

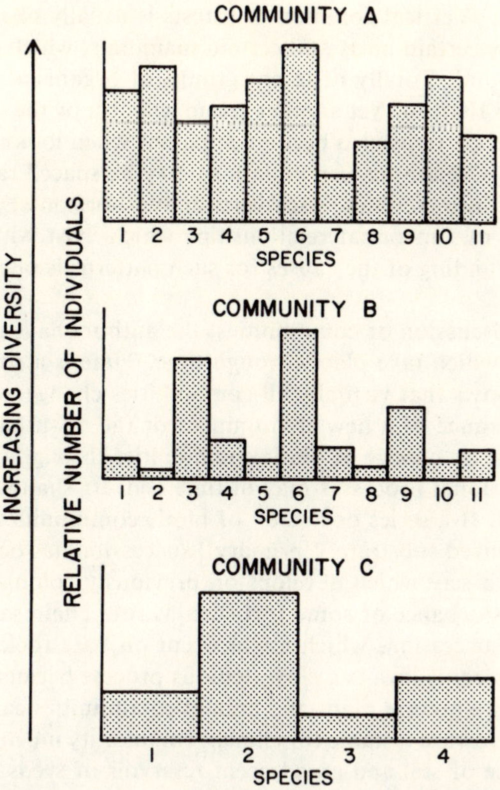

FIGURE 3. Community diversity as measured by the number of species and apportionment of individuals among the species. Communities A and B are more diverse than C because they contain more species. Community A is more diverse than B because the individuals are more evenly distributed among the same number of species.

As with other quantitative parameters used to describe communities, diversity can be expressed in various ways. One of the more frequently used indices of diversity ($\bar{H}$), is defined as

$$\bar{H} = -k \sum_{i=1}^{n} P_i \ln (P_i) \qquad (8)$$

where P_i is the probability of sampling the ith species, n is the total number of species encountered, and k is a constant.[10] It is important to note that this index does not depend upon which species are present, but only on the number and relative importance of each. Thus, structural changes in a community might change the CC or PS indices, but not $\bar{H}$; or, alternatively, diversity might change but, not the CC or PS. Usually, however, all such indices change across a stress gradient, but not necessarily in the same manner.

Another important concept with regard to biotic communities is spatial pattern and allocation of resources. Most communities for instance, have an organized vertical and horizontal structure.[3] Vertical zonation in forests is usually obvious, with a group of organisms, say trees, certain birds and certain mammals, which utilize the space 10 m or so above the ground. Totally different groups of organisms may utilize the space from ground level to 10 m and yet another group will occupy the space from the ground to 0.5 m. Horizontal patterns also become apparent when looking down upon a community from above. Organisms or populations may be spaced randomly, in a definite pattern, or clumped into colonies. Such vertical and horizontal patterns have evolved within the network of functional relationships which exist within communities and ecosystems. Understanding of the causes for such patterns is one of the keys to understanding ecosystems.

Thus far in our discussion of communities, the authors have given very little attention to the changes which take place through time. Time is a fundamental consideration because one knows that virtually all communities change, sometimes drastically, following the appearance of a new environment or the disturbance of an established community. The orderly change of biotic communities through time is called ''natural succession''. Successional processes may in time convert shallow lakes or bare rocky areas to lush forests. If a series or ''sere'' of biotic communities appear over time on a previously uncolonized substrate, ''primary'' succession has occurred. ''Secondary'' succession refers to a sere which develops on previously colonized substrate that has been denuded by disturbance of some sort such as fire, chain saws, plow, or ionizing radiation. Primary succession, which might occur on bare rock following the retreat of a glacier or on volcanic lava is a slow, tedious process because formation of fertile soil must occur before mature plant and animal communities can develop. Secondary succession proceeds toward a stable (or climax) community much more rapidly because of the usual presence of soil and an adjacent reservoir of seeds and other propagules which can colonize the damaged area.

Succession is the orderly, and usually predictable, progression of plant and animal communities toward a stable climax community which is in turn predictable on the basis of soil and climatic factors. A recognizable preclimax community may be referred to as a seral stage. Classic seral stages of secondary succession in portions of the southeastern U.S. might proceed from a bare field to early invading forbs, to grassland, to shrubs, to pine (*Pinus*) forest and finally to an oak-hickory (*Quercus-Carya*) forest climax.[1] It may take 150 years or more for the stable climax forest to become established. In the functional sense, succession proceeds until the total community respiration is equivalent to gross production, which is one way of defining climax. Prior to climax establishment, gross community production exceeds respiration and community biomass increases.[1]

The concept of natural succession is of prime importance to the radioecologist. Radiation stress may convert a climax community to a preclimax seral stage, which might be maintained under chronic exposure, or which will proceed toward climax again if the radiation stress is removed or reduced. In radioecological studies focused upon radionuclide distribution and cycling processes, recognition and understanding of the successional status of the system under investigation will aid in prediction of the long-term fate of the radioactive material. Climax communities and early seral stages show quantitative differences in mineral cycling, and by inference, in radionuclide cycling as well. A final thought along this line is that radioecologists are frequently involved in environmental impact assessments of activities that involve radioactive releases. Radiological impacts are frequently compared with the more tangible impacts of land surface disturbance. Analysis of surface disturbance impacts and reclamation efforts is heavily intertwined with the concepts of succession and community structure and it is important, therefore, to understand them.

C. Ecosystems

The development and succession of biotic communities cannot be separated from abiotic driving forces and constraints. Study of populations and communities cannot progress far without viewing such organizational entities in terms of the ecosystem. Energy transduction and mineral flows inherently involve abiotic forces and reservoirs and even purely behavioral aspects of organisms are closely linked to inorganic characteristics of the environment.

Ecosystems have attributes much like communities or populations, but these attributes include abiotic features of the environment. For instance, an ecosystem has a total inventory of essential minerals which is partitioned in a particular way between living and nonliving components. The ecosystem has quantifiable inputs of energy and minerals, and quantifiable losses. The ecosystem has composition, like communities, but its composition includes abiotic as well as biotic materials. Ecosystems have diversity, the measure of which is governed by nonliving as well as living things. Finally, ecosystems undergo development and succession, in concert with the communities they support.

Establishment and development of biotic communities is dependent upon inorganic resources and limitations. Such resources include solar radiation, water, and essential minerals. Limitations include lack of resources and presence of factors that limit growth such as cold temperatures, high winds, poor soils, and toxic elements or compounds. Resources and limiting factors vary tremendously throughout the biosphere, giving rise to an extremely diverse array of biotic communities.

Succession of biotic communities through time depends not only upon the abiotic environment already established in a particular locale, but also upon the changes in that environment which are brought about by the community per se. For instance, the death and decay of plant and animal matter provides organic material and nutrients essential to the development of productive soils. Plants and animals also assist in various ways the breakdown of rocks which provides additional inorganic nutrients. As the soil develops, larger-stature growth forms of vegetation can become established, and in turn, provide shelter needed by other populations of plants and animals. In this way, ecosystems as such undergo succession and longer-term evolution.

A convenient and often dramatic way to see how communities and their abiotic environments interact is to travel across environmental gradients, noting the composition of communities along the way.[3] Such gradients might include temperature, moisture, light intensity, salinity, etc. An environmental gradient usually is complicated by the simultaneous change in several important factors. For instance, the elevational gradient, which is very pronounced in mountainous areas, is confounded with gradients of temperature, precipitation, light, and wind. In any case, as one records, the frequency of occurrence of populations across an environmental gradient, a pattern not unlike that portrayed in Figure 4 is likely to be observed.

The patterns of population frequency in Figure 4 illustrate several important concepts. First, it is clear that each population has a definite range and an optimum position along the gradient. Secondly, specific regions within the gradient can be expected to contain certain populations, but not others. Third, some groups of populations, for example, populations 7 to 10 in Figure 4, respond to the environmental gradient in a similar fashion, perhaps implying some interspecific dependencies. Fourth, note that the population curves overlap each other, indicating the presence of transition zones, sometimes referred to as "ecotones". Fifth, some species have large tolerances and exist over wide ranges of the gradient, while others have narrow tolerance bands. Although this discussion illustrates several concepts, most real situations are more complex, owing to the usual presence of several environmental gradients. A more realistic

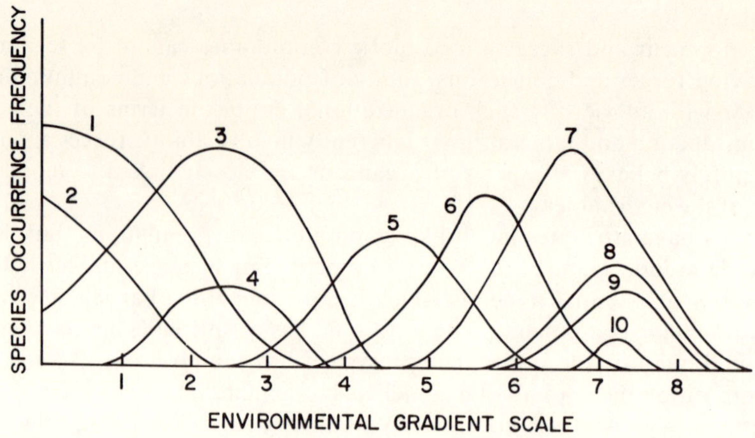

FIGURE 4. Occurrence frequencies of ten hypothetical populations across an environmental gradient.

method of portrayal would be multidimensional plots showing population response to two or more environmental factors.

As was mentioned earlier, ecosystem boundaries may not be sharp. It is clear from Figure 4 that if one defines an ecosystem from Position 2 to 3 on the environmental gradient, and compares it to the area between Positions 7 and 8, that these systems are very different, both in terms of species composition and abiotic factors. However, this contrast is purely arbitrary in terms of where one defines the boundaries. If one compares the region between gradient Positions 6 and 7 to that between 7 and 8, there is much less difference. The abiotic factors are more similar and 5 of the 6 species present are shared by both ecosystems. Thus, across environmental gradients, the pattern of a "continuum" of species as illustrated is more common than having sharp breaks between groups of species or communities.

Ecosystems can be categorized roughly according to broad geographic and environmental criteria. For instance, ecosystems may be terrestrial or aquatic, and the latter may be classified as marine or freshwater. Terrestrial ecosystems may be classified as tropical, temperate, or polar, according to latitude, and they may be classified on the basis of climate and elevation. Cold desert ecosystems of midtemperate latitudes are somewhat distinctive, as the descriptive terms imply. Aquatic ecosystems may be classified on the basis of salinity, latitude, climate, elevation, flowing or not flowing, and water depth.

One can observe rather large geographic areas within which climatic factors such as temperature and precipitation are similar when averaged over long-time periods. Such large geographic patterns of climate give rise to conspicuous, recognizable vegetative growth forms or distinctive physiognomic patterns. These larger areas which have similar climate and vegetation characteristics are called "biomes". A biome, although containing many distinct ecosystems in response to environmental gradients, is reasonably similar throughout. For instance, North America contains tropical rain forests, deserts, grasslands, deciduous forests, coniferous forests, and tundra.[1] Each biome extends over thousands of square miles. Each contains predictable species of plants and animals. Large research efforts in ecology were initiated in the 1960s under the auspices of the International Biological Program. These efforts were organized according to biomes.

The ecosystem concept can even be extended to the biosphere, or the spatial envelope surrounding the earth that contains life. This extension, while possibly seeming a bit far-fetched, is justified on the basis that biologically vital solids, liquids, and gases circulate on a global scale and thus are eventually shared by essentially all the earth's ecosystems. Migratory species cross many biomes at will and the activities of one species, man, can affect the entire biosphere. Despite the tremendous diversity and apparent separations within the biosphere, there is every reason to consider it in the context of one system. An excellent treatment of the biosphere is available.[11]

Although there is much more to say about ecosystems, the foregoing seems adequate for the purposes of this book. The important concept from the viewpoint of radioecology is that the fate and potential effects of radionuclides in the environment are determined not only by the amounts and properties of the radionuclides per se, but also by ecosystem properties. A good understanding of ecosystems, their parts and their network of functional relationships, is a crucial segment of the radioecologist's knowledge.

IV. FUNCTIONAL RELATIONSHIPS WITHIN ECOSYSTEMS

Now that the authors have considered some of the more obvious structural characteristics of ecosystems, let us turn to functional processes. The only reason for focusing our discussion first toward structure is that this concept seems more obvious, while function seems more subtle and more complex. Many authors, however, have good arguments for describing functional relationships first. In any event, it is important to realize that structure and function in ecosystems are so closely intertwined that it is difficult to consider them independently.

Ecosystem function refers to the things that the structural entities do, and how these things affect the other structural entities and the system as a whole. Our discussion of ecosystem function will be centered around energy flow, cycles of minerals and chemical compounds, and functional relationships of organisms. Energy flow is the principal driving force of ecosystems, mineral and chemical cycles provide necessary ingredients, and organisms evolve into niches and behavioral patterns which form the basis of ecosystem function.

A. Energy Flow in Ecosystems

It has already been stated that the ultimate energy source of the biosphere is the sun, which provides a constant energy input to the earth in the form of solar radiation of about 2 cal/cm^2/min. This amounts to a total biospheric energy income of about 13×10^{23} cal/year.[2] Of course, this energy input varies geographically and with time because of diurnal and seasonal cycles, so these values are annual averages. Somewhat less than half of this energy actually reaches the surface of the earth because of reflection by clouds and dust and absorption by constituents of the atmosphere. Of the solar energy that reaches the surface of the earth, a portion is reflected back into the atmosphere, a portion heats the water or ground surface and organisms upon it, and a small portion (usually less than 1%), is fixed by green plants in the form of chemical energy.

Fixation of solar energy by green plants is accomplished through photosynthesis. In the basic photosynthetic reactions, carbon dioxide and water are converted to oxygen, water, and energy rich sugar compounds, with solar energy and chlorophyll or other pigments acting as essential catalysts. The energy rich carbon compounds wind up largely as carbohydrates, proteins, and fats in nature, and oxidation of these compounds releases energy in the range of 4 to 10 cal/g. Since the amount of solar energy and other ingredients necessary for photosynthesis is finite and constant, there is a definite upper limit to primary production in the biosphere. By extension, this upper limit imposes a definite limit in turn to the quantity of consumer biomass that can be supported.

Although the fraction of the energy of the sun reaching the surface of the earth which is fixed by plants is small, up to several thousand grams of dry organic matter may be produced per square meter per year. The worldwide annual production of dry plant matter amounts to roughly 150 to 200 billion tons.[3,12] These figures represent "net primary production". Plants actually fix greater amounts of energy, but they also burn a portion of it through slow oxidation or respiration. It is the net production which appears as new tissue which can be transferred, potentially, to consumer organisms.

Net primary production varies dramatically from place to place in the biosphere, as might be expected from the variations in environments.[3] Extreme deserts or high mountains may produce less than 10 g/m^2/year, while swamps, tropical forests, and estuaries may produce over 2000 g/m^2/year. Some of the important temperate biomes in North America such as grassland, forest, and shrublands, produce roughly 500, 1300, and 600 g/m^2/year, respectively. Lakes and streams produce around 500 g/m^2/year, while the continental shelves and open oceans produce 350 and 125 g/m^2/year, respectively. Agricultural land may produce over 4000 g/m^2/year, but only because of manipulated inputs of fertilizers, pesticides, irrigation, and tillage. Surprisingly, modern agricultural systems require more energy than they produce![12]

Sometimes, net primary production and biomass or "standing crop" are confused. Biomass refers to the amount of living material per unit area and it is usually expressed as dry g/m^2. On the other hand, production is a measure of the amount of biomass produced per unit time. The total biosphere produces (net) about 320 g/m^2/year with a plant biomass which averages 3600 g/m^2. This suggests that about 9% of the earth's plant biomass is replaced annually. Herbaceous communities replace a larger fraction of their biomass annually than do woody communities. For instance, temperate grasslands replace about 30% of their biomass annually while temperate forests replace about 4% annually.[3]

Most plants are partially consumed by herbivores. Thus, a portion of the net primary production is consumed. The plant material consumed represents a given amount of energy. The energy consumed by herbivores is again partitioned, with part being burned through respiration and part being stored as consumer biomass. The respiratory losses are especially easily visualized for animals because they must expend energy to stay warm, capture food, reproduce, etc. The energy expended through respiration of all plants and animals is ultimately lost as heat to the environment.

Carnivores can now enter the picture because energy has been stored as herbivorous biomass and this is available for consumption. Therefore, a fraction of the herbivorous biomass provides energy input to carnivores. Again this energy must be partitioned into respiration and storage as carnivore biomass.

Through the processes of primary production and food chain energy conversions, a figure called "net ecosystem production" emerges. Net ecosystem production is the sum of the unconsumed biomass produced by plants and all consumers. The sum of energy consumed as respiration by all members of the biotic community is the total respiration of the ecosystem.

As a rule of thumb, roughly 10 to 20% of the net production at a given trophic level can be converted to net production at the next level.[12] Respiratory requirement is the major explanation for this figure seldom being much higher. This figure places definite limits on the growth of any population. It explains why high-density human populations are largely herbivorous. Human carnivory is expensive, both ecologically and monetarily.[13] Trophic level efficiency also is a basic reason why insects outnumber birds, which in turn outnumber mountain lions (*Felis concolor*) and bears (*Ursus* spp.). Our prized and rare top carnivores require large territories to provide their basic needs.

Let us now turn our attention to food chains for a moment. There are basically two kinds of food chains: the grazing food chains as the authors have been discussing with herbivores and carnivores, and the detritus or decay food chains. The detritus food chain utilizes dead plant or animal matter as its foundation, as opposed to green plants in the grazing food chain. Dead organic matter provides energy for myriad yet obscure populations of microorganisms (bacteria, fungi, actinomycetes), and soil/sediment-dwelling invertebrates (beetles, spiders, earthworms, mussels, crabs, etc.). These populations provide energy for higher order consumers, such as birds, fishes, and mammals. In many cases, higher order consumers can advantageously utilize energy originating from both food chains.

Food chains are descriptive of energy transfers between trophic (feeding) levels when species are lumped into categories such as producers, herbivores, carnivore 1, carnivore 2, etc. However, we sometimes choose to illustrate feeding relationships by species. In this case, alternative and multiple feeding possibilities result in a network of relationships which may be referred to as a food web (Figure 5). The diagram illustrates that consumers may have specific or broad food habits and that some are not by any means restricted to the preceding trophic level. All organisms provide dead organic matter which initiates the food web of decay.

The degree of complexity of food webs is a measure of functional diversity. The greater the number of species involved and the greater the number of alternative feeding relationships, the greater is the stability of the system. Short-term perturbations in specific populations can be offset by alternative food sources. Such complex naturally evolved networks show remarkable resiliency. This is one reason why ecologists worry over a greater and greater fraction of the energy of the biosphere being diverted to a single species, man. The growth of man is decreasing the structural and functional diversity of the biosphere and, quite possibly, reducing its stability.

Energy is the basic currency of the biosphere. Ecosystem function can be described in lucid detail on the basis of caloric flows. We look at a peaceful forest or pond and are largely unaware of the complex energy transformations which have made and which maintain that forest or pond. Two strongly recommended lucid articles on energy process of the earth and its biosphere, have been prepared by Oort[14] and Woodwell.[12]

Cycles which move nutrients and radioactive materials in the biosphere are driven by solar and earthbound energy. The transport of radioactive materials can be predicted in many cases on the basis of energy transformations.

B. Mineral Cycles of Ecosystems

The natural cycles of biologically important minerals, which occur on micro, local, and biospheric scales, are of fundamental importance to the behavior of radionuclides in the environment. There are numerous biological and geochemical processes which move materials of the earth from place to place. Radionuclides in the environment are moved by the same processes. Since the masses of radioactive material carried by these processes is negligible in comparison to the masses of stable material, one can think of the radionuclides as merely "going along for the ride" and as having little or no effect on the mechanisms of transport. This very brief review of the major mineral cycles of ecosystems should lay a basic cornerstone for understanding the environmental behavior of radionuclides.

All chemical elements which constitute the biosphere undergo cycling phenomena. Certain of those elements and their compounds play particularly significant roles in biological processes. Living matter consists mainly of H, O, C, and N, but other elements including P, S, Ca, K, Mg, Na, Cl, Fe, and Mn also play essential roles in the

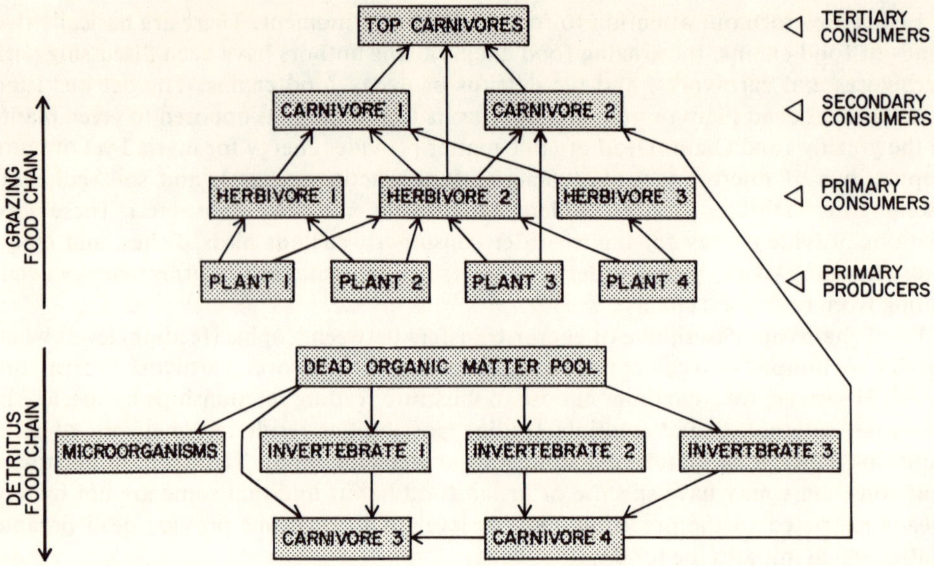

FIGURE 5. Grazing and detritus food chains organized by species to show the complex network of a hypothetical food web.

structure and function of biological systems. Cycling processes involve physical transport of elements from place to place, as well as complex chemical transformations. Elements may play key biological roles as single atoms or ions, or they may be involved in biological processes only by virtue of the need for certain chemical compounds of which they are a part. For illustrative purposes, the authors shall briefly consider the cycle of a most important compound, water, and a most important element, carbon.

It is known that water is one of the key compounds that is essential to life. It is a basic medium which surrounds life processes and it is a major source of hydrogen, an essential element in the synthesis of organic compounds. Water is also a major reservoir for oxygen. The vast majority of water in the biosphere is in the oceans, but significant quantities are also locked in ice and occur in bodies of fresh water. Although only a tiny fraction of the earth's water is in the form of atmospheric water vapor, this water is of vital importance in the development of weather systems.

The basic features of the water cycle include precipitation and evaporation. These processes provide continuous exchange between the earth's surface water bodies and the atmosphere. Generally, maintenance of stable water inventories on a biospheric scale requires that evaporation equals precipitation. Evaporation and precipitation vary widely in space and time according to myriad factors and circumstances. Precipitation and temperature probably affect the development of ecosystems more than any other factors. Also, these factors largely determine the geographic distribution of mankind. Water, flowing under the influence of gravity, is a major force which sculptures the land through erosion.

Tritium (^3H or T), an important radioisotope of hydrogen that is formed by natural processes as well as by the nuclear activities of man, follows the hydrologic cycle quite precisely because most of it combines with oxygen to form HTO, or tritiated water. The environmental behavior of HTO is nearly identical to that of H_2O, with only minor differences related to the slightly greater molecular weight of HTO and chemical bond energies which differ slightly between T-O and H-O. The transport of tritium in the biosphere can be predicted on the basis of knowledge of the hydrologic cycle. In return, many quantitative features of the hydrologic cycle can be measured using tritiated water as a tracer.

Precipitation and surface runoff patterns have a marked influence on the deposition and transport of most radionuclides in the environment. Airborne radioactivity, with the exception of the noble gases, is scavenged effectively from the atmosphere by precipitation through processes called "rainout" and "washout", as will be discussed in Volume I, Chapter 5 and Volume II, Chapter 1. Radionuclides scavenged by snowflakes are distributed according to snowfall patterns and redistributed by wind acting upon snow deposits. Radionuclides in soil and sediments can be redistributed by runoff events which cause scouring of water courses and sheet or gully erosion of susceptible terrain. The general trend on a biospheric scale is for radionuclides and other materials generated by man and nature to migrate toward the oceans under the persistence and power of flowing water. The oceans and their sediments serve as the ultimate waste receptacles of the biosphere.

The carbon cycle of the biosphere is comparatively simple in terms of general reservoirs and transport pathways, but highly intricate and complex in terms of the chemical transformations which take place in biological systems. The general cycle of carbon can begin with carbon dioxide (CO_2) which is present in the earth's atmosphere at a concentration of about 300 ppm. Through photosynthesis, atmospheric CO_2 is converted to carbon compounds in plant tissues on land. In the oceans and in freshwaters, dissolved CO_2 is assimilated by autotrophic organisms such as phytoplankton. Some carbon then finds its way into the tissues of consumer organisms through the processes of consumption and anabolism. Respiration of plants and animals results in the oxidation of carbon compounds and the release of CO_2 back into the atmospheric or aqueous pools. Carbon present in dead tissues enters the organic detritus pool, a portion of which is consumed by microorganisms and other decomposer organisms. This portion is ultimately converted to CO_2 as well through respiration of the decomposers, a process termed "soil respiration".

The carbon which circulates rather rapidly in the biosphere from CO_2 to organic carbon to CO_2, accounts for only a small fraction of the total biospheric inventory. Most of the biospheric carbon is locked in sedimentary deposits that have accumulated over hundreds of millions of years. The small fraction of organically bound carbon which escapes oxidation, forms deposits that can accumulate over geologic time periods. These fossil deposits include oil shale, coal, and petroleum. Other organically derived carbon deposits include calcium carbonates from the skeletons of plankton which settled to the ocean beds over eons of time. Inorganic carbon deposits include calcium carbonates (limestones) which can form in the ocean depths from carbonate and bicarbonate ions.

Our contemporary society is largely powered by the oxidation of fossil fuels. In utilizing these fuels, we are oxidizing carbon in a few hundred years that nature stored over millions of years. The biospheric ramifications of this are not known. The burning of fossil fuels at such a high rate releases tremendous quantities of CO_2 into the atmosphere and it has been documented that the atmospheric CO_2 concentration has been increasing by an average of some 3% per decade since 1850.[15] Some scientists believe that such a change in the quality of the atmosphere could trigger profound, worldwide climatic change.

The carbon cycle is of particular interest to the radioecologist because the energy flow in ecosystems involves the flow of carbon and because of the important natural and man-generated radionuclide, ^{14}C. As discussed earlier in this chapter, understanding ecosystem function largely involves the understanding of how energy in the form of carbon compounds is partitioned into separate flows. These energy flows describe what eats what, how much is eaten, and how minerals and radionuclides are transferred through food webs. Carbon-14, produced by cosmic rays, nuclear weapons, and nuclear reactors, enters biogeochemical cycles in the same manner as stable carbon. It results in radiation exposure to all living organisms and it also provides science with one of its most powerful investigative tools.

Any discussion of mineral cycles in ecosystems would be quite incomplete without considering at least oxygen, nitrogen, phosphorus, and sulfur, in addition to hydrogen and carbon. In addition, the importance of other essential elements such as Ca, K, Na, Cl, and Fe should be described. However, it is not our intent to treat all the mineral cycles and their roles in ecological systems. There are excellent reviews on these cycles by Bolin, Cloud and Gibor, Delwiche, and Deevey in one readily available volume.[11] However, the authors do not want to leave this topic until some of the natural processes which move the essential elements between and within ecosystems have been considered. Such processes are the key to radionuclide transport as well as stable mineral movements.

Mineral cycles can be visualized on the basis of those within local ecosystems with inputs from and exports to the surroundings,[16] on the scale of biomes, or at the level of the entire biosphere. The processes involved in the movement and conversion of elements involve biotic and abiotic phenomena. The latter may be considered on a local, short-term scale, or on a large, long-term geologic scale. Let us first consider some of the abiotic forces which move elements and their compounds.

The earth is a dynamic, restless ball of material surrounded by an even more restless atmosphere. On a time scale of hundreds of millions of years, land masses rise, mountains form, land masses erode and fill in the oceans, continents drift apart, glaciers come and go. Volcanoes erupt and spew gases and particles into the air, and lavas flow down hillsides covering everything in their path. Periodic earthquakes cause landslides, lakes, and tidal waves. These processes move astronomical quantities of the elements in the biosphere, but generally over periods of time which are too long to be fully appreciated by the transients of the surface of the earth.

More rapid processes, which also move materials about the biosphere, are much more noticeable and obvious. Such processes include erosion by wind and water and the flow of solids, gases, and liquids in the airstream or water courses. The deposition of snowflakes, sediments, and bits of windblown debris in the lee of physical structures can be seen by the most casual observer, and quantitatively measured by the research worker.

Erosional processes are of particular importance. Periodic high winds can move dry, unprotected topsoil across states, as during the dust bowl days of the 1930s, and they can disperse radioactive contamination from controlled areas to public lands. Winds prevent rich soil from accumulating on exposed mountain ridges by moving transitory deposits to protected areas of soil accumulation. Flowing water, especially when of high volume and velocity, has tremendous erosive power, as witnessed by the devastation wrought by occasional floods. The Big Thompson Canyon flood of 1976 in Colorado carried automobiles, boulders, and human bodies tens of miles, and silt and smaller debris hundreds of miles in just a few short hours and days. Tons of silt, driftwood, and dissolved minerals are moved many miles daily by large river systems, even under conditions of normal flow. Where reduced flow and turbulence occurs, gravity deposits suspended particulate materials. Large amounts of material are constantly being suspended and transported by wind and water and then deposited in a distant locale.

Minerals dissolved in water chemically adsorb to many kinds of surfaces. Later, the same minerals may be desorbed from a surface, once again becoming dissolved in the aqueous phase, where they are subject to transport and adsorption on a surface at some other location. Surface exchange phenomena can be used to predict the movement of ions, both stable and radioactive, through soils, sediments, and water courses. High in mountain watersheds, melting snows produce water that is initially cleared of its impurities by adsorption to vegetation and soil. Small rivulets and larger streams at high elevations are therefore usually low in dissolved minerals and sometimes approach the purity of distilled water. However, as water journeys to the sea, it encounters soils and sediments which contain larger quantities of exchangeable ions, and these are picked up by the flowing water and transported down the watershed. As water accumulates dissolved nutrients it becomes more capable of sustaining life until, in rare cases, it becomes so salty that life is depressed.

Minerals are also moved and cycled by biological processes. Plants absorb nutrients from the soil or water in which they grow and in some cases directly from the atmosphere. Nutrients assimilated by plants can be consumed by herbivores, enter the litter compartment through death of tissues, or be leached from the plant by water. Animals obtain minerals largely by ingestion of food and water, but inhalation and absorption through epidermal tissues from the environment may also occur in various groups of animals. Animals get rid of excess quantities of elements by excretions and secretions. Elements are returned to the environment through shedding, molting, and dying, and are transferred to other forms of life through predation, parasitism, and decay. Living animals can move over large distances through migrations, undirected wanderings, or passive movements in air or water and, in so doing, move elements from place to place. Plant parts can also move passively, as in the case of wind-transported pollens, blowing tumbleweeds, and drifting phytoplankton. The breakdown of dead organic matter by decomposer organisms results in the return of elements to soil and water in forms that can be assimilated by subsequent generations of life.

To accurately predict the fate of radioactive materials in the environment, a good knowledge of mineral cycling processes is essential. This knowledge must include the pathways and mechanisms of transport for specific elements, as well as quantitative understanding of the amounts which flow through the various pathways under specific conditions. This is a very tall order, and only in a few places on earth does one have such adequate knowledge of mineral cycling in whole ecosystems.

C. Interactions of Organisms

Ecosystem function can be viewed in terms of energy flow, mineral cycles, and the interactions of organisms. While these views can be made separately, they in fact are highly interwoven, and one should not lose sight of this. However, let us now consider some of the ways in which different species populations interact with each other and also how individual organisms interact, both within and between species. The interactions of organisms determine in large measure how energy is channeled through ecosystems and how minerals are biologically utilized. Therefore, such interactions are also of fundamental importance to the environmental behavior of radioactive materials. In addition, the response of biotic populations and communities to stress, such as ionizing radiation, is always tempered by interspecific relationships such that the effects of stress are frequently indirect and therefore are often unpredictable.

It should be obvious to any observant person who has spent time out-of-doors that populations are nurtured yet regulated by a mosaic of factors which include abiotic as well as biotic features of the environment. Each organism requires nourishment, space, and a certain range of temperature, moisture, and so on. Each organism depends upon other organisms of the same as well as different species for certain requirements, and in turn provides by its own presence for certain needs of other organisms. Each population occupies a rather specific physical territory and carries out specific functions within that territory. The physical territory occupied by a given population may be termed its "habitat". The functional role of that population may be termed its "niche".[1] However, since habitat and role are somewhat interdependent, confusion as to the meaning of niche has persisted over the years.[4]

Table 1
THE SCHEME OF INTERACTIONS BETWEEN TWO POPULATIONS

Type of interaction	Species 1	Species 2	General nature of interaction
Neutralism	0	0	Neither population affects the other
Competition	−	−	Direct inhibition of each species by the other/or/indirect inhibition when common resource is in short supply
Amensalism	−	0	Population 1 inhibited, 2 not affected
Parasitism	+	−	Population 1 (the parasitic) generally smaller than 2 (the host)
Predation	+	−	Population 1 (the predator) generally larger than 2 (the prey)
Commensalism	+	0	Population 1 (the commensal) benefits while 2 is not affected
Protocooperation	+	+	Interaction favorable to both, but not obligatory
Mutualism	+	+	Interaction favorable to both and obligatory

Note: 0 Indicates no significant interaction; + indicates growth, survival, or other population attribute benefited; and − indicates growth, survival, or other population attribute inhibited.

Adapted from Odum, E. P., *Fundamentals of Ecology,* 3rd ed., W. B. Saunders, Philadelphia, 1971. With permission.

A more satisfying way of looking at the requirements of niche and habitat for a particular population is to consider the "n-dimensional hypervolume" as postulated by Hutchinson.[17] This concept embodies the idea of a species occupying a position in the environment which is prescribed by the range of n-environmental gradients which it can tolerate. For instance, a particular bird species has a specific range of temperatures and moisture which it can tolerate. This defines a geographic area within which it may be found. However, this bird is a seed eater and can only effectively utilize a particular size range of seeds. This then prescribes a lesser geographic area within which the bird may be found, because the requirement of seed size has been superimposed upon temperature and moisture requirements. On top of this, these birds may require a certain stem size range in order to perch. This narrows the spatial volume that these birds can occupy even further. When all the tolerances and requirements of these birds are finally defined, a prescribed address within the n-dimensional hypervolume results.

Now then, the theoretical address of a particular species based upon requirements and tolerances is still not quite sufficient to describe precisely where it may be found. Other species may compete for the resources which are available within the address, there may exist predators, or there may be purely psychological reasons why the species in question may not be able to occupy its theoretical address. However, there does exist a "realized niche" within the address into which each species can fit. Therefore, the niche of a species essentially refers to the resources of an ecosystem which it is able to utilize; this set of resources is specified by physical tolerances, resource requirements, and interactions with other species.

Let us proceed to discuss how species interact with each other. The authors shall first consider the simplest case where they are concerned with the kinds of interactions between just two different populations. Basically, one population may affect another in a positive way (+), a negative way (−), or in a neutral way (0).[1] This scheme is frequently used to classify the ways in which two populations can interact (Table 1). These interaction classifications have several limitations because they may change with time, the intensity of the interaction is not indicated, and what may appear on the surface to be a specific type of interaction may in reality be another.

Neutralism is probably rarer than other classes of interactions. It usually seems that everything affects everything else in ecosystems, at least to some degree. For instance, each population provides input of organic detritus which alters the soil, which in turn affects the entire food base. The term, neutralism, might be used in practice when there is no obvious, direct interaction between two populations which occupy the same area.

Competition implies direct or indirect mutual inhibition between two species. The term can also be used to describe mutual inhibition between individuals of the same species. Direct competion implies immediate, mutual inhibition. For instance, two beetle populations might actually prey on one another, to the detriment of both populations. Chemicals secreted by some populations inhibit the growth of others. Two populations could limit one another by secretion of these "allelopathic" chemicals. Indirect competition usually implies mutual need for common resources which may be in short supply. Competition is frequently greatest between species which have similar or overlapping niches. Usually, when niches overlap too closely, competition becomes a matter of survival of the more fit population. One population, if it has a slight competitive edge over another, will survive in time while the other will either perish or be driven away. This concept is termed "competitive exclusion principle".[18]

Amensalism implies that one population suffers from the presence of another, but that the other population is not affected. Elephants and ants may provide a simple-minded example of this kind of interaction. As elephants walk about, they step on ants and ant hills, killing them by the thousands. On the other hand, the elephants are not likely affected one way or another by the ant population. However, this amensalistic interaction may only be apparent, for in reality, the ant population may have indirect effects on elephants.

Parasitism and predation are interactions which appear obvious. In the case of parasitism, the parasite usually feeds upon the tissues of the host. Usually, the parasite finds safety in its small size. It can find shelter in the hair, body openings or in the tissues of the host. The host usually suffers from this relationship while the parasite is usually dependent upon it. Predation usually involves the killing and consumption of a prey species by a predatory species. The predator always benefits to the expense of the prey, or so it seems. But does this +,− relationship always accurately portray the real situation? Probably not. For instance, while the individual prey organism always pays the price for being captured, the prey population may actually benefit from natural predation. Culling of the less fit members of a population and control of prey numbers through predation may actually enhance the fitness and stability of the prey population. So some predator-prey relationships probably deserve a +,+ rather than a +,−.

A commensal relationship is said to exist if one species benefits from the other, but the other is not affected. The oceans provide many examples of this type of interaction. For example, sponges, as well as many other marine organisms provide shelter for small organisms whose presence has neither a good nor a bad effect on the host organism. Most commensal relationships involve the provision of shelter, but some provide food as in the case of organisms which remain near a host obtaining surplus food of the host.

Mutually beneficial relationships include protocooperation in which the relationship is not obligatory and mutualism in which the relationship is obligatory. An example of protocooperation is the tick bird (*Buphagus* spp.) and the impala (*Aepyceros melampus*). This bird can and does feed on parasites which infect the impala. This relationship is mutually beneficial, but not obligatory, since the impala can survive without this "delousing" and the tick bird has other food sources. A good example of mutualism is the lichen, which is actually two organisms in one. Lichens are composed of an alga which synthesizes food through photosynthesis and a fungus which provides structure. This obligatory relationship involves two organisms living together, a situation sometimes referred to as "obligate symbiosis".

The authors have thus far considered the comparatively simple case of two-population interactions. There also exist higher-order interactions which involve two or more populations. In these, for example, the way in which species 1 relates to species 2 might depend upon species 3, which in turn depends in some way on species 4. Thus, natural communities consist of populations which find themselves in a complex network of interactions. These interactions are predicated upon tolerances, requirements, and social phenomena. Such interactions are frequently so complex that researchers are only at the fringes of understanding why organisms are where they are and why they do what they do.

Nevertheless, it seems that researchers should strive to understand natural relationships better if they are to avoid catastrophic events in the future of the human race, which is subject to many of the same environmental limitations as other populations. Organism interactions have definite, if sometimes subtle, relationships to human health, food production, energy procurement, and the social intricacies of man.

REFERENCES

1. **Odum, E. P.,** *Fundamentals of Ecology,* 3rd ed., W. B. Saunders, Philadelphia, 1971.
2. **Kormondy, E. J.,** *Concepts of Ecology,* Prentice-Hall, Englewood Cliffs, N.J., 1969.
3. **Whittaker, R. H.,** *Communities and Ecosystems,* 2nd ed., Macmillan, New York, 1975.
4. **Price, P. W.,** *Insect Ecology,* John Wiley & Sons, New York, 1975.
5. **Margalef, R.,** *Perspectives in Ecological Theory,* University of Chicago Press, Ill., 1968.
6. **Savage, J. M.,** *Evolution,* 2nd ed., Holt, Reinhart & Winston, New York, 1969.
7. **Cadwell, L. L. and Whicker, F. W.,** Radiation effects on a shortgrass plains ecosystem: the arthropod community, in 10th Annu. Progr. Rep. on Atomic Energy Commission, Contract AT(11-1)-1156, U.S. AEC Rep. COO-1156-54, Colorado State University, Fort Collins, 1972.
8. **Greig-Smith, P.,** *Quantitative Plant Ecology,* 2nd ed., Butterworths, London, 1964.
9. **Fraley, L., Jr. and Whicker, F. W.,** Response of shortgrass plains vegetation to gamma radiation. I. Chronic irradiation, *Radiat. Bot.,* 13(6), 331, 1973.
10. **Patten, B. C.,** Species diversity in net phytoplankton of Raritan Bay, *J. Mar. Res.,* 20(1), 57, 1962.
11. **Various authors,** *The Biosphere,* W. H. Freeman, San Francisco, 1970.
12. **Woodwell, G. M.,** The energy cycle of the biosphere, *Sci. Am.,* 223(3), 67, 1970.
13. **Ehrlich, P. R. and Ehrlich, A. H.,** *Population Resources Environment: Issues in Human Ecology,* 2nd ed., W. H. Freeman, San Francisco, 1972.
14. **Oort, A. H.,** The energy cycle of the earth, *Sci. Am.,* 223(3), 54, 1970.
15. **Bolin, B.,** The carbon cycle, *Sci. Am.,* 223(3), 125, 1970.
16. **Likens, G. E., Bormann, F. H., Pierce, R. S., Eaton, J. S., and Johnson, N. M.,** *Biogeochemistry of a Forested Ecosystem,* Springer-Verlag, New York, 1977.
17. **Hutchinson, G. E.,** Concluding remarks, *Cold Spring Harbor Symp. Quant. Biol.,* 22, 415, 1957.
18. **Hardin, G.,** The competitive exclusion principle, *Science,* 131(3409), 1292, 1960.
19. **Errington, P. L.,** *Of Predation and Life,* Iowa State University Press, Ames, 1967.

ADDITIONAL READINGS

Boughey, A. S., *Ecology of Populations,* Macmillan, New York, 1968.

Campbell, R., *Microbial Ecology,* Halsted Press, New York, 1977.

Cole, G. A., *Textbook of Limnology,* C. V. Mosby, Saint Louis, 1975.

Daubenmire, R., *Plant Communities: A Textbook of Plant Synecology,* Harper & Row, New York, 1968.

Daubenmire, R. F., *Plants and Environment: A Textbook of Plant Autecology,* 2nd ed., John Wiley & Sons, New York, 1974.

Davis, D. E., Ed., *Behavior as an Ecological Factor,* Dowden, Hutchinson & Ross, Stroudsburg, Pa., 1974.

Emlen, J. M., *Ecology: An Evolutionary Approach,* Addison-Wesley, Reading, Mass., 1973.

Gerking, S. D., Ed., *Ecology of Freshwater Fish Production,* Halsted Press, New York, 1978.

Gilbert, N., Gutierrez, A. P., Frazer, B. D., and Jones, R. E., *Ecological Relationships,* W. H. Freeman, San Francisco, 1976.

Golley, F. B., Ed., *Ecological Succession,* Dowden, Hutchinson & Ross, Stroudsburg, Pa., 1977.

Hutchinson, G. E., *An Introduction to Population Ecology,* Yale University Press, New Haven, Conn., 1978.

Krebs, C. J., *Ecology: The Experimental Analysis of Distribution and Abundance,* 2nd ed., Harper & Row, New York, 1978.

MacArthur, R. H. and Connell, J. H., *The Biology of Populations,* John Wiley & Sons, New York, 1966.

Meadows, P. S. and Campbell, J. I., *An Introduction to Marine Science,* Halsted Press, New York, 1978.

Pianka, E. R., *Evolutionary Ecology,* 2nd ed., Harper & Row, New York, 1978.

Pielou, E. C., *Population and Community Ecology: Principles and Methods,* Gordon & Breach, New York, 1974.

Pomeroy, L. R., Ed., *Cycles of Essential Elements,* Dowden, Hutchinson & Ross, Stroudsburg, Pa., 1974.

Reid, G. K., *Ecology of Inland Waters and Estuaries,* D Van Nostrand, New York, 1961.

Ricklefs, R. E., *Ecology,* Chiron Press, Newton, Mass., 1973.

Sadleir, R. M. F. S., *The Ecology of Reproduction in Wild and Domestic Mammals,* Methuen, London, 1969.

Singer, S. F., Ed., *Global Effects of Environmental Pollution,* Springer-Verlag, New York, 1970.

Smith, R. L., *Ecology and Field Biology,* 3rd ed., Harper & Row, New York, 1980.

Steele, J. H., *The Structure of Marine Ecosystems,* Harvard University Press, Cambridge, 1974.

Stoddart, D. M., Ed., *Ecology of Small Mammals,* Chapman & Hall, London, 1979.

Thomas, W. L., Jr., Ed., *Man's Role in Changing the Face of the Earth,* University of Chicago Press, Ill., 1956.

Whittaker, R. H. and Levin, S. A., Eds., *Niche: Theory and Application,* Dowden, Hutchinson & Ross, Stroudsburg, Pa., 1975.

Wiegert, R. G., Ed., *Ecological Energetics,* Dowden, Hutchinson & Ross, Stroudsburg, Pa., 1976.

Chapter 3

RADIOLOGICAL PRINCIPLES

I. INTRODUCTION

This chapter is intended for individuals who have little or no familiarity with the types, properties, and measurement of ionizing radiation. The understanding of subsequent chapters will be enhanced if the reader has a reasonable grasp of the terminology and concepts that are encountered in the radiation sciences. This section is intended as a survey of the subject rather than an in-depth treatment. Many excellent books are available for those who need more detailed information.[1-5]

II. THE RADIATIONS AND RADIOACTIVITY

The term radiation, broadly defined, refers to the propagation of energy through space. The energy may be in the form of either particles or electromagnetic waves. This book is concerned primarily with "ionizing" radiation, or those radiations having sufficient energy that their interactions with matter will produce an ejected electron and a positively charged ion. Ionizing radiation may originate spontaneously from the nucleus of an unstable atom or "radionuclide", or it may be produced by external stimulation of certain materials.

A. Types of Ionizing Radiations

The authors shall consider only the more common types of radiations: particles, namely alpha, beta, and neutrons; and electromagnetic waves (photons), the X- and gamma rays. Nearly all the work included within the scope of radiation ecology involves one of these types of radiations.

1. Alpha Particles

An alpha (α) particle, essentially a helium nucleus without orbital electrons, is composed of two protons and two neutrons with a charge of plus two. The theory of alpha emission is quite complex; however, simply stated, alpha particles are emitted by certain atoms, which results in a more stable nuclear configuration. Alpha particles originate primarily from nuclear decay of a number of the radionuclides with atomic numbers greater than 82 and they can be detected in samples containing uranium, thorium, or radium.

In order to present a specific example of a nuclear transformation which gives rise to the alpha particle, let us first consider a general nuclear equation. A radionuclide X may spontaneously decay to another nuclide Y with the formation of a smaller particle P (such as an alpha or beta particle) and energy as follows:

$$^{A}_{Z}X = {}^{A-a}_{Z-b}Y + {}^{a}_{b}P + \text{energy} \tag{1}$$

where: A is the atomic weight of X (number of neutrons plus protons), Z is the atomic number of X (number of protons), a is the atomic weight of P, and b is the atomic number of P. It is evident that the total equivalent number of neutrons and protons must be equal on each side of the equation. Now for example, the emission of an alpha particle from a radium-226 atom can be represented by

$$^{226}_{88}Ra \rightarrow ^{222}_{86}Rn + ^{4}_{2}He + Q \text{ (energy)} \qquad (2)$$

In this case, note the use of an arrow rather than the equality symbol, which implies a spontaneous, directional transformation.

The energy term Q in the above reaction is extremely important. Nearly all the energy liberated by this transformation is carried by the alpha particle ($^{4}_{2}He$) as kinetic energy. The kinetic energy of the alpha determines how far it can penetrate various types of matter and it is this energy which may cause biological damage.

How much energy is produced from alpha decay of ^{226}Ra? Before Q is calculated, a few terms should be defined. The unit of mass measurement of atomic particles is expressed as an atomic mass unit (amu), which is defined as 1/12 of the mass of an atom of ^{12}C. There are 1.6605×10^{-24} g/amu. Because of the very small quantities of energy involved in nuclear transformations, the unit electron volt (eV) is used. One eV $= 1.6 \times 10^{-12}$ erg, or only 3.8×10^{-20} cal. The value of Q can now be calculated because it is known that the energy is formed from part of the original mass of the radium atom and, since mass and energy must be conserved, one can use Einstein's mass-energy equation, $E = mC^2$, where E is energy (in erg), m is mass (in grams), and C is the velocity of light (2.998×10^{10} cm/sec). The calculation is as follows

$$\begin{aligned} \text{Mass of } ^{226}Ra &= 226.025438 \text{ amu} \\ \text{Mass of } (^{222}Rn + {}^{4}He) &= \underline{226.020213} \\ \text{Difference} &= 000.005225 \text{ amu} \end{aligned}$$

The mass difference of 0.005225 amu is equivalent to the energy value of Q. This value can be calculated from $E = mC^2$

$$\begin{aligned} Q \text{ (erg)} &= (0.005225 \text{ amu})(1.6605 \times 10^{-24} \text{ g/amu})(2.998 \times 10^{10} \text{ cm/sec})^2 \\ &= 7.80 \times 10^{-6} \text{ erg} \\ &= 4.87 \times 10^{6} \text{ eV or } 4.87 \text{ MeV} \qquad (3) \end{aligned}$$

Not quite all of the 4.87 million electron volt (MeV) is available as kinetic energy to the alpha particle, since part of the energy is associated with the radon atom ^{222}Rn. The partitioning of the energy between the alpha particle and the ^{222}Rn is inverse to their masses.[5] Taking this into account, the alpha particle will have a kinetic energy of 4.78 MeV, or 98% of the total.

Because alpha particles are comparatively heavy and have a double charge, they react strongly with matter, producing large numbers of ions per unit length of their path. As a result, they are not very penetrating and instrument detection is somewhat difficult. Special instrumentation and involved sample preparation must often be used. For example, a 5 MeV alpha particle will only travel about 3.6 cm in air and will not penetrate an ordinary piece of paper. Since alphas are low in penetration ability, they themselves are usually not hazardous, unless the alpha-emitting nuclide is deposited within the tissues of an organism. When internally deposited, however, alpha particles are often more damaging than most other types of radiations because comparatively large amounts of energy are deposited within a very small volume of tissue.

2. Beta Particles

Beta (β) particles are electrons which are spontaneously ejected from the nuclei of radioactive atoms during the decay process. These particles have the same mass as an electron (5.49×10^{-4} amu) and may either be positively or negatively charged. A positively charged beta, called the "positron", is less frequently encountered than its negative counterpart, the "negatron" or "beta particle".

For each element, there exists within the nucleus an optimum neutron/proton ratio for stability of the atom. Nuclides at the optimum ratio are usually stable, that is, they are highly unlikely to spontaneously disintegrate. The further the neutron/proton ratio from the optimum, the less stable the nuclide and the more probable that a given atom will disintegrate within a specified time interval.

If the neutron/proton ratio exceeds the optimum, then beta decay is extremely probable. Beta decay tends to relieve this departure from the optimum because, in essence, a neutron is converted to a proton, reducing the neutron/proton ratio. For example, in the decay of phosphorous-32, which has an extra neutron and, hence, is unstable,

$$^{32}_{15}P \rightarrow ^{32}_{16}S + ^{0}_{-1}\beta + \text{neutrino} + Q\ (\text{energy}) \qquad (4)$$

thus a stable isotope (^{32}S) is formed through the emission of a beta particle ($^{0}_{-1}\beta$).

Several important features should be noted in Equation 4. First, the proton number of the daughter atom has increased by one and the neutron number has decreased by one so that the atomic weight remains unchanged. Secondly, the sub- and superscripts on the beta symbol permit the reaction equation to balance. The atomic weight of the beta is symbolized as zero because its mass is very small compared to that of a proton or neutron. The atomic number is symbolized as -1 since in terms of electric charge, the presence of the beta is equivalent to the removal of one proton. The other feature is the neutrino, a small particle which always accompanies beta emission.

The neutrino has very little mass and is electrically neutral; therefore, it transmits little energy to the medium through which it travels. It does, however, carry off a variable part of the energy (Q) of the transformation. This helps to explain why beta particles are emitted from a given radionuclide with a spectrum of kinetic energies varying from zero up to a specific maximum. The energy of a neutrino is the difference between the maximum possible beta-particle energy and the actual energy of a particular particle. As a general rule, the average energy of beta particles is about one third their maximum energy.

Positively charged beta particles, positrons, are frequently emitted by radionuclides having a neutron/proton ratio less than optimum. For example, zinc-65, which is a radionuclide commonly used in biological research, can decay by positron emission to copper-65.

$$^{65}_{30}Zn \rightarrow ^{65}_{29}Cu + ^{0}_{+1}\beta + \text{neutrino} + Q \qquad (5)$$

Except for electrical charge, the properties of positrons are identical to beta particles and they may be detected with the same methods. One important difference, however, the ultimate fate of the positron, will be discussed later.

Radionuclides with a neutron/proton ratio less than optimum may also decay by a process called "electron capture". In this process, the parent nucleus may capture an orbital electron (usually from the K, or innermost, shell) and subsequently emit a neutrino and one or more photons, ridding itself of excess energy. For example, zinc-65 may also decay in this manner

$$^{65}_{30}Zn + ^{0}_{-1}e \rightarrow ^{65}_{29}Cu + \text{neutrino} + Q \qquad (6)$$

Notice that this process results in the same daughter nuclide (^{65}Cu) as positron emission. The only readily detectable radiations which accompany the process of electron capture are either gamma rays from the nucleus or X-rays from the electron shells.

Beta particles or positrons are emitted by many of the natural and artificially produced radionuclides and are readily detected by several types of instruments, including the ordinary Geiger-Mueller (G-M) counter. Betas are considerably more penetrating than alphas, but less penetrating than X- and gamma rays of equivalent energies. The ability of betas to penetrate matter increases with their kinetic energy.

3. Neutrons

Neutrons, although encountered less frequently in radiation ecology than the other radiations previously discussed, play an important role in the natural radiation environment. They are produced in the atmosphere by cosmic-ray interactions and combine with nitrogen and other gases to form carbon-14, tritium, and other radionuclides. In addition, since neutrons are formed in nuclear fission and fusion reactions, they may be prevalent around the core of nuclear reactors and at sites of nuclear detonations.

Once a neutron has been produced, it is not likely to exist autonomously for very long. If it were in free space, where the probability of its reactions with another particle or atom were remote, the neutron (^1_0n) would quickly undergo spontaneous decay to a proton (^1_1p), a beta particle, and a neutrino

$$^1_0n \rightarrow {}^1_1p + {}^0_{-1}\beta + \text{neutrino} + Q \qquad (7)$$

The mean lifetime of a free neutron is only about 19 min.

A neutron in close proximity to other matter is more likely to react with an atom before spontaneously decaying. Since the neutron is electrically neutral, it is not affected by the electrostatic forces of the atom's nucleus or orbital electrons. As a result, the neutron can easily penetrate the atom and if so, may lodge in the nucleus, causing a nuclear transformation. Neutrons can react with most elements, and, in many cases, the result is one or more radionuclides. This process is called "neutron activation". The atmospheric formation of radioactive carbon-14 from stable nitrogen-14 is but one of many types of transformations caused by neutrons

$$^{14}_7N + {}^1_0n \rightarrow {}^{14}_6C + {}^1_1p \qquad (8)$$

The probability of a neutron being captured by a nucleus varies both among elements and with the kinetic energy of the neutron. Newly produced neutrons are normally very energetic and it is usually not until the neutron has lost most of its kinetic energy that its capture is probable. The loss of energy is due to a process called "moderation". Moderation of a neutron is achieved through a series of collisions with other particles or atoms. Each time a neutron collides with an object, it imparts some of its energy to that object and it "bounces" off in a new direction. The loss of energy per collision is greatest when the mass of the object is equal to the mass of the neutron. Since neutron and proton masses are nearly equal, substances containing an abundance of hydrogen (protons) make the best moderators. This explains a curious fact: whereas very dense materials such as lead usually make the best shields for X- or gamma rays, very light materials such as paraffin make the best shields for neutrons.

Neutrons affect living matter by the process of moderation. A high-energy neutron encountering biological material is apt to collide with a proton with sufficient force to dislodge the proton from the molecule which held it. The proton (normally called the "recoil proton") may then have sufficient energy to travel some distance in the tissue, causing secondary damage through ionization and excitation of molecules along its path.

4. X- and Gamma Rays

Both X- and gamma (γ) rays (collectively termed photons) are electromagnetic waves having very short wavelengths in comparison to other members of the electromagnetic spectrum such as visible light, heat rays, and radio waves. X- and gamma rays are identical in terms of their properties, behavior, and effects. The only difference is in their origin. Gamma rays originate from atomic nuclei, whereas X-rays arise from the electron shells. Electromagnetic waves have no mass; essentially, an individual wave or photon consists of oscillating electric and magnetic fields which are propagated together through space. All photons travel at the velocity of light. Their energies, however, vary over a wide range.

The energy contained in an individual photon is inversely proportional to its wavelength. The weakest X-rays, for example, have wavelengths of the order of 10^{-6} cm, whereas the more energetic photons may reach wavelengths less than 10^{-12} cm. The energies of photons are normally expressed in terms of electron volts, usually millions of electron volts (MeV). In order to calculate the energy, E, of a photon having a wavelength of 10^{-10} cm, for example, one can use the relationship

$$E = h\nu \tag{9}$$

where E is in erg, h is Planck's constant (6.6×10^{-27} erg-sec), and υ is the frequency in sec^{-1}. Given that the frequency is the speed of light (C) divided by the wavelength (λ), one may express Equation 9 as

$$E = \frac{hC}{\lambda} \tag{10}$$

Substituting the appropriate values,

$$\begin{aligned} E(\text{erg}) &= \frac{(6.6 \times 10^{-27}\ \text{erg-sec})\ (3 \times 10^{10}\ \text{cm/sec})}{10^{-10}\ \text{cm}} \\ &= 1.98 \times 10^{-6}\ \text{erg} \end{aligned} \tag{11}$$

Since 1 erg is equivalent to 6.24×10^5 MeV,

$$\begin{aligned} E(\text{MeV}) &= (1.98 \times 10^{-6}\ \text{erg})\ (6.24 \times 10^5\ \text{MeV/erg}) \\ &= 1.23\ \text{MeV} \end{aligned} \tag{12}$$

Therefore, the energy of a photon with the wavelength 10^{-10} cm is equal to 1.23 MeV.

The energy of X- or gamma photons is very important because this is their only property which varies in such a way as to affect their behavior. The energy of a photon has tremendous influence on its ability to penetrate matter. In general, for X- and gamma photons of less than 10 MeV, the higher their energy, the more material they can penetrate. The authors will return to this subject in greater detail later in this chapter.

Many types of nuclear transformations are accompanied by the emission of gamma rays. For example, alpha and beta decay of many radionuclides is frequently accompanied by gamma photons. When a parent radionuclide decays to a daughter nuclide, the nucleus of the daughter frequently contains excess energy and is unstable. Stability is usually achieved very quickly through the release of energy in the form of one or more gamma photons, a process called "isomeric transition". The daughter nucleus decays from one energy state to another without a change in atomic weight or number. For example, in the beta decay of iodine-131, the nucleus of its daughter, xenon-131, is left in an unstable or "excited" state, represented by *,

$$^{131}_{53}\mathrm{I} \rightarrow {}^{131}_{54}\mathrm{Xe}^{*} + {}^{0}_{-1}\beta + \text{neutrino} + \mathrm{Q} \qquad (13)$$

However, the excited condition of ^{131}Xe* is quickly ($\sim 10^{-11}$ sec) relieved by the emission of a gamma photon (γ).

$$^{131}_{54}\mathrm{Xe}^{*} \longrightarrow {}^{131}_{54}\mathrm{Xe}\ (\text{stable}) + \gamma \qquad (14)$$

In the decay of ^{131}I, several gamma rays of different energies are actually observed. This is because the nucleus of ^{131}Xe can exist at several levels of excitation following its formation from ^{131}I. About 80% of the ^{131}I nuclei decay with the emission of a 0.61 MeV beta (maximum energy) and a 0.36 MeV gamma. The energy (Q) released by the entire process is 0.61 MeV + 0.36 MeV = 0.97 MeV.

The production of X-rays usually arises from external energy acting upon a material. The energy is usually in the form of energetic particles or rays and the material may be most any substance. In an X-ray machine, electrons are produced from a hot filament and accelerated to high energies within a vacuum tube by a strong electric potential. The high-speed electrons strike a target (usually a heavy metal such as tungsten), resulting in X-ray production. Different processes result in two types of X-rays, "bremsstrahlung" and "characteristic".

Bremsstrahlung, or "braking radiation", is produced when a charged particle encounters an electric field and is deflected. The deflection, or change in direction, of the charged particle results in a loss of kinetic energy, which is carried away by an X-ray. X-rays produced in this manner may vary in energy from zero up to the energy of the incident electron, depending upon the degree of interaction.

In contrast, characteristic X-rays may be produced in an X-ray machine when an incident electron interacts sufficiently with an electron in one of the target atoms to remove it from orbit. The electron "vacancy" becomes filled by a higher shell electron and an X-ray of a specific energy is released. The energy of the X-ray will depend upon the atomic species and upon the orbital shells involved in the exchange of electrons. The orbital shells usually involved are the K, or innermost, closest to the nucleus, and the L and M, the next distant shells. A very probable occurrence is the ejection of a K electron by the incident particle and the subsequent filling of the K shell with an electron from the L shell. As the electron drops from the L to the K shell, an X-ray will be emitted with an energy equal to the difference in electron binding energies between the two shells.

The nature of electromagnetic rays makes their detection relatively easy. Gamma rays originating from rocks and soil down to several feet beneath the earth's surface can be measured in air and contribute significantly to the total external radiation exposure. Gamma-emitting radionuclides can be detected and accurately measured in living organisms with strictly external instruments. Gamma-emitting tracers are quite simple to measure and time-consuming chemical separations are usually unnecessary. However, although gamma rays can be very useful, larger quantities of radionuclides emitting them require special techniques for safe handling.

B. Radioactivity

1. Radioactivity and Half-Life

Radioactivity is the result of the process in which a parent radionuclide undergoes spontaneous disintegration, releasing one or more radiations and forming a daughter nuclide. The number of nuclear disintegrations per unit time is proportional to the amount (mass or number of atoms) of radioactive material in a sample. Symbolic representation of radioactive decay is as follows:

$$\frac{dN}{dt} = -\lambda N \tag{15}$$

where N is the number of radioactive atoms, t is time, and λ is the decay constant. The expression dN/dt is the derivative of number of radioactive atoms with respect to time. The negative sign is used because the number of radioactive atoms decreases with time. The decay constant λ has the units of inverse time (time^{-1}) and indicates the fraction of N which can be expected to decay per unit time. Therefore, λ is a direct measure of the instability of the radionuclide. The decay constant is always the same for a given decay process for a specific radionuclide, but this parameter varies widely among different radionuclides. One may integrate Equation 15 to develop a simple expression for N as a function of time

$$\int \frac{dN}{N} = -\lambda \int dt \tag{16}$$

$$\ell n\, N = -\lambda t + C \tag{17}$$

To evaluate the constant of integration, C, one can set $N = N_o$ at $t = 0$,

$$\ell n\, N_o = -\lambda(0) + C \tag{18}$$

$$\ell n\, N_o = C \tag{19}$$

Substituting Equation 19 into Equation 17, one can obtain the expression

$$\ell n\, N - \ell n\, N_o = -\lambda t \tag{20}$$

which is equivalent to

$$N = N_o\, e^{-\lambda t} \tag{21}$$

The authors are now ready to examine the concept of half-life. Half-life is also a measure of the tendency of a nuclide to be radioactive. For example, a very short-lived radioactive substance is comparatively unstable and the atoms decay rapidly. Therefore, a given mass of this substance would be comparatively radioactive. The half-life ($T_{1/2}$) is that value of t in which the number of radioactive atoms N or the disintegration rate, dN/dt decreases by a factor of 2. This is represented symbolically by

$$N = \tfrac{1}{2} N_o \tag{22}$$

and

$$t = T_{1/2} \tag{23}$$

Substitution into Equation 20 gives

$$\ln \frac{N_o}{2} - \ln N_o = -\lambda T_{1/2} \tag{24}$$

$$\ln 2 = \lambda T_{1/2} \tag{25}$$

Since the natural logarithm of 2 is approximately 0.693,

$$T_{1/2} = \frac{0.693}{\lambda} \tag{26}$$

which defines the relationship between the decay constant and half-life of a radionuclide. Figure 1 portrays the relationship of time (expressed as number of half-lives and the percentage of the original number of radioactive atoms.

The half-life value of a radionuclide gives some immediate insight into its behavior and the nature of the hazard which might be associated with it. A small quantity of radionuclide with a half-life in the order of minutes will not persist long enough to present a significant hazard a few days later and it is not likely to become dispersed very far by natural forces. In contrast, a radionuclide with a half-life on the order of several years may represent a long-term hazard and become widely dispersed if not held in containment. These ideas do not imply, however, that short-lived isotopes are unlikely to be hazardous. A comparable quantity (N) of a short-lived substance may release as much radiation as a long-lived material, but it is released over a much shorter time period. This situation could be to the detriment of an organism because a given dose of radiation is more harmful if it is delivered very rapidly.

2. Specific Activity

The quantity of a radioactive substance may be expressed either as activity (number of disintegrations per unit time) or as mass of the substance. "Specific activity" is the ratio between the activity and the mass of material giving rise to the activity. It will become apparent numerous times throughout this volume that the environmental and physiological behavior of radionuclides, as well as the nature of their biological effects, is related to the specific activity of the radionuclide.

From Equation 15 it is seen that the instantaneous rate of change in the number of nuclear disintegrations is the product of the decay constant (λ) and the number of atoms (N). Using the notation for activity (A^*) instead of dN/dt, one can write

$$A^* = \lambda N \tag{27}$$

where the units of A^* are disintegrations/time. It is immediately clear that activity is proportional to the number of radioactive atoms. The number of radioactive atoms N in a sample of a specific radionuclide is given by

$$N = \frac{mN_A}{A} \tag{28}$$

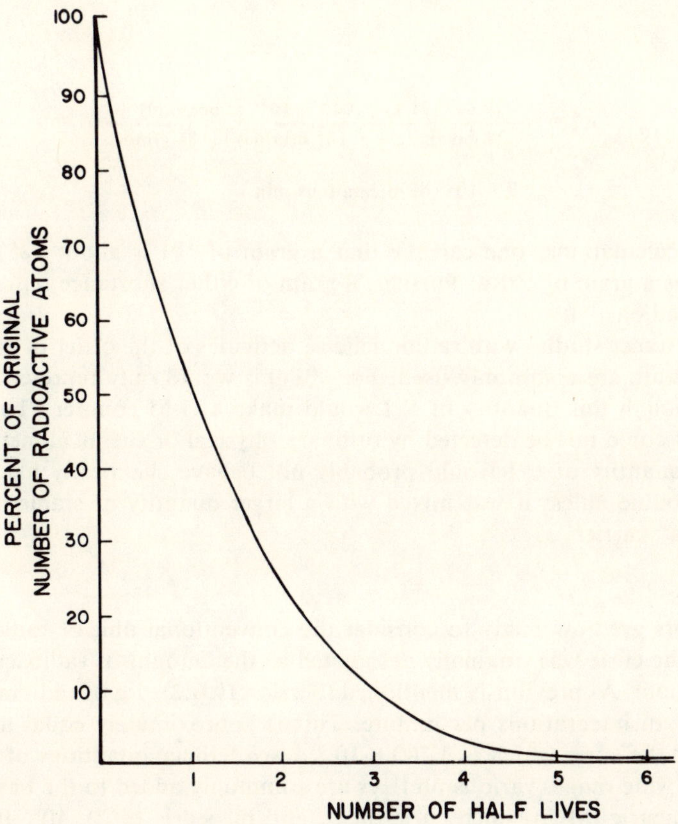

FIGURE 1. Relationship of time, expressed as number of half-lives, to the quantity of a radioactive substance.

where m is the mass of the substance in grams, N_A is Avogadro's number (6.025×10^{23} atoms/mol), and A is the atomic weight of the radionuclide in grams/mol. Substitution of Equations 28 and 26 into Equation 27 yields

$$A^* = \frac{0.693\ mN_A}{T_{1/2}\ A} \tag{29}$$

or

$$\frac{A^*}{m} = \frac{0.693\ N_A}{T_{1/2}\ A} \tag{30}$$

where A*/m is the specific activity of the radionuclide in question. This relationship indicates that radionuclides with low atomic weights and short half-lives have high specific activities. Let us consider a gram each of radium-226 and iodine-131 and calculate their relative activities:

For ^{226}Ra:

$$A^* = \frac{(0.693)\ (1\ g)\ (6.025 \times 10^{23}\ \text{atoms/mol})}{(1600\ \text{years})\ (5.26 \times 10^5\ \text{min/year})\ (226\ g/mol)}$$

$$= 2.2 \times 10^{12}\ \text{disintegrations/min}$$

For ^{131}I:

$$A^* = \frac{(0.693)\,(1\ g)\,(6.025 \times 10^{23}\ \text{atoms/mol})}{(8.08\ \text{days})\,(1.44 \times 10^3\ \text{min/day})\,(131\ g/\text{mol})}$$

$$= 2.7 \times 10^{17}\ \text{disintegrations/min}$$

From these calculations, one can see that a gram of ^{131}I is about 125,000 times as radioactive as a gram of ^{226}Ra. Further, a gram of either substance represents a large quantity of radioactivity.

For many tracer studies with radionuclides, activities of the order of 10^7 disintegrations per minute are commonly used. For ^{131}I this would only represent about 3.7×10^{-11} g. Although this quantity of ^{131}I would make a G-M counter "buzz" from its radiations, it could not be detected by ordinary physical or chemical methods. Such a small mass quantity of ^{131}I would probably not behave chemically like a weighable amount of iodine unless it was mixed with a larger quantity of stable iodine, which could act as a "carrier".

3. The Curie

The authors are now ready to consider the conventional unit of radioactivity, the curie (Ci). The curie was originally designated as the amount of radioactivity emitted by 1 g of radium. As previously mentioned (Section II.B.2), 1 g of radium corresponds to 2.2×10^{12} disintegrations per minute. This is approximately equal to the current definition of the curie which is 3.700×10^{10} d/sec.* Since quantities of radioactivity cover a very wide range, various prefixes are commonly added to the basic curie unit, such as megacurie (MCi), microcurie (μCi), and picocurie (pCi), 10^6, 10^{-6}, 10^{-12} Ci, respectively.

To give an idea of where some of the various units are useful, one can say that megacurie amounts of radionuclides have been produced by nuclear explosions; kilocurie (10^3 Ci) sources are used to irradiate tumors; microcurie quantities of tracers are often used in research; and a few picocuries of radionuclides from fallout may be found in a gram of plant or animal tissue.

It is important to note that the curie unit only specifies the number of nuclear disintegrations per second. It does not specify the number or types of radiations emitted per unit time. For this information it is necessary to know the number and kind of radiations emitted for each disintegration. An excellent handbook has been published which gives such information.[6]

III. INTERACTION OF RADIATION WITH MATTER

The detection, characterization, and effects of the radiations are almost entirely dependent upon their interaction with matter. In this section, the authors shall discuss the principal interactions and try to illustrate their significance.

A. Alpha Particle Absorption

The authors have already mentioned that alpha particles have very little penetrating ability, implying that they interact strongly with matter. Indeed, such is the case. Alpha particles can interact with either nuclei or orbital electrons in any absorbing medium

* In the International System of Units, the activity unit is the Becquerel (Bq), which is equal to 1 d/sec or about 2.7×10^{-11} Ci. Since there appears to be a definite though hesitant changeover from the curie to the Becquerel, it will be necessary for professionals and students to be familiar with both units.

such as air, water, tissue, or metal. An alpha passing in the vicinity of a nucleus may be deflected with no change in energy (Rutherford scattering), deflected with a small change in energy through bremsstrahlung (this process is negligible for alphas), or absorbed by the nucleus, causing a nuclear transformation.

The most probable processes involved in the absorption of alphas, however, are ionization and excitation of orbital electrons in the medium. Ionization occurs whenever the alpha particle is sufficiently near an electron to pull it by coulombic attraction from its orbit. Each time this occurs, the alpha loses kinetic energy and is thus slowed. The alpha also loses kinetic energy by exciting orbital electrons with interactions that are insufficient to cause ionization. As it becomes slowed, the alpha has a tendency to cause ionization at an increasing rate. As the alpha nears the end of its track, its rate of ionization peaks and within a very short distance, it stops, collects two electrons, and becomes a helium atom.

The range (average distance traveled) of alpha particles in air is roughly proportional to their initial kinetic energy. A 5 MeV alpha will travel about 4 cm in air while a 10 MeV particle will travel about 10.5 cm in air. The range of alphas in other materials with respect to air is approximately inversely proportional to the respective densities of each medium. For example, a 9 MeV alpha will travel about 9 cm in air. In mammalian tissue, an alpha of the same energy will travel approximately

$$(9\ \text{cm})\left(\frac{\text{density of air}}{\text{density of tissue}}\right) = (9\ \text{cm})\left(\frac{1.2 \times 10^{-3}\ \text{g/cm}^3}{1.1\ \text{g/cm}^3}\right)$$

$$= 9.7 \times 10^{-3}\ \text{cm}$$

$$\text{or } 97\mu\text{m}$$

B. Beta Particle Absorption

Like most other types of radiations, beta particles can interact with electrons as well as nuclei in the medium through which they are traveling. A beta particle passing near a nucleus will be deflected by the coulombic forces and a loss in kinetic energy of the beta may or may not occur. If no energy loss occurs, the interaction is said to be elastic, and the process is called "Rutherford scattering". In the case where energy is lost by the beta in an inelastic interaction, bremsstrahlung X-rays are produced. The probability of occurrence of either process increases as the atomic number of the absorbing medium increases.

Of greater significance are the interactions of beta particles with orbital electrons. Coulombic repulsion between a beta and an electron frequently results in ionization. In the ionization process, the beta loses an amount of energy equal to the kinetic energy of the electron plus the energy used to free it from the atom. A beta particle may produce 50 to 150 ion pairs per centimeter of air before its kinetic energy is completely dissipated. As mentioned earlier, characteristic X-rays are emitted as the vacant electron orbits are refilled with other electrons. Beta particles also cause excitation of orbital electrons which in turn leads to the emission of ultraviolet photons.

The ultimate fate of a beta particle depends upon its charge. A negatively charged beta particle, after its kinetic energy has been spent, either combines with a positively charged ion, or becomes a "free electron". Positrons, however, have a different fate. In spite of the fact that they dissipate their kinetic energy just like beta particles through ionization and excitation, they cannot exist at rest in the vicinity of electrons. When a positron has been slowed sufficiently, it will be attracted to the opposite charge of an electron. When the electron and positron collide, they are both annihilated and an amount of energy equal to the sum of the particle masses is released in the form of

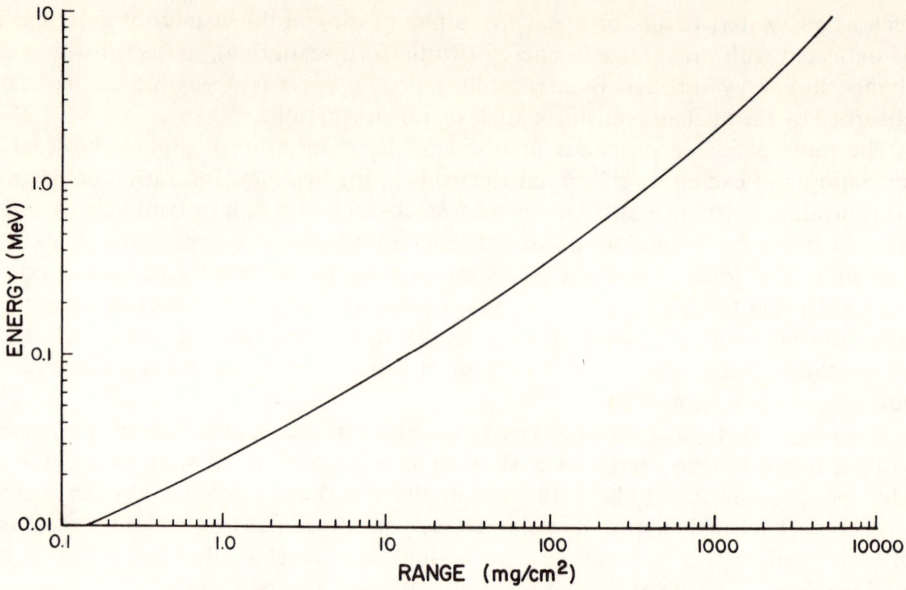

FIGURE 2. Range of beta particles of various energies in matter.

two photons. These photons are referred to as "annihilation radiations". Both annihilation photons carry an energy of 0.51 MeV, which is equivalent to the rest mass of the electron or the positron. Because of this phenomenon, 0.51 MeV photons often provide a convenient means for measurement of positron-emitting radionuclides.

Like alpha particles, betas have a characteristic range through matter that is dependent upon their initial kinetic energy. Beta particle range may be expressed as distance traveled in a certain medium. However, a more convenient way of expressing the range (r) of betas is the product of distance traveled (x) and the density of the absorbing medium (ρ)

$$r = x\rho \quad (31)$$

where r has the units of mg/cm² when x is expressed in cm and ρ in mg/cm³. Beta particle range expressed in this manner is fairly independent of the type of absorbing material, thus the range-energy plot (Figure 2) is reasonably applicable to most materials.

C. X- and Gamma Ray Interactions

The interaction of photons with matter involves several distinct processes. The relative importance and efficiency of each process is strongly dependent upon the energy of the photons and upon the density and atomic number of the absorbing medium. The authors shall first consider the general case of photon attenuation and then discuss some of the important processes separately.

1. Total Attenuation

Attenuation can be measured by placing various thicknesses of absorbing material in a narrow beam of photons that are emitted from a source and measured by a scintillation detection system (Figure 3). The intensity (I) of photons, measured by the counting rate of the detection system, can be determined without any absorber (I_o) and with various thicknesses of absorber (I_x). Note that the absorbing material may absorb

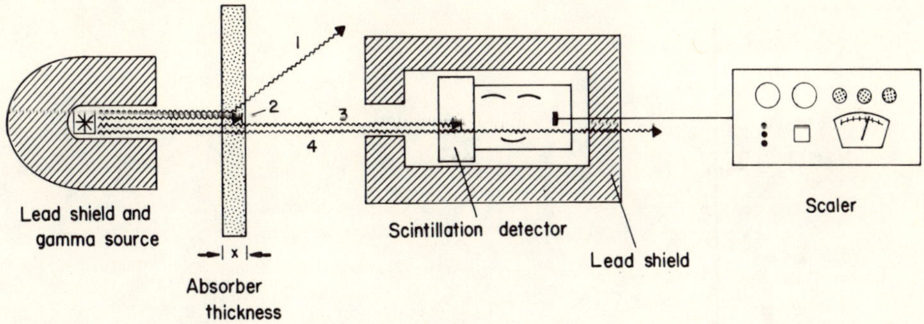

FIGURE 3. Apparatus used to measure photon attenuation by an absorbing material. Photons may be scattered (1) or absorbed (2) by the absorber. Photons not interacting with the absorber may be absorbed by the detector (3) or they may pass through the system without detection (4).

photons incident upon it (Case 2, Figure 3) or it may scatter photons in a new direction (Case 1, Figure 3). In both cases, photons that otherwise would have impinged upon the scintillation detector with a chance for absorption there, are prevented from doing so. Attenuation of a photon beam is the result of both absorption and scattering. A semilogarithmic plot of I_x vs. absorber thickness (x), is called an "attenuation curve" (Figure 4). The curve is exponential, providing that the photon beam is monoenergetic. The algebraic form of the curve is

$$\ell n\, I_x = \ell n\, I_o - \mu x \quad (32)$$

or

$$I_x = I_o\, e^{-\mu x} \quad (33)$$

where μ is called the "attenuation coefficient" and has the units cm^{-1}. The value of μ is dependent upon photon energy, and atomic number and density of the absorbing material.

Photon attenuation is the sum of several processes. The attenuation coefficient may be expressed as

$$\mu = \sigma_r + \sigma_a + \sigma_s + \tau + \kappa \quad (34)$$

where σ_r is the Rayleigh scattering coefficient, σ_a is the Compton absorption coefficient, σ_s is the Compton scattering coefficient, τ is the photoelectric absorption coefficient, and κ is the pair production coefficient. Each coefficient has the units of cm^{-1} and may be visualized as the probability of a specific type of photon interaction per centimeter of absorber traversed by the photon. Whereas σ_r and σ_s represent photon scattering processes, σ_a, τ, and κ denote photon absorption processes.

2. Rayleigh Scattering

When a photon encounters an orbital electron, it may or may not impart some energy to it. If the photon energy is less than the binding energy of the electron, the photon may be deflected with no energy transfer. This process is called Rayleigh scattering and is most probable for very low-energy photons.

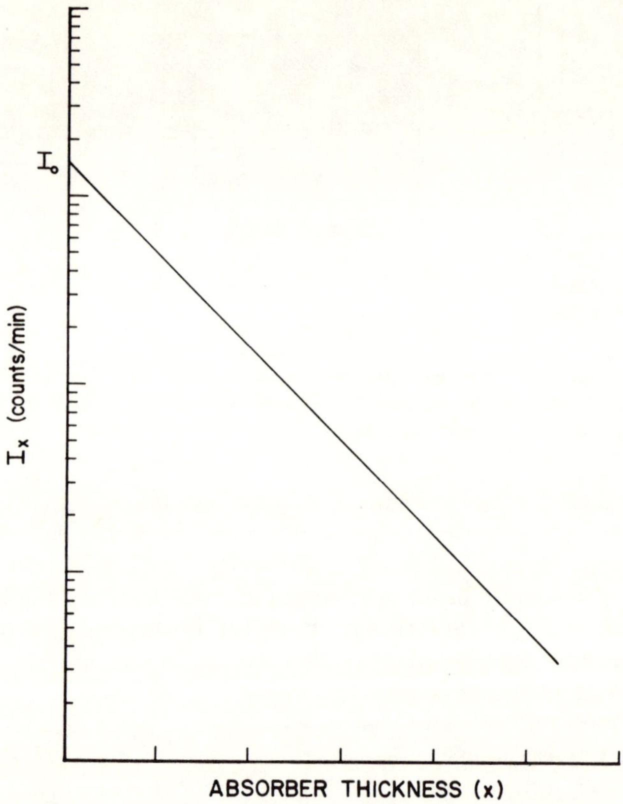

FIGURE 4. Attenuation curve for a beam of monoenergetic photons.

3. Compton Effect

The Compton effect is usually the predominant type of interaction for medium energy photons (0.3 to 3 MeV). In this process the photon interacts with an electron sufficiently to eject it from orbit, yet the photon retains a portion of its original energy and continues in a new direction. Thus, the Compton effect has an absorption component (σ_a) and a scattering component (σ_s). The amount of energy lost by the photon can be related to the angle at which the scattered photon travels relative to the original direction of travel.

The scattered photon will interact again, but since its energy has decreased, it becomes more probable that it will enter into a photoelectric or Rayleigh interaction. The free electron produced by the Compton process may be quite energetic and behave like a beta particle of similar energy, producing secondary ionization and excitation before coming to rest.

4. Photoelectric Absorption

The most probable fate of a photon having an energy slightly higher than the binding energy of an encountered electron is photoelectric absorption. In this process, the photon transfers all of its energy to the electron and its own existence terminates. The electron will escape its orbit with a kinetic energy equal to the difference between the photon energy and its own binding energy. Photoelectric absorption is most important

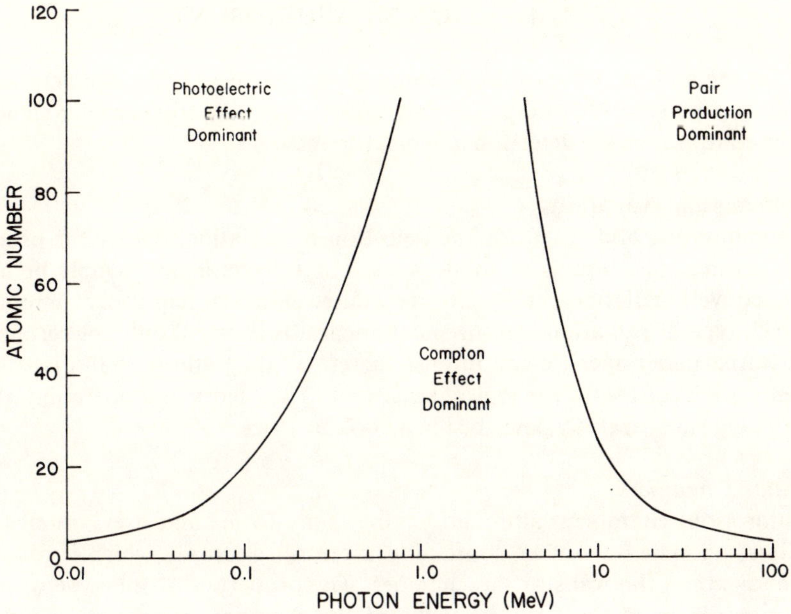

FIGURE 5. Effects of photon energy and atomic number of absorbing medium on the dominant type of photon attenuation processes.

for photons below 0.1 MeV if the absorbing medium is water or biological tissue. However, in high Z (atomic number) materials such as lead, this process is relatively important for photons up to about 1 MeV.

As with ionization produced by any process, secondary radiations are initiated, in this case, by the photoelectron which may have sufficient energy to produce additional ionization and excitation of orbital electrons. Also, filling of the vacancy left by the photoelectron results in characteristic X-rays.

5. Pair Production

Photons of energies greater than 1.02 MeV, under the influence of the electromagnetic field of a nucleus, may be converted into an electron and a positron. At least 1.02 MeV are required because the energy equivalent of the rest mass of the electron and positron is 0.51 MeV each. Pair production is not very probable, however, until the photon energy exceeds about 5 MeV. The available kinetic energy to be shared by the electron and the positron, is the photon energy minus 1.02 MeV, or that energy needed to create the pair. The probability of pair production increases with Z of the absorber and with the photon energy.

6. Relative Importance of Photon Attenuation Processes

The various processes of photon attenuation can now be considered by examining the effects of photon energy and atomic number of the absorber on their relative importance (Figure 5). The lines in the figure indicate the values of the photon energy and Z where the probability of occurrence of two major processes are equal. Photon attenuation coefficients for various materials are available from a number of sources.[5-7]

IV. RADIATION MEASUREMENT

Now that the authors have explored some of the properties and interactions of the commonly encountered radiations, they are now ready to briefly consider some of the methods used for radiation detection and measurement.

A. Instruments and Apparatus

Most common methods used for the detection of radiation involve the production of ionization in a gas, liquid, or solid. A radiation detector may simply be a device which can convert radiation energy into recordable electrical impulses or into a visible image. Each type of radiation measurement apparatus is useful only for certain radiations measured under specific conditions; therefore, the authors shall discuss which instruments are most useful for typical situations. The theory and operation of radiation instruments are carefully described in a book by Price.[8]

1. Ionization Chambers

Ionization (ion) chambers, although having many forms and uses, usually consist of a hollow, gas-filled chamber containing a central electrode which is placed at a higher voltage than the walls of the chamber. The properties of this type of detector depend primarily upon the size and shape of the chamber, the amount of electrical potential applied between the chamber and the central electrode, and the type and amount of gas within the chamber. In principle, ionizing radiation which passes through the chamber may produce ion pairs within the gas. If the electric field is sufficiently strong, electrons will migrate toward the positive central electrode and cations will drift toward the negative chamber walls. This process either reduces the voltage between the chamber and the electrode or, if the electrode and chamber are connected with an appropriate circuit, generates a current.

Pocket dosimeters carried by radiation workers are frequently ion chambers (Figure 6). They are charged by an external battery, and if the chamber is exposed to any type of ionizing radiation that can penetrate its walls, the initial charge is reduced. The amount of reduction in charge is a measure of the radiation received by the dosimeter. Radiation exposure can be related to voltage change through calibration or calculation. The voltage of a pocket dosimeter may be measured with an external meter or by direct reading of those dosimeters equipped with an internal indicating fiber.

Most ionization chambers used for careful measurement of radiation fields are connected to an electronic circuit which maintains a certain voltage across the chamber and measures the current produced in the chamber. The current flow is proportional to the rate of ionization in the chamber, which in turn is proportional to the radiation field intensity. For precise external background radiation measurements, large ionization chambers with electronic circuits designed to accurately measure very small currents may be used (Figure 7).

The detection of X- and gamma radiation can usually be accomplished easily with ion chambers since photons can readily penetrate the chamber walls. However, very low quantities of gamma radiation may be difficult to measure because photons may pass completely through the ion chamber without causing ionization. This problem can be alleviated by using a large-volume ionization chamber filled with gas under pressure. Alternatively, a more efficient solid scintillation detector may be used to measure low-intensity X- or gamma ray fields.

Unless an ion chamber has thin walls, it is not suitable for measurement of alpha or beta particles; however, the problem of absorption of alphas or betas by the chamber walls can be circumvented in the laboratory by placing the radioactive sample within the chamber itself. The radioactive sample may be prepared on a small metal

FIGURE 6. Dr. Les Fraley checks a ^{137}Cs source used to irradiate a natural grassland area to be sure that the source is safely within its shield. Note the use of an ion chamber pocket dosimeter and film badge to record radiation exposure. The hand-held instrument is a portable Geiger-Mueller survey meter.

pan (planchet) which can be inserted into the ion chamber or, in some cases, the sample can be vaporized and mixed with the gas in the chamber. Carbon-14, which emits a weak beta particle, for example, is counted very efficiently by combusting samples, collecting the CO_2, and mixing the ^{14}C-containing CO_2 with the counting gas. Nearly every beta particle emitted within the gas would produce ionization, which in turn would produce a pulse for counting.

2. Proportional and Geiger-Mueller Detectors

Proportional detectors are ionization chambers used primarily for counting alpha-

FIGURE 7. Jerry Martin records the natural background exposure rate on Loveland Pass, Colo., using a large, pressurized ionization chamber and a sensitive electrometer.

or beta-emitting samples in the laboratory. Ordinarily, the sample is placed within the ionization chamber on a metal planchet and a continuous supply of gas is provided for the chamber since the chamber is not permanently sealed. These detectors are termed ''proportional'' because they operate in a voltage range where the magnitude of the electrical impulse produced by ionization within the chamber is proportional to the amount of ionization and also the voltage applied across the chamber. Advantages of this type of detector are that alpha and weak beta particles can be counted since they originate inside the chamber and particles of various energies can be distinguished, enabling simultaneous measurement of more than one radionuclide.

The Geiger-Mueller (G-M) detector usually consists of a sealed, gas-filled chamber, fitted with a thin window to permit entry of beta particles from outside the chamber. These detectors are operated at considerably higher potentials than ionization chambers or proportional detectors. Potentials across G-M detectors are great enough that any amount of ionization produced within the chamber leads to some upper limit of electrons which are collected at the anode to form an electrical impulse that can be registered by an electronic scaler or counting device. G-M counting systems are comparatively simple, reliable, and inexpensive. They frequently are used as portable survey instruments (Figure 6). They are most useful for detection of beta particles, but they are also used for detection of X- and gamma radiation, although their efficiency for photons is low. A disadvantage of G-M detectors is that different radiation energies cannot be distinguished.

3. Scintillation Detectors

Scintillation detectors operate on the principle that ionization and excitation of certain substances (fluors) lead to excited electron states and the subsequent return of excited electrons to normal states is accompanied by the emission of light. The light flashes or "scintillations" from a fluor can be converted to countable electrical pulses with appropriate equipment and electronic circuitry.

Scintillation detectors have certain advantages over other types of detectors. Since the fluors commonly used are either in the solid or liquid state, they are much denser and have a higher detection efficiency than gases. Furthermore, with the proper electronic equipment, the output pulse from the scintillation detector is proportional to the energy dissipated in the detector. Therefore, the scintillation counter can be used to determine the amount and energy of various kinds of radiations, even in very small quantities. Scintillation detectors of the solid type are usually crystals of anthracene or naphthalene for the measurement of beta particles, and sodium iodide (NaI) activated with thallium for the measurement of photons.

With the exception of a few radionuclides such as ^{3}H, ^{14}C, ^{32}P, ^{35}S, ^{45}Ca, and ^{90}Sr, most radioactive substances of ecological importance emit X- or gamma rays which can be measured easily with solid scintillation detectors. Gamma ray-emitting nuclides can usually be measured in living or nonliving matter with NaI detectors without the chemical or physical separations that are often necessary for measurement of alpha or beta particles (Figure 8). A major advantage of many solid scintillation counting systems is that with electronic equipment capable of storing impulses from the detector according to energy,* several gamma-emitting radionuclides in the same sample can be simultaneously identified and quantified.

Application of the scintillation principle to counting of alpha and beta particles may be accomplished by dissolving or dispersing the sample into a liquid containing a scintillator.[9] A commonly used liquid scintillator consists of toluene (the solvent), 2,5-diphenyloxazole (PPO, the primary fluor), and 2,2′-paraphenylene-*bis*-5-phenyloxazole (POPOP, the secondary fluor). The function of PPO is to convert the energy of the radiations into light through fluorescence and POPOP is a substance which converts the light from PPO to a longer wavelength which more readily activates the photomultiplier tube, a device which converts light into countable electrical impulses. Major advantages of liquid scintillation counters are that alpha- and beta-emitting radionuclides can be measured with high efficiency and adequate energy resolution and that such counters are usually equipped with automatic sample changers to permit unattended counting of many individual samples. These counting systems are relatively expensive, however.

4. Semiconductor Detectors

In a semiconductor radiation detector, an electric field is established across a solid material having semiconductor properties. Ionizing events caused within the semiconductor by a radiation will produce "free" electrons and electron vacancies, or "holes". The electrons and the holes are separated by the electric field, collected at the electrodes, and produce a pulse which can be amplified, sorted, and recorded.

Within the last 2 decades, the state-of-the-art of applying semiconductor detectors to radionuclide measurement has shown steady improvement and increased usage.

* Equipment having such capability is referred to as a "multichannel pulse-height analyzer".

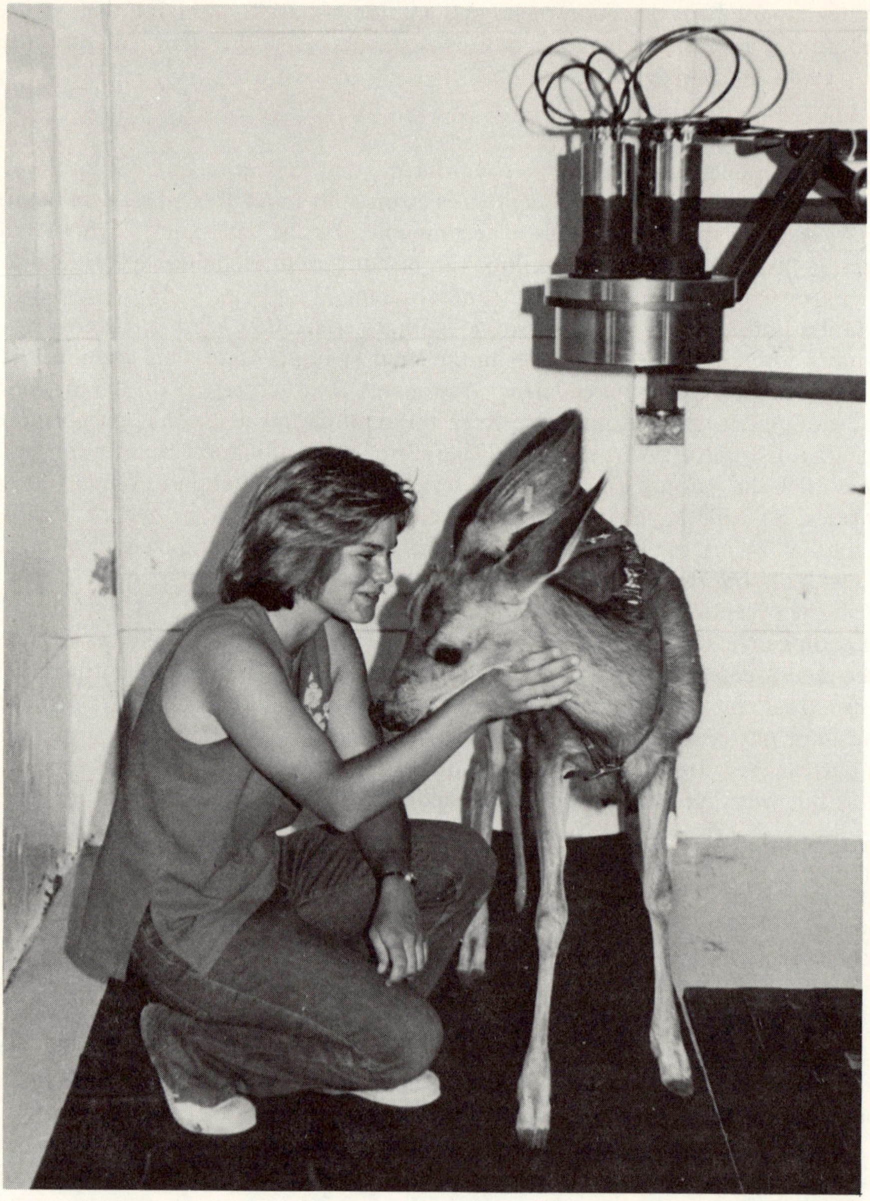

FIGURE 8. Graduate student Lynn Alexander positions a tame mule deer fawn beneath a large sodium-iodide scintillation detector. Such detectors are useful for in vivo measurement of gamma-emitting radionuclides.

These detectors, particularly lithium-drifted germanium [Ge(Li)] and silicon [Si(Li)] diodes, appear to be particularly useful in the measurement of complex gamma-ray spectra often found in environmental samples.[10] The primary advantage of these detectors is the extremely high gamma ray energy resolution that can be obtained as compared to NaI detectors. However, these detectors are less efficient than NaI because of smaller size and lower density.

5. Photographic Emulsions

Photographic emulsions, in their various forms, provide investigators with another means of measuring radiation.[11] For certain types of applications, emulsions may be far superior to the electronic devices which the authors have considered so far; often emulsions provide the only suitable means of radionuclide detection consistent with the needs of the experiment.

A photographic emulsion consists of silver halide (usually AgBr) crystals suspended in a medium of gelatin and various other ingredients of secondary importance. The emulsion may be spread thinly over a celluloid base to form a semirigid film, mounted on glass plates, or handled as a gel. When the emulsion is exposed to radiation, some of the AgBr crystals become sensitized through ionization and a "latent image" forms on the film.[12] The latent image consists of crystals that will form visible specks of metallic silver upon development with certain chemicals. The degree of blackening of an emulsion depends upon the number of AgBr crystals that have been sensitized, which in turn depends upon the amount of radiation absorbed by the crystals. Emulsions are sensitive to visible light, as well as X- and gamma rays, alpha particles, and beta particles.

The developed emulsion which has been prepared for viewing and storage is called a "radiograph". Medical radiographs used for diagnostic purposes are prepared by exposing a patient to a point source of X-rays and recording the rays which pass through the patient on a sheet of film. The differential density and chemical composition of internal structures of the patient cause differential absorption of the X-rays which forms an image of such structures on the film. The ecologist is likewise able to study internal structure (shape, density, and composition) of biotic and abiotic components of natural ecosystems using radiographic techniques.

A radiograph which is exposed from radiations emanating from the object itself is called an "autoradiograph". Autoradiographs are extremely useful for studying the uptake and distribution of radionuclides or radiolabeled substances in cells, organs, and organisms. In principle, the object is prepared so that it can be placed in close proximity to the film and held there long enough for sufficient exposure of the emulsion. Regions of the object containing sufficient radioactivity will project a blackening on the nearest part of the film.

6. Thermoluminescent Detectors

In recent years, thermoluminescent detectors (TLD) have been applied to many problems in radiation ecology, especially dosimetry. Their main advantages are small size, sensitivity, ruggedness, ability to measure various types of radiations, and the ability to be reused. Thermoluminescent detectors work on the principle that radiation raises electrons in certain materials to higher energy states.[13] The electrons remain "frozen" in the excited states until the TLD is heated. As the temperature increases, the electrons return to their normal states and light is emitted. The amount of light emitted is proportional to the radiation dose received by the TLD. Although several substances are suitable as TLD's, lithium fluoride (LiF) is most widely used because its effective Z and radiation absorption characteristics are similar to biological tissues. The small size and ruggedness of the TLD makes it particularly suitable for insertion into the tissues of living organisms and for general field use.[14] TLDs must be calibrated by exposure to known radiation fields. Since LiF can be impregnated into thin disks of Teflon® that require essentially no protective covering, these dosimeters are extremely useful for alpha and beta as well as gamma ray dosimetry. LiF dosimeters are also sensitive to neutrons, thus useful for their measurement.

B. Radiation Counting

Interpretation of many data in radioecology, as well as the design and execution radiological surveys or tracer experiments, requires understanding of some basic principles of radiation counting and sampling. This section provides a brief survey of some of these principles.

1. Sample Collection and Preparation

Some general remarks concerning sample collection and preparation are warranted because the degree of success in obtaining meaningful radiological data is dependent upon the way in which samples have been obtained and handled. In terms of sample collection, for example, suppose that a research objective is to measure the ^{226}Ra concentration in a plant community that has invaded an abandoned uranium mill tailings pile. The investigator must sample vegetation from the plant community and assay the samples for ^{226}Ra. However, questions such as where to sample, how many samples, how much volume per sample, and how to handle the samples after collection must be addressed if credible data are to be obtained.

The question of where to sample is not necessarily unique to radionuclides and will involve the size and shape of the area to which inferences will be made. It may also require some prior knowledge of gradients or spatial patterns of ^{226}Ra in the area of concern. A normal practice is to impose a grid system over the area of concern and to employ an unbiased method of choosing specific sampling locations. The overall objective is to sample from a set of locations which will give an unbiased estimate of the average ^{226}Ra concentrations over the entire plot, as well as a measure of the variability in the concentrations.

The question of number of samples may be difficult to answer without some prior knowledge about the variability to be expected and the precision desired. The greater the spatial variability, the greater the number of independent samples needed to determine the mean ^{226}Ra concentration with a specified degree of confidence. Frequently, the limiting factor with respect to the number of samples is the budget for radioanalytical work. A common practice in answering the question of sample number is to conduct a pilot survey, in which a relatively small number of samples is randomly collected and analyzed, to get a preliminary estimate of the mean and variance. Then, based upon statistical, budgetary, and other considerations, a more appropriate sampling scheme can be designed, likely with the help of a knowledgeable statistician.

The decision of sample volume to be taken is based mainly upon the expected concentration of the radionuclide sought. It will be seen later in this section that the precision with which the quantity of radionuclide in a sample can be measured increases with the activity present. Thus, it is highly desirable to collect a sufficient volume of material so that the activity contained within it is large enough for an adequate measurement. Of course, the larger each sample becomes, the more time and effort is required to collect and process it. Therefore, some practical compromise between measurement precision and cost or effort must be made.

Major concerns with regard to handling of samples that contain radioactivity are safety, preservation of sample integrity, and prevention of cross-contamination. Most radioecological investigations should be evaluated by someone qualified in health physics to assure that the amounts of radioactivity involved, as well as handling procedures, are adequately considered to assure the safety of the investigator. Samples should be handled so as to prevent deterioration of the sample itself and to prevent the escape of radioactivity from the sample. If the radionuclide sought is short-lived, time becomes an important consideration. Samples should always be packaged carefully to prevent cross-contamination between samples, or between samples and laboratory or field equipment. This usually requires at least two barriers (double-containment).

Specific sample preparation techniques for radionuclide assay are too numerous, and usually too involved, to discuss in detail here, but a few general considerations are offered. Preparation of a sample for radiation counting varies with the sample matrix (soil, water, vegetation, bone, etc.) and with the radionuclide (chemistry, radiations emitted, etc.). If the sample is to be assayed for a radionuclide which emits only alpha or beta particles, some sort of sample volume reduction is necessary because these radiations are not sufficiently penetrating to get out of most types of environmental samples. This problem is termed "sample self-absorption." If the sample contains a reasonably small mass of inorganic mineral, such as water or soft biological tissue, it may be adequate to evaporate or ash the material to produce a small quantity of residue for counting. A reasonably large fraction of the more energetic beta particles (say E_{max} >0.1 Mev) could emanate from a few milligrams and be adequately counted with a G-M or proportional gas-flow detector. On the other hand, this may not be an adequate procedure for efficient measurement of alpha or weaker beta particles. The procedure may also be inadequate if radionuclides in addition to the one sought are present. In this case, chemical separation of the element of interest may be necessary prior to counting, unless some other means can be used to account for their presence.

The case of radionuclides which emit photons of energies greater than 100 keV is usually simpler in terms of sample preparation, since the photons emanate from the sample and can be detected and counted with a solid-scintillation detector, such as a NaI crystal, or a semiconductor detector such as a Ge(Li) diode. As long as a sufficient number of photons emanate from the sample per unit time, and the sample geometry is consistent with respect to the detector, quantitative determination of the photon emitter is feasible without sample volume reduction and without chemical or physical separation of the radionuclide. In addition, if a multichannel pulse-height analyzer is available, a photon-emitting radionuclide can usually be measured in the presence of other radionuclides, if the detection system has adequate energy resolution.

A general summary of sample preparation methods and radiation detection systems in relation to type of radiation and environmental matrix is given in Table 1. Note that this is a very general guide and there are many special circumstances which would require more elaborate sample preparation and specific counting systems. Common problems that may require special procedures include low activity and presence of interfering radionuclides.

As alluded to earlier, some samples may contain a mixture of two or more different radionuclides. In many cases in radioecology, it is necessary to determine the kinds and amounts of the radionuclides in a sample. Identification of individual radionuclides can usually be accomplished by determination of the types and energies of their radiations and half lives.[3] Sometimes additional work to elucidate the chemical properties of the radionuclides is necessary. Determination of the types and energies of radiations usually involves absorption criteria and spectroscopy.[10,15] Direct half-life determinations through repeated counting are feasible for half-lives in the range of minutes to a few years. Half-lives of extremely long-lived radionuclides can be estimated from their measured specific activity (Equation 30).

2. Counting Yield

Radiation counting systems used to quantitatively assay the amount or activity of a radionuclide in a prepared sample normally record electrical impulses from the detector for some period of time. The number of electrical impulses recorded per unit time is called the "count rate". The count rate is related but seldom equal to the nuclear disintegration rate of the sample because only a certain fraction of the disintegrations produce radiations which activate the detector. The ratio of count rate (R^*) to disintegration rate or activity (A^*) is the counting yield (Y)

$$Y = \frac{R^*}{A^*} \tag{35}$$

Table 1
TYPICAL SAMPLE PREPARATION METHODS AND RADIATION DETECTION SYSTEMS USED FOR ASSAY OF RADIONUCLIDES IN ENVIRONMENTAL MATRICES

Principle radiation of radionuclide	Type of sample matrix	Typical minimum degree of sample preparation[a]	Recommended radiation detection systems[b]
Alpha	Water	Evaporation	GFP, LS
	Soil	Chemical separation	GFP, SB, LS
	Soft tissues	Ashing	GFP, LS
	Bone	Chemical separation	GFP, SB, LS
Beta ($E_{max} < 0.1$ MeV)	Water	Evaporation	GFP, LS
	Soil	Chemical separation	GFP, LS
	Soft tissues	Ashing	GFP, LS
	Bone	Chemical separation	GFP, LS
Beta ($E_{max} > 0.1$ MeV)	Water	Evaporation	GFP, LS, G-M
	Soil	Ashing or chemical separation	GFP, LS, G-M
	Soft tissues	Ashing	GFP, LS, G-M
	Bone	Chemical separation	GFP, LS, G-M
Photon (E < 0.1 MeV)	Water	Evaporation	NaI, Ge(Li)
	Soil	Uniform geometry	NaI, Ge(Li)
	Soft tissue	Ashing	NaI, Ge(Li)
	Bone	Chemical separation	NaI, Ge(Li)
Photon (E > 0.1 MeV)	All types	Uniform geometry	NaI, Ge(Li)

[a] Special circumstances such as the presence of other radionuclides may require more elaborate sample preparation.

[b] The relative merits of the alternative detection systems vary with type of sample preparation, presence of other radionuclides, activity present, and other factors. Legend: GFP = gas flow proportional; LS = liquid scintillation; SB = surface barrier (α spectrometry); G-M = Geiger-Mueller; NaI = sodium iodide crystal; and Ge(Li) = Germanium (Lithium) diode.

where Y has the units of counts/disintegration (c/d). A major objective of most sample preparation procedures is to arrange the sample such that Y is maximized to produce the highest counting rate feasible. The higher the counting rate the more precision is inherent in the measurement for a fixed counting time, as will be shown in the subsequent section. In order to maximize Y, it is essential to understand the factors affecting it. A relationship which can be applied to most types of radiations and counting systems is

$$Y = \frac{f_r f_s f_g f_w f_e}{f_b} \tag{36}$$

where f_r = fraction of disintegrations that produce detectable radiations, f_s = fraction of detectable radiations that emanate from the sample, f_g = fraction of radiations emanating from the sample that travel toward the detector, f_w = fraction of radiations traveling toward the detector that reach its sensitive volume, f_e = fraction of radiations reaching sensitive volume which produce a count, f_b = ratio of count rate without backscatter to count rate with backscatter.

The value of f_r can be less than or greater than 1.0 and can be determined for the radionuclide of interest from handbooks.[6,7] The value of f_s, termed the "sample self-absorption factor", increases as the sample mass decreases and as the radiation energy increases. The geometry factor, f_g, is governed mainly by the distance between the sample and detector, and the relative volumes and configurations of the sample and detector. In the case of G-M, solid-scintillation, and semiconductor detectors, f_g is maximized by reducing sample mass as much as possible and by placing it as close as possible to the detector. Also, f_g increases with the size of the detector. In liquid-scintillation detectors, f_g approaches 1.0, since the sample is mixed throughout the detector. Attenuation of radiations between the sample and the sensitive volume of the detector is expressed by f_w. This factor is usually more important for alpha and beta particles than for photons. The detector efficiency for the radiation of interest is denoted by f_e, which increases with the interaction probability between the radiation and detector. Radiations which emanate from the sample that are initially directed away from the detector, but due to scattering from surrounding material find their way into the detector to produce a count, affect f_b. This process is mainly of concern in counting beta particles with a G-M detector, but it may also occur with photons.

Ordinarily, the various factors which affect counting yield are not quantitatively determined individually, even though they are considered when setting up a counting system. In the process of standardization of a counting procedure with a sample containing a known activity, the value of Y is measured directly. Standardization usually involves placement of a specified quantity of radionuclide into the sample matrix in such a way that it is uniformly distributed and the matrix is in the same form and configuration as subsequent samples containing unknown amounts of activity. The count rate from the standardized sample is then measured and Y is calculated (Equation 35). Accurately calibrated radionuclide standards can be obtained from commercial laboratories and from the U.S. National Bureau of Standards.

3. Counting Statistics

Radioactive decay is a probabilistic or random process. Consequently, the instantaneous disintegration rate of a radioactive sample fluctuates through time and repeated counts made on the same sample for the same length of time will vary. This introduces a source of error (called the "counting error") into each measurement, which must be quantified if counting data are to be meaningful. This section provides but a capsule summary of the subject since it is treated in detail elsewhere.[1,2]

Frequency distributions of counting data closely approximate the Poisson distribution and the standard deviation of the distribution of counts recorded may be estimated by

$$s = \sqrt{c} \tag{37}$$

where s = the estimated standard deviation of c counts recorded. The number of recorded counts c is an estimate of the "true" number of counts that could be expected to be recorded. It can be shown that the "true" number of counts will have a 68% chance of lying within the interval c ± s. It will have a 95 and 99% chance of falling within the intervals c ± 2 s and c ± 3 s, respectively.

Somewhat more useful is the standard deviation of a count rate, $s_R{}^*$, since counting data are usually expressed in terms of recorded count rate (R*) in c/time. The expression is

$$s_R{}^* = \sqrt{\frac{R^*}{t}} \tag{38}$$

where t is the counting time and R^* and t must employ the same time units (minutes, seconds, etc.). Notice in this case that $s_R{}^*$ may be reduced by increasing the counting time t. In this case, the "true" counting rate has a 68% chance of falling within the interval $R^* \pm s_R{}^*$ and a 95% chance of falling within $R^* \pm 2\, s_R{}^*$.

Finally, one must be concerned about the standard deviation of an observed net sample counting rate. The net counting rate R_s^* is given by

$$R_s^* = R_{s+b}^* - R_b^* \tag{39}$$

where R_{s+b}^* = gross count rate (includes sample and background) and R_b^* = background count rate. The standard deviation of R_s^* is

$$s_s = \sqrt{\frac{R_{s+b}^*}{t_{s+b}} + \frac{R_b^*}{t_b}} \tag{40}$$

where s_s = estimated standard deviation of the observed net count rate R_s^*, t_{s+b} = time of recording of the gross count rate R_{s+b}^*, and t_b = time of recording of the background count rate R_b^*. Again all time units involved in Equation 40 must be the same.

To illustrate the use of Equation 40, suppose that 105 counts were recorded in 5 min with no sample present, then with a sample present, 324 counts are recorded in 3 min. Calculation of the net sample counting rate and its standard deviation is as follows

$$R_b^* = \frac{105\ \text{c}}{5\ \text{min}} = 21\ \text{c/min}$$

$$R_{s+b}^* = \frac{324\ \text{c}}{3\ \text{min}} = 108\ \text{c/min}$$

then

$$R_s^* = 108\ \text{c/min} - 21\ \text{c/min} = 87\ \text{c/min}$$

and

$$s_s = \sqrt{\frac{108\ \text{c/min}}{3\ \text{min}} + \frac{21\ \text{c/min}}{5\ \text{min}}} = \pm\ 6.3\ \text{c/min}$$

Therefore, the "true" net sample counting rate has a 68% probability of being within the interval 87 − 6.3 c/min and 87 + 6.3 c/min. It has a 95% probability of being within the interval 87 − 12.6 c/min and 87 + 12.6 c/min.

The foregoing equations enable the investigator to evaluate the counting error relative to other sources of error, such as that which results from sampling, weighing, chemical processing, etc. Ordinarily, the counting error will be comparatively small if $s_s/R_s^* < 0.1$. If the relative counting error is larger than desired, one can increase the counting time, count larger samples, try to improve the counting yield, try to reduce the background count rate, and any combination of these efforts. A large relative counting error (defined as s_s/R_s^*) may be unavoidable in the case of very low activity samples.

C. Radiation Dosimetry

An ultimate concern in many radioecological investigations is the biological effects that might result from radiation in the environment. The most common "yardstick" used for relating the amount of radiation to specific biological effects is "radiation dose", which is a measure of the amount of radiation energy that is dissipated in a given volume of living tissue. This section represents our attempt to present, in a simplified way, the concept of radiation dose, as well as methods used for measurement or calculation of dose. As with the other sections in this chapter, the reader is again referred to other works for a more complete treatment of the subject.[4,16-18]

1. Concept of Radiation Dose and Exposure

Radiation-induced damage to biological tissue results from the absorption of energy in or around the tissue. The amount of energy absorbed in a given volume of tissue is related to the type, energy, and number of radiations traversing the tissue volume, and the interactions which occur between the radiations and the atoms and molecules of the tissue. The fundamental unit of radiation absorbed dose is the rad, defined* as

$$1 \text{ rad} = \frac{100 \text{ erg (absorbed)}}{1 \text{ g material}} \tag{41}$$

The definition of the rad states that 100 erg of energy are absorbed or dissipated in a gram of tissue or other substance. While this definition is simple in concept, its measurement or calculation is not.

Frequently, it is not convenient to measure dose in rads. In the particular case of irradiation of organisms or tissues from an external X- or gamma ray source, it is more convenient to measure the radiation exposure at the surface of the tissue of interest using an appropriate instrument or device. The dose to the tissues can be estimated from the measured exposure from theoretical considerations. The fundamental unit of radiation exposure, which is applicable only to X- or gamma radiation, is the roentgen (R), which is defined as

$$1 \text{ R} = 2.58 \times 10^{-4} \text{ coul/kg air} \tag{42}$$

This definition of the roentgen is equivalent to the production by X- or gamma rays of 1 electrostatic unit of charge of either sign per cubic centimeter of dry air at 0°C and 760 mm mercury. The roentgen quantity is defined on the basis of production of ionization in air because it is normally measured with an air-filled ionization chamber.[4]

The careful measurement of W, the amount of energy absorbed in air per ion pair produced, enables one to relate the roentgen unit to the rad unit for air. The quantity W is 33.7 eV per ion pair for X- and gamma rays in air. Therefore,

$$1 \text{ R} = \frac{(2.58 \times 10^{-4} \text{ coul/kg})\ (33.7 \text{ eV/ion pair})\ (1.6 \times 10^{-12} \text{ erg/eV})}{(1.6 \times 10^{-19} \text{ coul/ion pair})\ (1000 \text{ g/kg})}$$

* In the International System of Units, the absorbed dose unit is the grey (Gy), which is equal to 1 J/kg or 100 rad. Since this unit is in the process of replacing the rad, both are encountered in the literature and it is, therefore, important to be familiar with each.

$$= 87 \text{ erg/g}$$

$$= 0.87 \text{ rad in air} \quad (43)$$

Consideration of Equation 43 and the absorption of X- and gamma radiation by materials in addition to air enables us to relate exposure measured in air to dose absorbed by tissue or other medium.

For any medium, the dose to the medium D_{med} in rads is

$$D_{med} = D_{air} \frac{(\mu_{en}/\rho)_{med}}{(\mu_{en}/\rho)_{air}} = 0.87 \text{ R} \frac{(\mu_{en}/\rho)_{med}}{(\mu_{en}/\rho)_{air}} \quad (44)$$

where D_{air} = 0.87 R, R = exposure in air in roentgens, μ_{en} = energy absorption coefficients for the medium or air ($\mu_{en} \approx \tau + \sigma_a + K$), and ρ = density of the medium or air. Values of μ_{en} depend upon photon energy and "effective atomic number" of the medium in question and may be obtained from several sources.[4-6] The relationship between dose (rads) and exposure (roentgens), calculated with Equation 44 as a function of photon energy for various kinds of tissue, is shown in Figure 9.

Biologic damage from radiation is dependent mainly upon the quantity of energy deposited in tissue. However, it is also dependent upon the microdistribution of energy dissipation. Because of the latter dependence, different types of radiation elicit varying biologic responses even though the energy absorbed per gram of tissue is equivalent. The relative biological effectiveness for a given type of radiation is termed the "RBE". The RBE is determined from experimentation and

$$\text{RBE} = \frac{\text{dose (rad) of medium energy X-rays to produce a given effect}}{\text{dose (rad) of radiation in question to produce the same effect}} \quad (45)$$

The effective radiation dose in rems may be expressed in terms of the dose in rads and the RBE

$$\text{Equivalent RBE dose (rem)} = \text{D (rad)} \times \text{RBE} \quad (46)$$

The concept of RBE dose is now limited to studies in radiation biology. A factor related to the RBE, but which has preferred usage in the field of radiation protection is the quality factor (QF). The quality factor varies with the density of energy transfer per length of track of the radiation. This quantity is termed the "linear energy transfer" (LET), which is usually expressed in terms of kilo electronvolts per micrometer of track length. The QF is used to express dose equivalent (DE), also in units of rems:

$$\text{DE (rem)} = \text{D (rad)} \times \text{QF} \quad (47)$$

Table 2 shows the dependence of QF on LET, as well as the QF for various types of radiation. Dose, when used for purposes of radiation protection, is usually expressed in units of rems. On the other hand, basic investigations in radiation biology and radioecology most frequently use the more fundamental unit of absorbed dose, the rad.

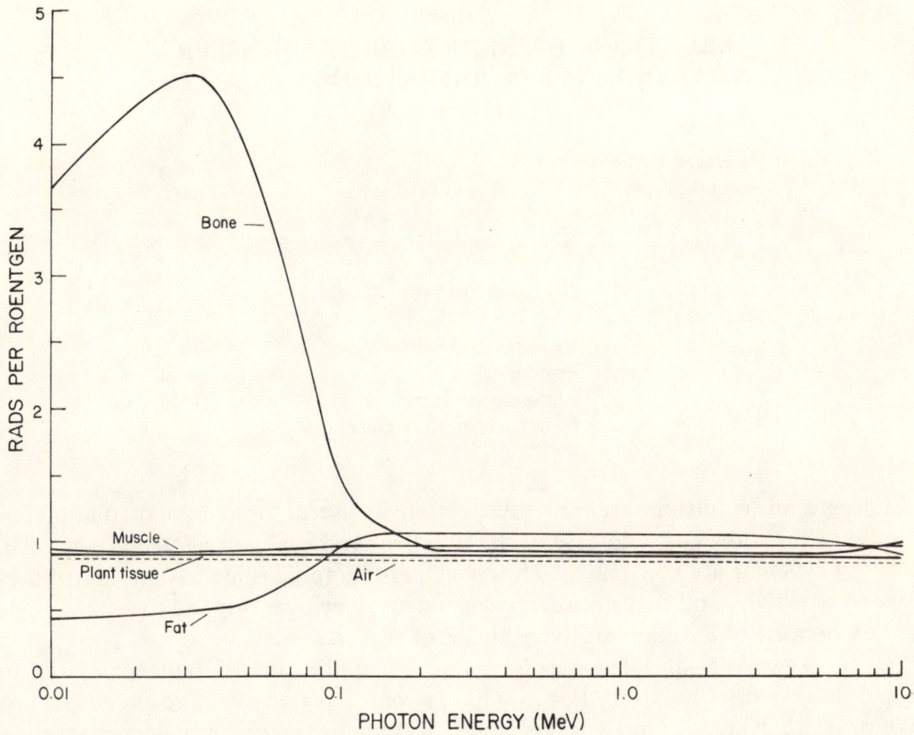

FIGURE 9. Relationship of absorbed dose in rads to exposure in roentgens as a function of photon energy for various absorbing materials.

2. Dosimeters

The more common types of dosimeters used in radioecology include ionization chambers, thermoluminescent detectors, and photographic emulsions. Over the years, ionization chambers have probably received the greatest usage in measurement of dose from external X- or gamma ray sources. Instruments such as the Victoreen "condenser R-meter"[4] and the Landsverk roentgen meter are accurate, sensitive, and simple to use. Under the appropriate conditions, exposures measured in roentgens by these instruments can be converted to absorbed dose in rads using the principles embodied in Equation 44. Perhaps the main disadvantages of the use of ion chambers in field dosimetry are that they are generally too large to insert into plant or animal tissues and some are sensitive to climatological changes. Further, the cost of individual chambers precludes numerous simultaneous dose measurements within an exposure field. For measurement of extremely low exposures, such as from natural background radiation, large ion chambers coupled to sensitive electrometers are practical.[19]

Thermoluminescent dosimeters probably have the greatest overall utility and versatility in radioecology. They are very practical for mapping radiation fields because individual detectors can be placed at many locations for simultaneous measurement of dose. These detectors are relatively inexpensive, small, and are usually not affected by weather conditions. Lithium fluoride dosimeters, having a low effective atomic number, are energy independent, that is, they absorb radiation over a wide energy range at an efficiency similar to water or most soft biological tissues. In addition, TLDs are usually accurate over a large dose range. When embedded in a suitable material such as Teflon®, TLDs may be surgically placed within tissues for the measurement of internal dose.

Table 2
RELATION OF LINEAR ENERGY TRANSFER (LET) AND TYPE OF RADIATION TO QUALITY FACTOR (QF)

Average LET in water (keV/μm)	Type of radiation	Quality factor
3.5 or less	X-rays, gamma rays, beta particles	1
7		2
	Neutrons (<10 keV)	3
23		5
53	Neutrons (>10 keV)	10
	Protons	1—10
	Alpha particles	1—20
175	Fission fragments, recoil nuclei	20

Photographic emulsions are sometimes extremely useful for mapping radiation fields because of their low cost and ease of development and evaluation. Dental-sized X-ray films in particular are convenient. However, photographic films have a limited useful range of sensitivity and they are energy dependent (they over respond to photons below 200 keV because of the comparatively high Z of silver).

Several other types of dosimeters have possible applications in radioecological studies and deserve brief mention. For example, silver-activated metaphosphate glass rods have proven advantages in field dosimetry, so long as their energy dependence is accounted for.[20] Portable detectors, especially scintillation detectors, are useful in obtaining comparative external exposure rates very rapidly. These instruments can be quite sensitive to photons and can be read instantaneously; however, they are usually not adequate for accurate dose measurement.

3. Calculation of Dose

In many situations, radiation dose may be calculated from theoretical considerations. This is true when the type and amount of radioactivity is known, as well as its spatial distribution with respect to the tissues of interest. Theoretical considerations involve the types, energies and distributions of the radiations and their interactions with matter. Such theoretical calculations can be applied to external radiation sources or fields, as well as to internally deposited radionuclides. Since this subject is rather complex, only some of the simpler calculation possibilities are pointed out here. Actual calculations will require additional reference material.[4,6,16,18]

Radiations of external origin that may be absorbed in living tissues include natural or man-generated rays from the atmosphere, earth, water, and other tissues or organisms. Also, objects such as television sets, X-ray machines, and sealed gamma ray sources emit radiations which may be absorbed by tissues. The type of external exposure that is probably the simplest to evaluate is that resulting from a "point source" of photons. Photon emitters have a gamma ray exposure constant, termed I_γ, which has the units of roentgen per hour per curie at an air distance of 1 m. The value of I_γ may be calculated for any gamma-emitting radionuclide and it is dependent upon the number of photons emitted per disintegration and the energies of the photons. Values of I_γ are, for example, 0.33 and 1.32 R/hr/Ci for ^{137}Cs and ^{60}Co, respectively. Values for other radionuclides are tabulated elsewhere.[6] The exposure rate (E) in roentgen per hour at any distance (d) in meters can be calculated for a point source containing A* curies by

$$E = \frac{I_\gamma A^*}{d^2} \tag{48}$$

Nonpoint sources and cases in which several gamma emitters are present are more difficult to evaluate.[18] The case in which the radionuclide is distributed within a mass of substance is even more difficult to evaluate because of absorption and scattering of photons prior to their exit from the substance. If such complexities are great enough, direct measurement of exposure rate may be necessary.

Dose evaluation from external charged particles is considerably more complex than gamma-dose calculations. The increased complexity arises from several factors, including significant absorption by media between the source and point of interest, rapid decrease in absorbed dose with tissue depth, distribution of particle energies, and varying LET with particle energies. Beta particles may contribute significantly to the total external dose that small or thin organisms receive from environmental radioactivity. Hine and Brownell,[16] Brown,[21] and Brown and Yu[22] have developed models for the prediction of external beta doses from many radionuclides for certain geometric arrangements. The models have been tested by actual measurements and the effects of varying complicating factors such as surface roughness have been evaluated.[23] The problem of external exposures from alpha particles is generally insignificant due to their very low penetration power and has received very little attention.

An important aspect of external radiation dosimetry is depth dose. Depth dose generally refers to the dose received at various depths in tissue, relative to the surface dose. In general (but not always), the dose from an external source at some depth within a block of tissue is less than the surface dose. The primary factors causing a decreasing dose with depth are absorption of radiations by the tissue and the distance (inverse square) effect. Factors which add to the dose with depth include the buildup of scattered radiation and an increasing LET as charged particles become degraded in energy. The buildup factor from scattered radiations may exert sufficient influence at shallow depths to raise the depth dose above the surface dose.

Situations frequently exist in which organisms receive a far greater dose of radiation from internal sources than from the external environment. Some radionuclides enter biological tissues because they are isotopes of essential nutrient elements, others enter because they are chemically similar to nutrient elements. Some radionuclides may be insoluble and not enter the tissues at all, but, if ingested, may reside in the digestive tract long enough to deliver a significant dose. Adding considerably to the potential dose from internally deposited radionuclides is the fact that most or essentially all of the energy of alpha and beta particles is deposited within the organism (unless the organism is small compared with the range of the particles). In many calculations, it can be assumed that nearly all the particulate radiation will be absorbed by the organism and that part of the gamma radiation will be absorbed. The fraction of the total energy released that is absorbed by the organism depends upon the size and shape of the organism, the type and energy of the radiations, and upon the distribution of the radionuclide within the tissues. When practical, doses from internal radionuclides can be measured directly by inserting small, luminescent dosimeters. However, radionuclides in biological samples are usually expressed in terms of activity per gram of tissue and dose can often be calculated more easily than it can be measured. A situation which can complicate considerably the estimation and significance of dose is nonuniform distribution of activity within the tissue. If activity is localized in "hot spots" within the tissue in question, then doses in and near these deposition sites may be much higher than the average for the entire tissue volume. Damage inflicted within such hot spots may then become manifest in the condition of the whole block of tissue. Autoradiography is often used to study the localization of activity in hot spots. Morgan and Turner present an excellent summary of dose calculations for internally deposited radionuclides.[18]

4. Field Dosimetry

Many of the problems of field dosimetry, although qualitatively predictable from theory, are somewhat unique to the radiation ecologist. The relatively simple design of placing a large point source in a natural ecosystem leads to a complex radiation field due to perturbations caused by living and nonliving components of the system. Even more complex is the distribution of dose within an ecosystem exposed to fallout particles because nonuniform distribution of activity and heterogeniety of radiations is compounded with the shielding and scattering effects of the system's components.

A number of studies on the effects of radiation on natural plant communities have been conducted by placing single point sources of gamma radiation in vegetative stands.[24-30] Extensive dosimetric surveys were undertaken in each study by exposing dosimeters at various locations relative to the source. Of interest in these studies were the effects of distance from the source, height above (and below) the ground, and vegetation and terrain on the relative exposure rate. Factors tending to decrease the exposure rate from a point source are distance (inverse square law) and absorption of radiation by air, vegetation, and terrain. Scattering of radiation, on the other hand, tends to increase with distance from the source. Cowan and Meinhold[28] treated these effects for Woodwell's study in an Oak-pine *(Quercus-Pinus)* forest on Long Island, N.Y. The net effect of these factors is always a decrease in exposure rate with distance. The decrease with distance is usually somewhat greater than that which might be predicted from the inverse square law alone.

McCormick and Golley[25] found no detectable differences in vertical distribution of exposure rates in old field vegetation. Some vertical exposure rate differences were found near the source in a pine forest, but the effects could be related either to actual distance from source or to scattering effects. Exposure to below-ground structures relative to surface exposures drops off rapidly with depth and somewhat more slowly with distance from the source.[27]

The shielding effects, particularly from tree trunks, are noteworthy. Woodwell[24] found twofold differences in exposure rates between open areas and areas shielded by trees. At greater distances from the source, however, the radiation field becomes more uniform because it consists less of primary photons and more of diffusely scattered rays. Exposure rates at the ground surface would be similarly affected, particularly near the source, by rocks, litter, and microterrain as well as by vegetation.

The photon energy degrading process involved in Compton scattering causes a definite decrease in average photon energy with distance from the source. Studies in Colorado shortgrass plains indicate that the average energy of photons in a ^{137}Cs field decreases from about 660 keV at the source to about 250 keV at 50 m. Thus, it is not possible to apply the same quality of radiation to a large field from a single, stationary point source.

Field dosimetry of animals is complicated by their behavioral mobility. French et al.[31] found considerable seasonal variation in the dose received by desert rodents exposed to an external gamma source. When the animals entered subterranean burrows, radiation exposures decreased markedly. Thus, the time spent in burrows, which varied seasonally, affected their exposures. Doses received by animals from contaminated ground may vary within the animal tissues, and seasonally due to variations such as ground water level.[20]

The most practical approach to the dosimetry of ecosystems appears to be direct measurement of the dose to the critical tissues using reliable dosimeters. Through this approach, empirical relationships between the quantity of radiation applied and the dose to critical biological tissues can be developed.

V. FISSION AND FUSION

Production of large amounts of radioactive material accompanies the processes of nuclear fission and fusion. A brief description of these phenomena, therefore, appears warranted. While most laypersons recognize the enormous potential for energy production through these processes, few understand in detail how radionuclides come into being through fission or thermonuclear fusion transitions.

Fission, as the name implies, involves the splitting of nuclei and the release of energy, free neutrons, and radioactive fragments called "fission products". Although bombardment of various nuclei with high energy particles can cause fission, we are principally concerned with those nuclei which have a high probability of fissioning upon the entry of thermal (slow) neutrons. Naturally occurring ^{235}U and the man-produced nuclides ^{233}U and ^{239}Pu all exhibit high fission probabilities in the presence of thermal neutrons. These three fissile materials are therefore suitable for use as nuclear fuels.

The "chain reaction", which may result in controlled burning of nuclear fuel in a reactor or an uncontrolled explosion in a nuclear weapon, results from the fact that two or three neutrons are released per fission. These neutrons can cause additional fissile nuclei in the vicinity to fission, which produces still more neutrons, in turn producing still more fissions. Roughly 200 MeV of energy is released for each fission. The speed with which the chain reaction takes place is governed by the density and geometry of fissile nuclei and the presence of materials which will slow (moderate) the neutrons or capture them.

In a nuclear reactor, the fuel density and geometry are designed to promote the fission process, but at a controllable rate. The fission rate, which determines the amount of heat produced, is controlled by neutron-absorbing rods that can be inserted to various degrees into the reactor core. In a fission weapon, on the other hand, much higher concentrations of fissile material are required. If subcritical masses can be forced rapidly together to form a critical mass, and contained in such a geometry long enough, and providing there is appropriate geometry and lack of extraneous materials to absorb neutrons, the chain reaction can proceed explosively. A nuclear explosion is not physically possible in a reactor because of fuel density, geometry, and other considerations.

In the fission process, a single neutron entering a fissile nucleus causes an unstable situation which usually causes the nucleus to split into two fragments of roughly equal mass. The vast majority of fission fragments range in mass number from 72 to 158 and include some 200 radionuclides of 34 different elements.[1] The most probable mass partitioning results in fragments having masses roughly in the ranges of 90 to 106 and 134 to 144. Symmetrical fission is considerably less probable with thermal neutrons. A small fraction of the fissions are ternary and produce light nuclei, including ^{3}H.

Radionuclides, produced in conjunction with the fission process, include fission products and activation products, the latter resulting from reactions between neutrons and elements in the vicinity of the reaction. Fission products for the most part contain an excessive number of neutrons and therefore are unstable and radioactive, most decaying by beta particle emission. Many of the primary fission products decay to radioactive daughter products. Radionuclides as light as tritium to as heavy as plutonium, americium, and curium, can be produced by neutron activation. Plutonium is of special significance because it is produced purposely for use as a nuclear fuel or warhead material. It arises from neutron capture by ^{238}U and successive beta decays of ^{239}U and ^{239}Np. Reactor cores may also be loaded with naturally occurring ^{232}Th, which is activated by neutrons to form ^{233}Th. Successive beta decays of ^{233}Th and ^{233}Pa lead to fissile ^{233}U.

Fusion reactions involve the coalescence of light nuclei, the formation of heavier nuclei, and the release of energy. Such reactions are termed "thermonuclear" because they require very high temperatures, of the order several million degrees, to initiate the reaction. Most thermonuclear reactions employ isotopes of hydrogen or helium. Thermonuclear fusion reactions account in large measure for the tremendous releases of energy in the sun and stars; their potential for explosive energy release has been amply demonstrated by man's megaton-equivalent warheads.

We owe our very existence on earth to light- and heat-producing reactions which occur in the sun, such as

$$^{1}H + {}^{1}H \rightarrow {}^{2}H + e^{+}$$
$$^{2}H + {}^{1}H \rightarrow {}^{3}He + \gamma$$
$$^{3}He + {}^{3}He \rightarrow {}^{4}He + 2\,{}^{1}H$$

Other solar and stellar thermonuclear reactions involving carbon, nitrogen, and oxygen have also been proposed.

Artificial production of a thermonuclear reaction has been accomplished mainly by the use of deuterium (^{2}H) and tritium (^{3}H). Reaction possibilities include

$$^{2}H + {}^{2}H \rightarrow {}^{3}He + n + 3.26\ \text{MeV}$$
$$^{3}H + {}^{2}H \rightarrow {}^{4}He + n + 17.6\ \text{MeV}$$
$$^{3}H + {}^{3}H \rightarrow {}^{4}He + 2n + 11.3\ \text{MeV}$$

These reactions can be triggered to produce a cataclysmic explosion by a fission-type detonation which can produce the required temperature. Theoretically, it should be possible to control the thermonuclear reaction process in such a way that constructive energy could be obtained. The natural abundance of deuterium is sufficient to supply our most fantastic conceivable energy requirements. Apparently, one of the major technical problems standing in the way of controlled thermonuclear fusion is that of containing the reaction plasma.

It is clear from the thermonuclear reactions listed above, that the major sources of associated radioactivity are residual (unreacted) tritium and neutron activation of materials in and around the reaction. With the exception of residue from a detonation-triggering advice, no fission products are produced in the fusion process.

REFERENCES

1. **Lapp, R. E. and Andrews, H. L.,** *Nuclear Radiation Physics,* 4th ed., Prentice-Hall, Englewood Cliffs, N.J., 1972.
2. **Overman, R. T. and Clark, H. M.,** *Radioisotope Techniques,* McGraw-Hill, New York, 1960.
3. **Chase, G. D. and Rabinowitz, J. L.,** *Principles of Radioisotope Methodology,* 3rd ed., Burgess, Minneapolis, 1967.
4. **Johns, H. E.,** *The Physics of Radiology,* 3rd ed., Charles C Thomas, Springfield, Ill., 1969.
5. **Evans, R. D.,** *The Atomic Nucleus,* McGraw-Hill, New York, 1955.
6. **Bureau of Radiological Health and Training Institute,** Radiological Health Handbook, Public Health Service Publ. 2016, U.S. Department of Health, Education and Welfare, Washington, D.C., 1970.
7. **Blatz, H., Ed.,** *Radiation Hygiene Handbook,* McGraw-Hill, New York, 1959.
8. **Price, W. J.,** *Nuclear Radiation Detection,* 2nd ed., McGraw-Hill, New York, 1964.
9. **Bell, C. G., Jr. and Hayes, F. N., Eds.,** *Liquid Scintillation Counting,* Pergamon Press, Elmsford, N.Y., 1958.
10. **Cooper, J. A., Wogman, N. A., Palmer, H. E., and Perkins, R. W.,** The application of solid state detectors to environmental and biological problems, *Health Phys.,* 15(5), 419, 1968.

11. **Boyd, G. A.,** *Autoradiography in Biology and Medicine,* Academic Press, New York, 1955.
12. **Fuchs, A. W.,** *Principles of Radiographic Exposure and Processing,* 2nd ed., Charles C Thomas, Springfield, Ill., 1958.
13. **Cameron, J. R., Suntharalingam, N., and Kenney, G. N.,** *Thermoluminescent Dosimetry,* University of Wisconsin Press, Madison, 1968.
14. **Lucas, A. C. and French, N. R.,** A miniature thermoluminescent dosimeter and its application in radioecology, in Luminescence Dosimetry, *(AEC Symp. Ser.),* Attix, F. H., Ed., U.S. Atomic Energy Commission, Washington, D.C., 1967, 402.
15. **Siegbahn, K., Ed.,** *Alpha- , Beta- and Gamma-ray Spectroscopy.,* Vol. 1, North-Holland, Amsterdam, 1965.
16. **Hine, G. J. and Brownell, G. L., Eds.,** *Radiation Dosimetry,* Academic Press, New York, 1956.
17. **Whyte, G. N.,** *Principles of Radiation Dosimetry,* John Wiley & Sons, New York, 1959.
18. **Morgan, K. Z. and Turner, J. E., Eds.,** *Principles of Radiation Protection,* John Wiley & Sons, New York, 1967.
19. **Kastner, J., Shonka, F. R., and Rose, J. E.,** A muscle-equivalent environmental radiation meter of extreme sensitivity, in *The Natural Radiation Environment,* Adams, J. A. S. and Lowder, W. M., Eds., University of Chicago Press, Chicago, 1964, 655.
20. **Kaye, S. V.,** Use of minature glass rod dosimeters in radiation ecology, *Ecology,* 46(1/2), 201, 1965.
21. **Brown, S. L.,** Disintegration Rate Multipliers in Beta Emitter Dose Calculations, Rep. for Contract N228(62479)65702, Stanford Research Institute, Menlo Park, Calif., 1965.
22. **Brown, S. L. and Yu, O. S.,** Computational Techniques in Beta Dose Problems, Rep. TRC-68-13, Stanford Research Institute, Menlo Park, Calif., 1968.
23. **Mackin, J. L., Brown, S. L., and Lane, W. B.,** Beta Radiation Dosimetry for Fallout Exposure Estimates: Comparison of Theory and Experiment, Rep. TRC-69-26, Stanford Research Institute, Menlo Park, Calif., 1969.
24. **Woodwell, G. M.,** Design of the Brookhaven experiment on the effects of ionizing radiation on a terrestrial ecosystem, *Radiat. Bot.,* 3(2), 125, 1963.
25. **McCormick, J. F. and Golley, F. B.,** Irradiation of natural vegetation: an experimental facility, procedures and dosimetry, *Health Phys.,* 12(10), 1467, 1966.
26. **Odum, H. T. and Pigeon, R. F., Eds.,** A Tropical Rain Forest: A Study of Irradiation and Ecology at El Verde, Puerto Rico, U.S. AEC Rep. TID-24270, U.S. Atomic Energy Commission, Washington, D.C., 1970.
27. **Fraley, L., Jr. and Whicker, F. W.,** Response of shortgrass plains vegetation to gamma radiation. I. Chronic irradiation, *Radiat. Bot.,* 13(6), 331, 1973; II. Short-term seasonal irradiation, *Radiat. Bot.,* 13(6), 343, 1973.
28. **Cowan, F. P. and Meinhold, C. B.,** Radiation dosimetry for Co^{60} and Cs^{137} gamma ray field irradiation facilities, *Radiat. Bot.,* 2(3/4), 241, 1962.
29. **Zavitkovski, J., Ed.,** The Enterprise, Wisconsin, Radiation Forest. Radioecological Studies, U.S. ERDA Rep. TID-26113-P2, U.S. Energy Research and Development Administration, Washington, D.C., 1977.
30. **Fabries, M., Grauby, A., and Trochain, J. L.,** Study of a Mediterranean type phytocenose subjected to chronic gamma radiation, *Radiat. Bot.,* 12(3), 125, 1972.
31. **French, N. R., Maza, B. G., and Aschwanden, A. P.,** Periodicity of desert rodent activity, *Science,* 154(3753), 1194, 1966.

ADDITIONAL READINGS

Blaylock, B. G. and Witherspoon, J. P., Jr., Environmental factors affecting glass rod dosimetry, *Health Phys.,* 11(6), 549, 1965.

Crook, M. A. and Johnson, P., *Liquid Scintillation.* Vol. 4, Heyden, New York, 1977.

Godbold, B. C. and Jones, J. K., Eds., *Radiological Monitoring of the Environment,* Pergamon Press, Elmsford, N.Y., 1965.

Guthrie, J. E. and Scott, A. G., Measurement of radiation dose distribution in a pond habitat by lithium fluoride dosimetry, *Can. J. Zool.,* 47(1), 17, 1969.

Hale, J., *The Fundamentals of Radiological Science,* Charles C Thomas, Springfield, Ill., 1974.

Halford, D. K. and Markham, O. D., Radiation dosimetry of small mammals inhabiting a liquid radioactive waste disposal area, *Ecology,* 59(5), 1047, 1978.

Hall, R. M., Jr., Thermoluminescent dosimetry, in A Tropical Rain Forest: A Study of Irradiation and Ecology at El Verde, Puerto Rico, Odum, H. T. and Pigeon, R. F., Eds., U.S. AEC Rep. TID-24270, U.S. Atomic Energy Commission, Washington, D.C., 1970, C37.

Harley, J. H., Ed., HASL procedures manual (plus supplements), U.S. AEC Rep. HASL-300, Health and Safety Laboratory, U.S. Atomic Energy Commission, Washington, D.C., 1972.

International Atomic Energy Agency, Rapid Methods for Measuring Radioactivity in the Environment, Publ. STI/PUB/289, International Atomic Energy Agency, Vienna, Austria, 1971.

International Atomic Energy Agency, Solid State and Chemical Radiation Dosimetry in Medicine and Biology, Publ. STI/PUB/138, International Atomic Energy Agency, Vienna, Austria, 1967.

Knoll, G. F., *Radiation Detection and Measurement,* John Wiley & Sons, New York, 1979.

Lecleric, J. C., Cornu, A., and Ginier-Gillet, A., *Neutron Activation Analysis Tables,* Heyden, New York, 1974.

Metcalfe, T. B., *Radiation Spectra of Radionuclides,* Noyes Data Corp., Park Ridge, N.J., 1976.

Miller, D. G., *Radioactivity and Radiation Detection,* Gordon & Breach, New York, 1972.

National Council on Radiation Protection and Measurements, Environmental Radiation Measurements, NCRP Rep. No. 50, National Council on Radiation Protection, Washington, D.C., 1977.

Parchevskaya, D. S., Estimation of absorbed doses of beta-radiation in spherical hydrobionts of small size, in *Marine Radioecology,* Polikarpov, G. G., Ed., Publishing House Academy of Science, Ukraine, SSR, Kiev, 1970, 135; AEC-tr-7299, U.S. Atomic Energy Commission, Washington, D.C., 1972; (English transl.).

Parchevskaya, D. S., Estimation of absorbed doses of alpha-radiation in spherical hydrobionts of small sizes, *Radiobiologiya,* 9(2), 281, 1969, AEC-tr-7025, U.S. Atomic Energy Commission, Washington, D.C., (English transl.).

Platt, R. B., McGinnis, J. T., and Cowan, J. J., An automatically controlled gamma irradiation facility at Emory University, *Bull. Ga. Acad. Sci.,* 22(3/4), 75, 1964.

Rogers, A. W., *Techniques of Autoradiography,* 2nd ed., Elsevier, New York, 1973.

Strettan, J. S., *Ionizing Radiations,* Pergamon Press, Elmsford, N.Y., 1965.

Wang, Y., Ed., *Handbook of Radioactive Nuclides,* CRC Press, Boca Raton, Fla., 1969.

Watson, D. G. and Templeton, W. L., Thermoluminescent dosimetry of aquatic organisms, in Radionuclides in Ecosystems, Nelson, D. J., Ed., U.S. AEC Rep. CONF-710501-P2, U.S. Atomic Energy Commission, Washington, D.C., 1973, 1125.

Woodhead, D. S., Methods of dosimetry for aquatic organisms, in Methodology for Assessing Impacts of Radioactivity on Aquatic Ecosystems, Tech. Rep. Ser. No. 190, International Atomic Energy Agency, Vienna, Austria, 1979, 43.

Chapter 4

ENVIRONMENTAL RADIOACTIVITY

I. INTRODUCTION

In contemporary societies, questions of radioactivity in the environment are frequently raised. Often the manner in which the questions are raised leads the public to believe that radiation is necessarily harmful, regardless of its intensity. If it is detectable, it must be harmful. Although scientific research has demonstrated time and again that the greatest factors affecting biological effects or risk of some effect are the duration and intensity of exposure, the layman in general has a very poor appreciation for the quantitative relationships between radiation dose and effect or risk. To make intelligent decisions, the layman needs to understand more clearly that life on earth has evolved in an environment of ionizing radiation from natural sources and that life can and does prosper within a certain range of chronic radiation exposures.

Throughout the history of life on earth, organisms continuously have been exposed to cosmic rays, radionuclides produced by cosmic ray interactions in the atmosphere, and radiations from naturally occurring substances which are ubiquitously distributed in all living and nonliving components of the biosphere. It is clear that contemporary life forms have adjusted or are doing so to all features and limitations of the environment, including the natural radiation background. Although higher levels of radiation are definitely harmful to organisms, some environmental radiation is of importance to life as we know it. For example, background radiations have contributed, though we do not know how much, to the fundamental processes of chemical and biological evolution. Of clearer importance is the fact that the earth's heat content is principally provided and maintained by the heat of decay of primordial, naturally occurring radionuclides. Were it not for this, the earth would be a totally different place.

These facts should be considered when one attempts to evaluate the implications of raising the total radiation exposure by allowing anthropogenic radioactivity into the biosphere. This is done, of course, by the individuals who are formally responsible for setting standards and guidelines for radiation exposure. In general, there is little, if any, scientific evidence that the normal geographical variations in the intensity of natural radiation cause measurable biological perturbations or limits the geographical occurrence of life forms. If one can believe this, then it is difficult to believe that an anthropogenic increment less than the normal geographic variations in natural radiation exposure will cause undue risk.

In this chapter, the authors attempt to present a general outline of the natural radiation environment and of the incremental radiation exposure that has resulted from the activities of modern man.

II. NATURAL SOURCES OF RADIATION

A. Cosmic Radiation

Radiations of extraterrestrial origin which rain continuously upon the earth are termed "cosmic rays". These radiations are also referred to as "galactic radiation". The effects of cosmic rays were first noticed around 1910 by investigators who were unable to completely eliminate ionization events in radiation detectors by adding lead shielding. The first few centimeters of lead shielding reduced the rate of ionization substantially, but a point was soon reached where additional shielding had little effect on the readings. This suggested the presence of a highly penetrating form of ionizing radiation, which we now term the "hard component" of cosmic radiation.

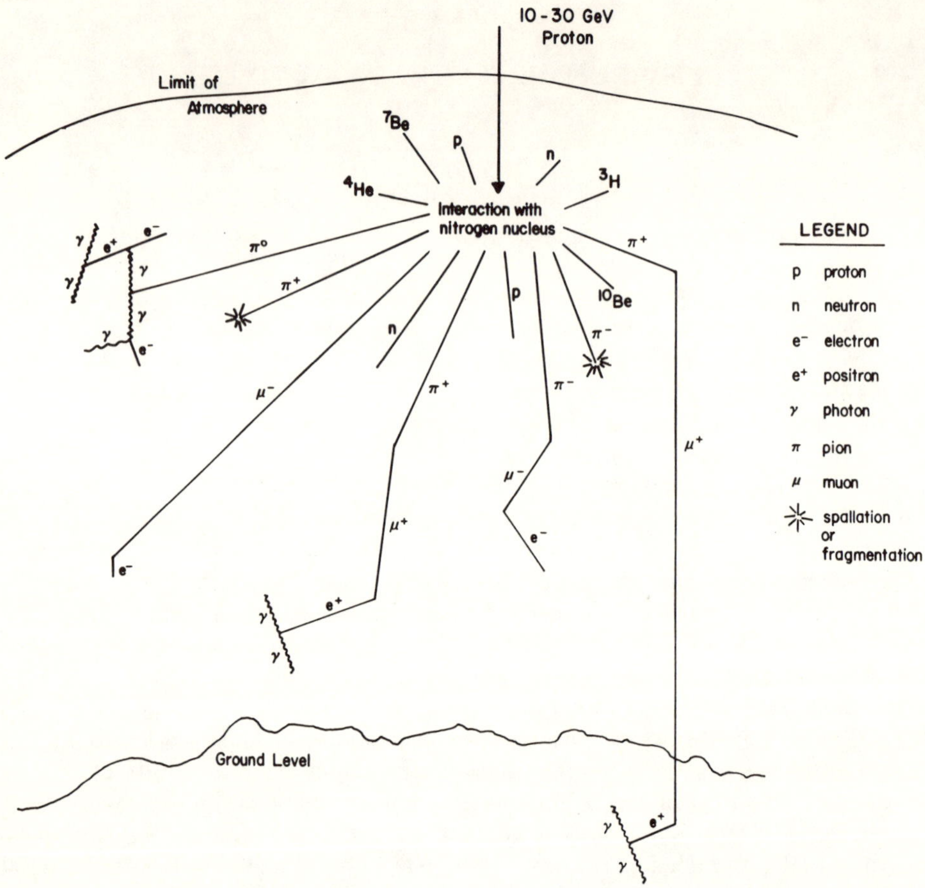

FIGURE 1. Schematic diagram showing the interaction of a primary cosmic proton with an atom in the earth's atmosphere to form numerous secondary particles. Decay products and some interaction possibilities of secondaries are also shown.

The fact that this highly penetrating radiation was impinging upon the earth from space, rather than emanating from the earth, was deduced from balloon experiments in which ionization measurements were made at various altitudes from sea level to 30,000 ft. It was found that the ionization rate decreased for some 2300 ft and from that point increased quite rapidly with elevation. The initial decrease could be explained by a decreased intensity of terrestrial gamma rays, while the increasing component was due to cosmic rays. The likely origin of cosmic rays is the almost infinite number of stars in the universe. One evidence for this is the increased cosmic ray intensity observed on earth following solar flares. However, it is clear that the sun is not normally a major contributor to the total cosmic flux since diurnal variations are very small.

Cosmic rays may be termed "primary" or "secondary". Those which have not yet interacted with matter in the earth's atmosphere, lithosphere, or hydrosphere, are termed primary. These consist principally of protons ($\sim$85%) and alpha particles ($\sim$14%), with much smaller fluxes ($<$1%) of heavier nuclei. Secondary cosmic rays, which are produced by interactions of the primary rays and matter, consist largely of subatomic particles such as pions, muons, and electrons. At sea level, nearly all the observed cosmic radiation consists of secondaries, with some 70% of the flux ac-

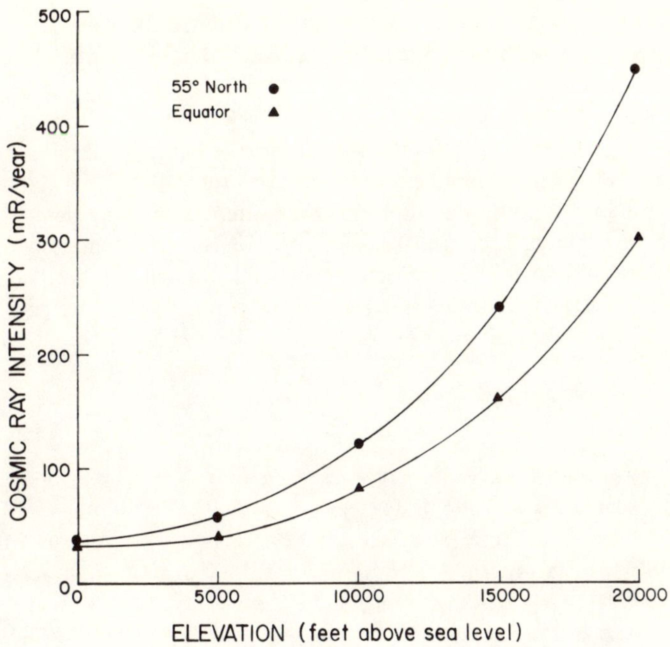

FIGURE 2. Relation of elevation and latitude to cosmic ray intensity.

counted for by muons and 30% by electrons. Less than 1% of the flux at sea level consists of protons.

Primary cosmic rays usually possess tremendous kinetic energy. These rays are positively charged and gain energy by acceleration within the magnetic fields of space. In the vacuum of outer space, charged particles may exist for long periods of time and travel millions of light years. During this flight, they gain high kinetic energies, on the order of 2 to 30 GeV (1 GeV = 10^9 eV). Occasional particles having energies up to 10^{10} GeV have been observed.

The high energies of primary cosmic rays enables them to literally blast apart atoms in the earth's atmosphere upon collision (Figure 1). Such high-energy reactions are termed "spallation" or "fragmentation", depending upon the size of the fragments and of the residual nucleus.[1] Notice (Figure 1) that lighter elements, such as hydrogen, helium, and beryllium, may be formed by the cataclysmic reaction, as well as neutrons, protons, and subatomic particles. Pions, which may be charged or uncharged, decay rapidly ($<10^{-7}$ sec) to form either photons or muons. Muons are always charged and also decay rapidly ($<10^{-5}$ sec) to electrons or positrons. The high energy and small mass ($\sim$1/9 mass of a proton) of muons enables them to penetrate matter readily. For instance, muons can be detected in deep mines and they readily penetrate several inches of lead shielding to produce "background" counts in radiation detectors. Other interaction possibilities of cosmic rays are also symbolized in Figure 1.

Cosmic ray intensity increases sharply with elevation until a maximum is reached at an altitude of about 12 miles.[2] From 12 miles to the limit of the atmosphere ($\sim$30 miles), the intensity decreases. This pattern is explained by increased production of secondaries resulting from the increasing atmospheric density as one moves toward earth from an altitude of 30 miles. At 12 miles, most of the primaries have interacted and the decreasing intensity from 12 miles altitude to earth reflects absorption of secondaries by the atmosphere, which is equivalent to some 30 ft of water. Cosmic ray exposure rates over the range of elevations of most interest from a biological viewpoint are shown in Figure 2.

Cosmic ray intensity is also related to latitude (Figure 2). At a given altitude, the cosmic flux increases from the equator to a latitude of 50 to 60°.[2-4] The flux remains roughly constant from 50 to 60° to the poles. The latitude effect can be explained by the shape of the earth's magnetic field and the energy distribution of primaries. Magnetic lines of force extending beyond the atmosphere are generally parallel to the earth's surface at the equator, and perpendicular at the poles. Charged primary particles can easily penetrate the magnetic field when aligned with the lines of force. However, only the more energetic primaries can penetrate the field when not aligned. From the poles to 60 deg, virtually all the primaries reach the earth's atmosphere, whereas at the equator, only those primaries exceeding about 15 GeV penetrate the magnetic shield. From empirical data, it has been found that

$$E_{min} = 15\,(\mathrm{GeV})\ \cos^2 \lambda \tag{1}$$

where E_{min} is the minimum cosmic ray energy in GeV needed for penetration of the magnetic field and λ is the latitude in degrees.

As a result of the elevational and latitudinal effects on cosmic ray intensities, one finds considerable variation throughout the world. Much of the contiguous U.S. is below 1000 ft elevation at latitudes between 30° and 50°N. In this large region, cosmic ray exposure rates average close to 35 mR/year. In mile-high Denver, Colo., the cosmic ray exposure is about 55 mR/year and on 14,000 ft mountain peaks near Denver it is about 190 mR/year. The cosmic ray exposure on 20,000 foot Mt. McKinley in Alaska is of the order of 450 mR/year. Passengers in commercial jetliners receive exposure rates of the order of 500 to 2000 mR/year while at cruising altitudes. Cosmic radiation intensities that confront astronauts are orders of magnitude higher, particularly in the Van Allen belts[5] and anywhere in space if solar flares occur.

Time variations occur in cosmic ray intensities.[2] Comparatively modest cyclic variations occur which seem related to solar activity and to the interplanetary magnetic field. More dramatic cosmic ray variations have been observed in connection with solar flares. For instance, the large flare of February 23, 1956, resulted in cosmic intensities that exceeded by more than tenfold the normal flux. The particle shower peaked about an hour after the flare was observed visually, and tailed off over a period of several hours. Fairly recent scientific evidence suggests that the earth's magnetic field has reversed polarity some 9 times over the past 4 million years. During such reversals, it is conceivable the magnetic shield was missing for periods of time of the order of 1000 years. If this theory is correct, it is likely that cosmic intensities on earth were substantially higher during such polarity reversals.

B. Cosmogenic Radionuclides

A considerable number of radionuclides are continuously produced in the atmosphere by cosmic ray interactions with matter (Table 1). Most of these radionuclides are produced as spallation fragments, but some are formed by activation of stable atoms with neutrons or muons. The natural production of radionuclides in the atmosphere shows elevational and latitudinal patterns similar to those of cosmic ray intensities. About 70% of the spallation-produced nuclides arise in the stratosphere, while about 30% are formed in the troposphere.

With the possible exceptions of ^3H and ^{14}C, the radionuclides in Table 1 are normally found in very minute concentrations and require very sensitive methods for detection. Tritium is diluted and mixed with the earth's water and H_2 gas reservoirs, while ^{14}C combines with oxygen to form $^{14}CO_2$, which mixes with the atmospheric CO_2 pool. Carbon-14 enters plants through the process of photosynthesis and prior to the indus-

Table 1
RADIONUCLIDES PRODUCED FROM COSMIC RAYS

Radionuclide	Half-life	Primary production mode	Atmospheric production rate (atoms/cm²-sec)	Detected and measured in
^{10}Be	2.7×10^6 year	Spallation	4.5×10^{-2}	Deep sea sediments
^{36}Cl	3.1×10^5 year	$^{35}Cl(n,\gamma)^{36}Cl$	1.1×10^{-3}	Rocks, rain
^{14}C	5568 year	$^{14}N(n,p)^{14}C$	1.8	Organic material, CO_2
^{32}Si	500 year	Spallation	1.6×10^{-4}	Marine sponges, sea water
^{3}H	12.3 year	Spallation $^{14}N(n,^3H)^{12}C$	0.25	Water, air
^{22}Na	2.6 year	Spallation	5.6×10^{-5}	Rain, air, organic material
^{35}S	88 day	Spallation	1.4×10^{-3}	Rain, air, organic material
^{7}Be	53 day	Spallation	8.1×10^{-2}	Rain, air
^{33}P	25 day	Spallation	6.8×10^{-4}	Rain, air, organic material
^{32}P	14.3 day	Spallation	8.1×10^{-4}	Rain, air, organic material
^{27}Na	15.1 hr	Spallation		Rain
^{38}S	2.9 hr	Spallation		Rain
^{39}Cl	55 min	$^{40}A(\mu^-,n)\,^{39}Cl$	1.6×10^{-3}	Rain
^{38}Cl	37 min	Spallation		Rain

Reproduced from *Health Physics,* Volume 11, Issue 2, 1297, 1965. By permission of the Health Physics Society.

trial revolution was found in the ratio of about 15 dpm/g of carbon. Very old biological specimens can be dated by measurement of the ^{14}C activity of carbonaceous material. A number of the radionuclides listed in Table 1, including ^{3}H, ^{14}C, ^{7}Be, ^{22}Na, and ^{32}P, can also be produced by nuclear explosions and geochemical research on these radionuclides must take this fact into account.

Although cosmogenic radionuclides are not sufficiently abundant to add significantly to the total radiation dose from natural background, they should not be dismissed because several of them have been extremely useful to science. For example, in addition to dating of various materials using ^{14}C or ^{3}H, some of the nuclides have been used as atmospheric tracers.[6] After their formation in the atmosphere, these tracers move with surrounding air masses and most eventually find their way to Earth where they accumulate in soil and sediment. The behavior of air masses can be studied by careful, systematic measurements of such natural tracers. A similar approach using $^{12}C/^{14}C$ measurements has been applied to the study of ocean currents and their mixing rates.[7] Meteoritic dust, which rains upon the Earth at the rate of about 10^4 tons/day,[6] also contains several cosmic ray-induced radionuclides. Two of the longer-lived radionuclides in extraterrestrial dust are ^{53}Mn and ^{26}Al. The measurement of these and other radionuclides in the ocean sediments give considerable insight into ages and rates of accumulation of these sediments.

C. Primordial Radionuclides

Radionuclides which appeared at the time of formation of the earth are termed "primordial". Of the many radionuclides that must have been formed with the earth, only a few have half-lives sufficiently long to explain their current existence. If the earth was formed about 6×10^9 years ago, a primordial radionuclide would need a half-life of at least 10^8 years to still be present in measurable quantities. Of the primordial radionuclides that are still detectable, three are of overwhelming significance.[8] These are ^{40}K, ^{238}U, and ^{232}Th. Uranium and thorium each initiate a chain of radioactive progeny which are nearly always found in the presence of the parent nuclides (Table

Table 2
PRIMARY DECAY SCHEMES OF ^{238}U and ^{232}Th

Uranium-238			Thorium-232		
Radionuclide	Half-life	Radiation	Radionuclide	Half-life	Radiation
^{238}U	4.5×10^9 year	α,γ	^{232}Th	1.4×10^{10} year	α,γ
^{234}Th	24 day	β,γ	^{228}Ra	6.7 year	β,γ
^{234}Pa	1.2 min	β,γ	^{228}Ac	6.1 hr	β,γ
^{234}U	2.5×10^5 year	α,γ	^{228}Th	1.9 year	α,γ
^{230}Th	8×10^4 year	α,γ	^{224}Ra	3.6 day	α,γ
^{226}Ra	1620 year	α,γ	^{220}Rn	55 sec	α,γ
^{222}Rn	3.8 day	α,γ	^{216}Po[a]	0.16 sec	α,β
^{218}Po[a]	3.1 min	α,β	^{212}Pb	11 hr	β,γ
^{214}Pb	27 min	β,γ	^{212}Bi[a]	61 min	α,β,γ
^{214}Bi[a]	20 min	α,β,γ	^{212}Po	3×10^{-7} sec	α
^{214}Po	1.6×10^{-4} sec	α	^{208}Pb	Stable	None
^{210}Pb	19 year	β,γ			
^{210}Bi[a]	5.0 day	α,β,γ			
^{210}Po	138 day	α,γ			
^{206}Pb	Stable	None			

[a] Alternate, less frequent branching decays not shown.

Table 3
SINGLY OCCURRING PRIMORDIAL RADIONUCLIDES[a]

Radionuclide	Half-life (year)	Radiation
^{40}K	1.26×10^9	β,γ
^{50}V	6×10^{15}	β,γ
^{87}Rb	4.8×10^{10}	β
^{115}In	6×10^{14}	β
^{123}Te	1.2×10^{13}	EC[a]
^{138}La	1.1×10^{11}	β,γ
^{142}Ce	$>5 \times 10^{16}$	α
^{144}Nd	2.4×10^{15}	α
^{147}Sm	1.1×10^{11}	α
^{149}Sm	$>1 \times 10^{15}$	α
^{152}Gd	1.1×10^{14}	α
^{174}Hf	2×10^{15}	α
^{176}Lu	2.2×10^{10}	β,γ
^{180}Ta	$>1 \times 10^{12}$	β
^{187}Re	4.3×10^{10}	β
^{190}Pt	6.9×10^{11}	α

[a] Electron capture.

From Eisenbud, M., *Environmental Radioactivity*, 2nd ed., Academic Press, New York, 1973. With permission.

2). Although many of the daughter radionuclides are short-lived, they are ubiquitously distributed in the environment because they are continually being formed from long-lived precursors.

A list of 16 other primordial, long-lived radionuclides which occur in nature is provided in Table 3. With the exception of ^{40}K, these radionuclides are generally found

Table 4

CALCULATED TOTAL EXPOSURE RATE AT 1 M ABOVE GROUND FOR NATURAL EMITTERS UNIFORMLY DISTRIBUTED IN THE SOIL[13,14]

Radionuclide	Exposure rate/radionuclide concentration: $\mu R\ hr^{-1}/pCi\ g^{-1}$	$\mu R\ hr^{-1}$/indicated concentration
^{40}K	0.179	1.49/% K
^{226}Ra + daughters	1.80	$0.61/0.358 \times 10^{-6}\ \mu g\ g^{-1}\ Ra^{a}$
^{214}Pb	0.20	$0.07/0.358 \times 10^{-6}\ \mu g\ g^{-1}\ Ra^{a}$
^{214}Bi	1.60	$0.54/0.358 \times 10^{-6}\ \mu g\ g^{-1}\ Ra^{a}$
^{238}U + daughters	1.82	$0.62/\mu g\ g^{-1}\ ^{238}U$
^{232}Th + daughters	2.82	$0.31/\mu g\ g^{-1}\ ^{232}Th$
^{228}Ac	1.18	$0.13/\mu g\ g^{-1}\ ^{232}Th$
^{208}Tl	1.36	$0.15/\mu g\ g^{-1}\ ^{232}Th$
^{212}Bi	0.09	$0.01/\mu g\ g^{-1}\ ^{232}Th$
^{212}Pb	0.09	$0.01/\mu g\ g^{-1}\ ^{232}Th$

[a] Concentration of ^{226}Ra in equilibrium with $1\ \mu g\ g^{-1}\ ^{238}U$.

in very low concentrations and very careful methods are required for their detection and measurement. Potassium-40, however, is abundant and ubiquitously distributed in living and nonliving matter.

Naturally occurring neutrons form many other radionuclides, but their abundance is usually very low. There is fascinating evidence that a "natural reactor" produced fission and activation products in the Oklo quarry in Gabon, Africa. Depletion of fissile ^{235}U and the presence of fission products within this rich uranium deposit indicate that spontaneously triggered fission chain reactions occurred some 2 billion years ago.[9]

Even though the primordial radionuclides are ubiquitous, their concentrations vary substantially with location. The main reservoir of natural radioactivity is the lithosphere. However, considerable variation in radioactivity exists within the lithosphere. Some variations appear associated with specific types of formations and certain minerals, while other variations appear to be strictly regional, with little correlation to types of rocks and minerals. The geological processes responsible for the distribution of primordial radionuclides are discussed by Adams.[10] Considerable information on the distribution of natural radionuclides in the earth's crust is available.[7,8,11,12]

Because of the penetrating gamma rays emitted by most primordial radionuclides and progeny, their presence in soil and rock results in external exposures at the earth's surface. Table 4 indicates the exposure rates at one meter above ground for naturally occurring gamma emitters uniformly distributed in soil.

Radiation exposure levels from natural radionuclides show wide geographic variations. Table 5 lists several locations and their exposure rates from terrestrial gamma rays. Exposure levels over the vast majority of the earth's surface from terrestrial gamma rays probably range between 40 and 90 mR/year. The value for Clallam Bay, Wash. is unusually low, and that for Denver, Colo., relatively high. Denver is at the edge of the Colorado Front Range, much of which contains relatively high concentrations of uranium and thorium.[21] Some local areas of the world contain very high concentrations of natural radioactivity (Table 5). The readings indicated are mostly maximum intensities from highly localized deposits. The high radiation regions in India and Brazil have received considerable attention because they are inhabited by people.[7,22,23]

Table 5
RADIATION EXPOSURE LEVELS FROM NATURAL TERRESTRIAL GAMMA RAYS AT VARIOUS LOCATIONS

Location	Exposure rate (mR/year)	Ref.
Clallam Bay, Wash.	24	15
"Typical" for U.S.A	60	15
Denver, Colo.	114	15
Atypically high local sites		
Kerala, India	1,600	16
Black Forest, Germany	1,800	17
Central City, Colo.	2,200	18
Guarapari, Brazil	17,000	19
U.S.S.R.	70,000	20

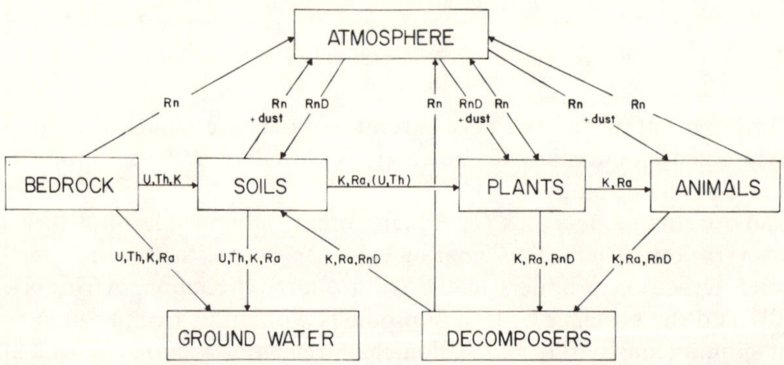

FIGURE 3. Major pathways of primordial radionuclides and important progeny in a terrestrial ecosystem. Symbols: U = uranium isotopes; Th = thorium isotopes; K = ^{40}K; Ra = radium isotopes; Rn = radon isotopes; and RnD = progeny of radon decay.

It is evident that plants and animals indigenous to certain areas receive considerable external gamma radiation. Added to the external exposures are doses which arise from internal deposition of the natural radionuclides. Intimate contact of organisms with soil and rock results in additional external alpha and beta exposure. The latter type of exposure may be relatively important for root tips and shoots of plants just emerging from the soil and also for microorganisms and soil invertebrates.

The distribution of the primordial radionuclides and their progeny in natural ecosystems is influenced by many factors, including chemical properties of the nuclides, physical factors of the ecosystem, and physiological and ecological attributes of the biota. Since quantitative generalizations would be very difficult to make, the authors shall settle, for the time being, on a qualitative model of natural radionuclide behavior. A simple compartmental diagram of a terrestrial ecosystem, showing pathways along which some of the more abundant natural radionuclides have been shown to move, is given in Figure 3.

Weathering of bedrock, the main reservoir of the primordial radionuclides, releases U, Th, and K to the soil. Progeny of U and Th also take part in this transfer. From soil, K, Ra, and small amounts of U and extremely small quantities of Th are taken up by plants. Potassium-40 is utilized by the plant in the same manner as it uses the essential element, stable K. Radium, an important longer-lived progeny of ^{238}U, is utilized by the plant, not because it is an isotope of an essential element, but because it is chemically similar to calcium, which is essential. The uptake of U and Th by plants is usually small or negligible since these nuclides normally are relatively insoluble.

Soils may slowly become depleted in radioactivity due to leaching losses. This may partially account for the observation that with the possible exception of ^{226}Ra, soils generally contain less of the nuclides under discussion than their parent materials.[11] Recycling of ^{226}Ra by the biota may explain its persistence in soil relative to the bedrock. Radionuclides attached to small soil particles at the ground surface may be transferred as a dust or aerosol through the atmosphere to plants or animals by wind action. The importance of this mechanism varies with climate, soil, and vegetation cover.

Herbivores ingest all radionuclides in or on consumed vegetation, but only K and Ra isotopes are readily assimilated. Most other natural radionuclides are poorly absorbed. Within the animal, K accumulates mainly in soft tissues, while Ra accumulates in bone and in any other structures which require calcium. Excretory products and death of plants and animals result in the return of radionuclides to the soil. Radioactivity incorporated within dead tissues reaches the soil either through the decomposer organisms involved in the breakdown and decay of organic material or through leaching of litter.

Bedrock, soil, plant, animal, and decomposer compartments all release radon to the atmosphere. Radon is the decay product of radium and is produced in any material containing radium. Since radon is one of the inert gases, it can escape from surfaces which are in contact with the atmosphere. The amount of radon which emanates from a given mass of rock depends upon the quantity of radium present and upon the amount of surface area presented by the mass. The more finely broken a given mass of rock, the more radon it can release. The concentration of radon in the air adjacent to radium-bearing material also depends upon the rate of fresh air movement into the space in question. In basements, caves, and mine shafts that have poor air circulation, the radon concentrations can build up to very significant levels. Efficient ventilation in mines is often necessary to maintain radon concentrations below those which would be hazardous for workers. Radon escapes from plants through the stomates and from animals via exhaled gases. Analysis of expired breath samples is sometimes used to study radium burdens in humans.

Radon decays quite rapidly to form a series of daughter radionuclides (Table 2). Once formed in the atmosphere, the progeny of radon attach to small dust particles which are subject to deposition on soil and plants and to inhalation by animals. Rainfall is particularly efficient in scrubbing radon daughters from the atmosphere, but impaction and sedimentation of aerosol particles also contribute to deposition. The major portion of the total dose to plants and animals from natural airborne radioactivity is usually from radon progeny. Two progeny of radon which are particularly significant for herbivorous animals and also for humans are ^{210}Pb and ^{210}Po.[24,25] These comparatively long-lived nuclides find their way to plant tissues through uptake from the roots and surface deposition. When consumed by herbivores, the assimilated portion of ^{210}Pb locates primarily in bone and that of ^{210}Po locates in bone, kidney, liver, and other soft tissues.[26] Although most of the intake of ^{210}Po and ^{210}Pb in animals is from ingestion, some is inhaled. This is apparently particularly true in humans who smoke cigarettes. One study showed that smokers contain over twice the levels of these radionuclides as nonsmokers.[27]

Leaching of bedrock and soil may lead to detectable concentrations of U, Th, and Ra in ground water. In local areas, concentrations of natural radioactivity may exceed normal concentrations by orders of magnitude.[7] In some areas of the world, health spas have advertised radium springs as having beneficial health effects. Visitors to such spas have paid for the "privilege" of drinking and bathing in water containing radium well in excess of present maximum permissible concentrations and for breathing radon emanation. In the U.S., some individual states, the Nuclear Regulatory Commission (NRC), and the Environmental Protection Agency (EPA) now have authority to inspect public and commercial establishments to insure compliance with applicable health standards. On the other hand, government agencies have no jurisdiction over private use of privately owned springs and mineral baths.

The redistribution of primordial radionuclides by the uranium and thorium industries leads to significant contact of these materials by humans and local environments. Uranium, used principally as a nuclear fuel, is mined in the U.S. largely in New Mexico, Wyoming, Utah, and Colorado.[28] Large quantities of uranium are also mined in concert with phosphate. Thorium is mined for the purpose of breeding fissile ^{233}U in nuclear reactors, and for other uses such as welding rods and mantles in gas lanterns.

Underground uranium mine atmospheres have high radon concentrations, and usually require ventilation to maintain permissible concentrations for mine workers. In some mines, respirators are required for workers in addition to ventilation. The ventilation of mines releases radon and progeny to the aboveground atmosphere but dispersion processes significantly reduce the concentrations within a short distance. Open pit mines release large quantities of radon to the atmosphere, but the release area is usually diffuse and respirators are seldom required in such mines (Figure 4).

Both underground and open pit mines usually require dewatering. Mine water ordinarily contains measurable concentrations of Ra, U, Th, Pb, and other natural radionuclides. Mine water is typically pumped to a retention pond or series of ponds to allow settling of suspended solids and sorption of dissolved minerals to algae and sediments. The outflow from the retention pond system is sometimes allowed to enter a natural drainage where measurable but reduced concentrations of radioactivity can be found.

Ore and mill tailings that contain significant quantities of uranium and/or progeny are measurable sources of environmental contamination. A major mechanism of transport, especially in arid regions, is wind erosion. Some ores and most tailings are comprised of erodible-sized particles that can be easily resuspended by strong winds, especially when in a dry, loose state. Mill tailings can be reasonably well-protected by pumping them as a slurry to a retention pond where the solids settle out. If the evaporation and percolation rate equals the liquid input to the pond, surface water discharge from the pond can be eliminated. A small portion of the solid tailings can settle out on a beach where wave and wind action can move the material away from the retention pond. Minimizing wind erosion of ore and tailings is a matter of good engineering and land-management practice.

D. Summary of Dose Rates from Natural Radiation

A brief discussion of the relative contributions of various natural radiation sources is appropriate at this point to place each source into general perspective. Annual doses from natural radiation sources, which are generally applicable to humans and most other terrestrial vertebrates, are listed in Table 6. It is apparent that terrestrial gamma rays make the largest contribution to the total dose. Also, terrestrial gammas are generally the most variable of the natural radiation components from place to place in the biosphere. The internal sources, together accounting for about 19% of the total dose, also vary with location because of variable amounts of radionuclides in local environments.

Terrestrial vegetation probably receives somewhat higher natural doses than terrestrial vertebrates because of a higher ^{40}K content, because of dust and debris that cling to foliage surfaces, and because of the intimate contact of roots and emergent shoots

FIGURE 4. A typical open-pit uranium mine in Wyoming. This pit, which covers some 500 acres, was opened to reach an ore body some 300 ft below the surface.

Table 6
TYPICAL ANNUAL RADIATION DOSES RECEIVED BY HUMANS AND OTHER TERRESTRIAL VERTEBRATES FROM NATURAL SOURCES

Source of irradiation	Dose rates (mrad/year)	% of total	Ref.
External sources			
Cosmic rays (1000 ft, 50° latitude)	35	30	4
Terrestrial γ rays	60	51	15
Internal sources			
^{40}K	19	16	29
^{14}C	1	1	4
$^{226,228}Ra$	1	1	29
^{3}H, ^{87}Rb, ^{210}Po ^{220}Rn, ^{222}Rn, ^{238}U	1	1	29
Total	117		

with the soil. One study found that trunks of oak trees received exposures up to 3.5 mR/week during periods of growth from ^{40}K and other internal radionuclides.[30] The intimate contact of many invertebrate animals and microorganisms with the soil probably results in higher doses in these species also.

According to compiled data,[31] aquatic organisms generally receive radiation exposure from natural sources of roughly the same order of magnitude as terrestrial organisms (Table 7), although the dose distribution varies by source. It is clear that internal radioactivity and sediments are the largest dose contributors for marine species, whereas cosmic rays and sediments provide the largest doses for freshwater organisms. Substantial variation in dose occurs between and within taxonomic groups as a result of differing life habits.

III. MAN-GENERATED RADIOACTIVITY

Man has produced a vast array and quantity of radioactive nuclides from naturally occurring raw materials. Some of these radionuclides are produced to serve a particular purpose, while some are unwanted waste products. Some radionuclides are primarily associated with peaceful applications of nuclear energy, while others are largely associated with potential weapons of war. All radionuclides, at certain times and places, and to varying degrees, find their way into the biosphere or local ecosystems.

A. The Nuclear Fuel Cycle

The staggering growth in the world demand for energy has forced a careful examination of all forms of energy production. Nuclear power production, despite not having grown as fast as it might have, still promises to grow significantly in comparison to other forms of power. When the alternative energy sources are analyzed for safety, reliability, economics, and environmental impact to the extent that nuclear has, the authors believe that the nuclear option will clearly emerge as one of the more feasible options. In a practical sense, there is need to develop all reasonable energy alternatives simultaneously, to the extent dictated by economic and environmental limitations, and political realities.

Nuclear reactors, and the entire nuclear fuel cycle which revolves around them, are still not well-understood by the public. In fact, the public has been misled and misinformed about their safety by individuals and special interest groups.[32] In this section,

Table 7
ESTIMATES OF ANNUAL DOSES (MRAD/YEAR) RECEIVED BY MARINE AND FRESHWATER ORGANISMS FROM NATURAL SOURCES OF RADIATION

Source	Taxonomic group	Marine (20 m depth)	Freshwater (1—2 m depth)
Cosmic	Phytoplankton	4.4	24
	Zooplankton	4.4	24
	Mollusca	4.4	19
	Crustacea	4.4	19
	Fish	4.4	19—24
Water	Phytoplankton	3.5	0.06—54
	Zooplankton	1.8	0.009—7.4
	Mollusca	0.9	0.004—3.1
	Crustacea	0.9	0.004—3.1
	Fish	0.9	0.004—6.1
Sediment ($\beta + \gamma$)	Phytoplankton	0	0
	Zooplankton	0	0
	Mollusca	27—324	27—324
	Crustacea	27—324	27—324
	Fish	0—324	0—324
Internal	Phytoplankton	17—64	—
	Zooplankton	23—138	—
	Mollusca	65—131	—
	Crustacea	69—188	—
	Fish	24—37	32—42
Sum of natural sources	Phytoplankton	25—72	24—78
	Zooplankton	29—168	24—31
	Mollusca	97—460	46—346
	Crustacea	101—517	46—346
	Fish	29—366	51—396

Data adapted from IAEA, Effects of Ionizing Radiation on Aquatic Organisms and Ecosystems, Tech, Rep. Ser. No. 172, International Atomic Energy Agency, Vienna, 1976.

the authors will briefly discuss nuclear reactors and the nuclear fuel cycle, mainly from the standpoint of production of radioactive materials which may be released to the environment. The nuclear power option will be explored in Volume II, Chapter 3 from the standpoint of ecological consequences.

There is much more than meets the eye when one sees a nuclear power reactor with its cooling towers, transformers, and transmission lines leading into a power grid network (Figure 5). In order to support a nuclear reactor, new fuel must be supplied and spent fuel must be handled and disposed of. The nuclear fuel cycle involves mining of uranium ore, milling it down to a concentration where it can be economically transported, chemical conversion, purification, enrichment of the fissile isotope of uranium, ^{235}U, and fuel fabrication before the reactor can operate. When the fuel elements have been used or "spent" to a certain degree in the reactor core, they must be stored for a period to allow decay of short-lived fission products, and then either disposed of, as is, or reprocessed to separate unused fissile material from waste products that have no practical use. The basic elements and options within the nuclear fuel cycle are shown in Figure 6.

It is evident from the diagram that the flow of natural uranium through the cycle is all that is really needed to supply the reactor. However, although natural uranium is ubiquitous, high grade ore reserves are limited and the isotopic abundance of fissile

FIGURE 5. The St. Vrain nuclear power station near Platteville, Colo.

^{235}U is only about 0.72%. The latter problem can be handled through isotopic enrichment, but this process is difficult and expensive. Therefore, there is economic incentive to reclaim unused ^{235}U, as well as fissile ^{239}Pu that is formed in the reactor from neutron capture of ^{238}U, the abundant isotope of uranium. This reaction sequence is

$$^{238}U + n \longrightarrow {}^{239}U \xrightarrow[\text{decay}]{\text{beta}} {}^{239}Np \xrightarrow[\text{decay}]{\text{beta}} {}^{239}Pu$$

An alternative addition to the cycle is natural thorium. Purified thorium pellets, placed in the reactor core, lead to the production of another fissile uranium isotope, ^{233}U, through the reaction sequence

$$^{232}Th + n \longrightarrow {}^{233}Th \xrightarrow[\text{decay}]{\text{beta}} {}^{233}Pa \xrightarrow[\text{decay}]{\text{beta}} {}^{233}U$$

Production of ^{239}Pu and/or ^{233}U is termed "breeding" and the precursors of these fissile substances, ^{238}U and ^{232}Th, are termed "fertile". Breeding increases the efficiency of nuclear reactors and "breeder reactors", those designed to actually produce more nuclear fuel than they consume, offer tremendous energy implications for the future. Successful employment of the breeder reactor concept could substantially reduce the need for uranium mines in the future, and possibly the total cost per megawatt-hour of electricity.

1. Mining

As mentioned earlier, New Mexico, Wyoming, Utah, and Colorado produce the majority of uranium in the U.S. Several other western states, Texas, and Florida also produce uranium (Figure 7). In Florida, uranium is produced as a by-product from phosphate mining. Uranium ore above 0.01% U_3O_8 is presently considered economically feasible to mine and most ore actually mined averages between 0.04 and 0.42%

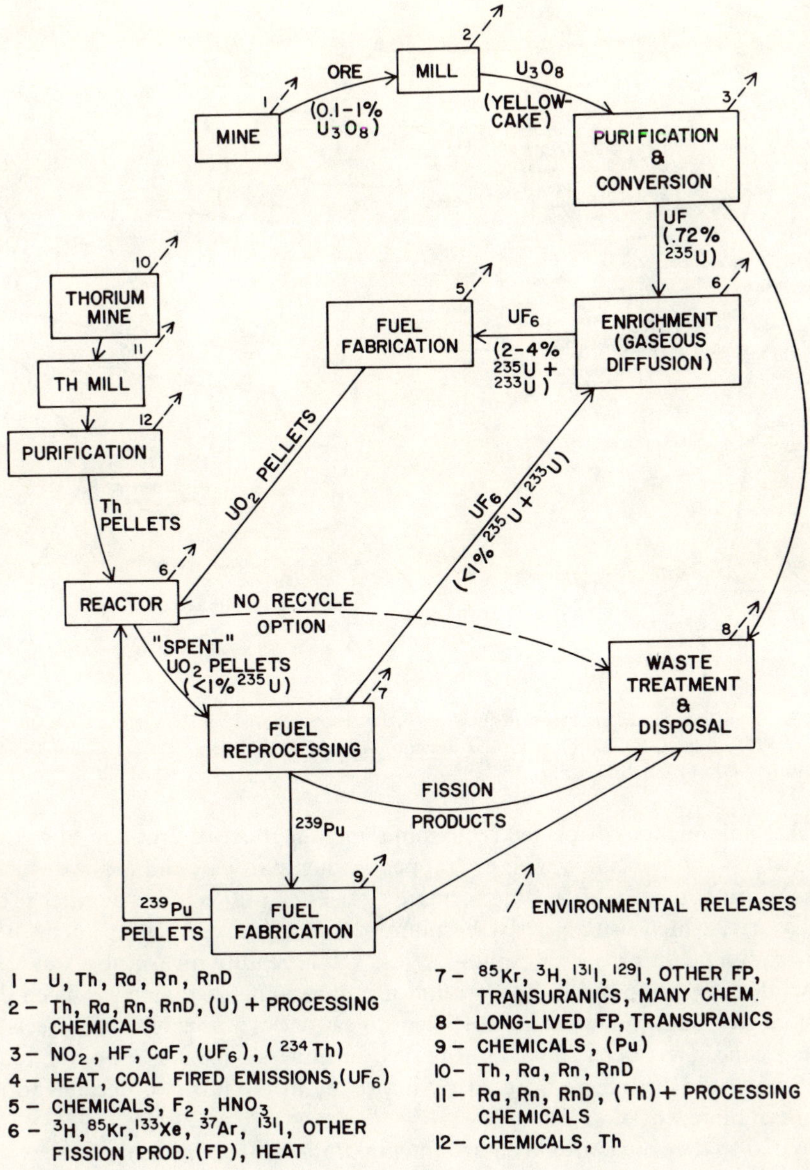

FIGURE 6. Basic elements and options within the nuclear fuel cycle, showing the kinds of materials that can be released to the environment.

U_3O_8.[28] Purchases of U_3O_8 by the U.S. Atomic Energy Commission peaked at 34,410 ton/year in 1960, but fell to 1295 ton/year by 1971. However, domestic commercial purchases in 1977 amount to roughly 12,300 ton U_3O_8/year and by 1990, this may grow to 45,000 ton/year. The average price for U_3O_8 was around \$15/lb in 1976 and the forecast is for an average price of \$25/lb by 1982.[28]

Mining methods primarily involve deep shafts or large open pits, but other methods, such as *in situ* leaching have also been used. Open pits are used for ore up to several hundred feet deep when the overburden is easy to move (Figure 4). Underground mines are usually more feasible for ore bodies deeper than 400 ft or those lying beneath hard rock strata. The principal environmental impacts of uranium mining involve landscape alterations, human and heavy equipment activities, dust generation, and radionuclide

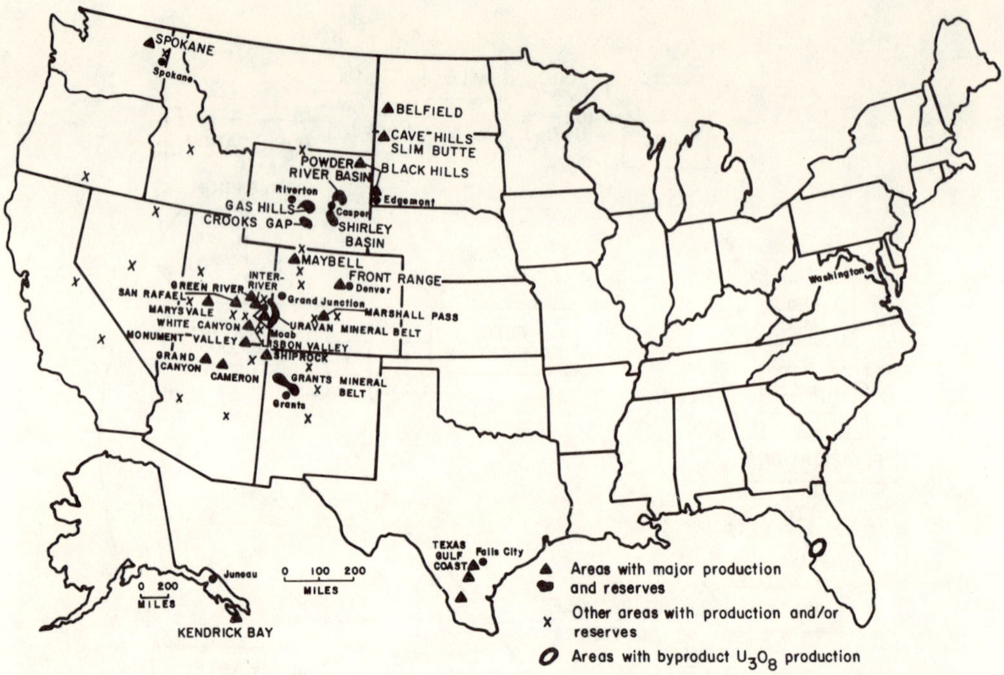

FIGURE 7. Significant uranium areas of the U.S. (From **Anon.**, Statistical Data of the Uranium Industry, January 1, 1977, Rep. GJO-100 (77), Grand Junction Office, U.S. Energy Research and Development Administration, Grand Junction, Colo., 1977.)

dispersal. Radionuclides dispersed from mines include those of uranium, thorium, and radium, which are associated with or trapped in dust particles and radon, which emanates from ore as a gas and decays to create a series of radioactive daughter products. Ground water, which usually must be pumped out of mines to keep them dry, also contains radionuclides of the uranium series. While readily measurable quantities of radionuclides can be found in and around uranium mines, it has never been demonstrated that the radiation levels are sufficient to cause obvious ecological effects. However, lung cancer has been caused in the past in miners working in poorly ventilated shafts. Safety measures have been vastly improved in the last two decades to protect the health of mine workers.

Compared to the conventional environmental impacts from people and heavy equipment, radiation-caused impacts are negligible around uranium mines. And, although landscape disturbance at a typical open pit uranium mine involves hundreds of acres and looks very impressive, the impact is quite small in comparison to that from an open-pit coal mine which can provide the same amount of useable energy. This comparison is a result of the fact that a pound of unrefined low-grade (0.01%) uranium ore contains at least 100 times the energy as a pound of coal.

2. Milling

The purpose of a uranium mill is to extract U_3O_8 from the raw ore so that it can be shipped in reasonably pure form to a conversion and purification plant. As of January 1977, there were 18 uranium mills in operation in the U.S. Essentially all of these are located close to major mining centers in the western states. The collective capacity of these mills is roughly 30,000 ton of ore per day. The smallest mill processes 400, the largest 7000 ton of ore daily.[28] The milling process is 90 to 95% efficient in extracting uranium from the ore.

FIGURE 8. A typical uranium mill and impoundment which receives tailings in slurry form from the mill. The solids settle to the bottom of the pond, gradually filling it, while the water evaporates or percolates into the ground. Foreground shows overburden poles from the adjacent open pit mine. These will be graded, covered with topsoil, and seeded.

Mills generally extract uranium using acid or alkali leach methods. Processing steps involve crushing and grinding of the ore; leaching with sulfuric acid or sodium carbonate; clarification to remove solid residues; recovery using chemical precipitation, ion exchange, or solvent extraction; and filtration and drying of the final product. Uranium is barreled at the mill for shipment as uranium oxide or salt concentrates, generally termed "yellowcake". Yellowcake contains 70 to 90% U_3O_8 (equivalent) by weight.

Residues of the uranium extraction process, termed "tailings" or "spoils" are usually pumped as a slurry to a liquid retention impoundment (Figure 8). The concept of such a tailings pond is to create a system which permits the solids to settle out and the liquids to evaporate. The liquids largely prevent the solids from drying out and becoming airborne dust. Because of the presence of acids or bases and other chemicals,[33] tailings ponds seldom support life. They generally are sterile systems which contain substantial quantities of radioactivity.

Uranium ores contain all members of the uranium series (Table 2). In most cases, it can be assumed that activities of most members of the series are in secular equilibrium with the ^{238}U parent. In practice, this means that for every nuclear disintegration of a ^{238}U nucleus, there is a nuclear disintegration of each radioactive daughter product. Making this assumption and also the assumption that 30,000 ton of 0.2% ore is processed daily in the U.S., one can calculate that about 15 Ci of ^{230}Th and ^{226}Ra enter tailings each day. Not only are these radionuclides long-lived and potentially hazardous but, in addition, ^{226}Ra produces gaseous ^{222}Rn, some of which escapes the tailings system. Daughter products of ^{222}Rn, such as 19-year ^{210}Pb, cause measurable radiation exposure to the surrounding biota. Radium forms soluble compounds and some may find its way into ground water, depending on geological and hydrological factors.

Despite the large quantities of radioactivity in tailings, proper management practices can reduce the escape of radionuclides to the extent that they do not pose a measurable hazard to humans or other organisms. Land use necessary for milling undoubtedly has a far greater environmental impact than radioactivity. The question of chemical effects of mill residues on the environment has received only limited investigation.

3. Purification and Conversion

Yellowcake from the mill is shipped to commercial refineries at Gore, Okla. or Metropolis, Ill., which convert uranium to high purity gaseous UF_6, used in the enrichment process. Two processes are used to convert yellowcake to UF_6 — the dry hydrofluor method, which processes the concentrates directly, and a wet solvent extraction method.[33] Intermediate chemical forms of uranium in the processes may include UO_2, UO_3, and UF_4.

The refining steps involve the handling of dry powder forms of uranium and the generation of fine dusts. Modern filtration systems are capable of preventing all but a tiny fraction of the uranium-bearing particles from reaching the environment. The value of high purity uranium presents economic incentive for maintenance of a high integrity operation in addition to regulatory requirements. Chemical effluents from the refineries are likely more significant than uranium discharges. Such effluents may contain NO_2, HF, H_2S, and organic solvents.[33] Radionuclides of uranium, thorium, and protactinium are present in solid wastes generated during UF_6 production. The total activity involved is some 6 Ci/year. The solid wastes consist largely of ash residue which is packaged for waste burial.[33]

4. Isotopic Enrichment

A crucial and costly step in the fuel cycle involves isotopic enrichment of fissile ^{235}U. Having a natural abundance of only 0.7%, the fissile atoms of ^{235}U in high purity unenriched uranium is insufficient to sustain a nuclear chain reaction. Uranium fuel suitable for nuclear reactors must contain about 3% ^{235}U, that suitable for weapons must be enriched to over 90%. Incidentally, the difference in enrichment is one of the fundamental reasons why a nuclear explosion in a nuclear reactor cannot physically happen.

The enrichment procedure involves gaseous diffusion of uranium in the form of UF_6 in large diffusion plants at Oak Ridge, Tenn., Paducah, Ky., and Portsmouth, Ohio. Enrichment by gaseous diffusion is based on the principal that molecules of $^{235}UF_6$ are roughly 0.85% lighter than molecules of $^{238}UF_6$, and therefore, diffuse slightly more rapidly. As UF_6 molecules follow a tortuous path in passing through multiple stages of the diffusion plant, the ratio of ^{235}U to ^{238}U gradually increases.

Isotopic enrichment by gaseous diffusion is expensive and requires considerable amounts of energy to compress and pump the gas through the thousands of porous barriers required to achieve adequate separation. More efficient methods of isotopic separation are being investigated, one of the more promising being centrifugation. When commercially realized, centrifugation will be far more efficient in terms of energy consumption than gaseous diffusion.

The major emissions resulting from gaseous diffusion operations are nonradiological. Coal-fired emissions, including solid particulates and heat, are of principal environmental importance. Radiological emissions include small amounts of UF_6 that escape containment and also small amounts of ^{234}Th and ^{234}Pa which are products from the decay of ^{238}U. These radioactive emissions are of little significance in comparison to those from other steps in the nuclear fuel cycle.

5. Fuel Element Fabrication

Following enrichment, uranium hexafluoride is shipped to a fuel fabrication plant where it is hydrolized to uranyl fluoride, converted to ammonium diuranate, and calcined to the dioxide (UO_2). The dioxide powder is pelletized, sintered, and loaded into stainless steel or Zircaloy tubing which is then capped and welded. Actually, the geometry of fuel elements varies as rods, pins, plates, or tubes may be used. The main purposes of the fuel element cladding are to prevent the escape of fission products and to protect the fuel from the environment of the fuel element. As contrasted to only three gaseous diffusion plants in the U.S. there are a considerable number of fuel fabrication plants. The EPA[34] lists ten commercial plants in the U.S.: two in Connecticut, and one each in Missouri, North Carolina, Oklahoma, Pennsylvania, South Carolina, Tennessee, Virginia, and Washington. During fabrication, some of the enriched uranium is released, and very small quantities may be detected in the environment. Chemical effluents are produced in gaseous, liquid, and solid forms.[33]

6. Radionuclides in a Reactor

The reactor is the dominant feature of the nuclear fuel cycle. This is where the energy used to produce and dispose of the fuel is more than compensated for. The ratio of usable energy produced by reactors to the energy required to operate them (including the entire fuel cycle) is about 3.6 to 1.[32] This ratio can be improved considerably if more efficient methods of enrichment, such as centrifugation, can be successfully developed. Space does not permit, and ample literature[1,7] obviates, the need for a lengthy discussion here of the principle of reactor operation. However, a very brief review of reactor fundamentals and the associated production of radionuclides is useful at this point.

As discussed in Chapter 3, Section V the fission process in the core of a nuclear reactor leads to the production of a large spectrum of radioactive fission fragments, large numbers of neutrons, and large amounts of energy, most of which appear as heat within the core. The neutrons are necessary for maintenance of the nuclear chain reaction, and the heat is converted in various ways, depending upon the reactor design, into steam, which is used to spin a turbine and electrical generator. Figure 9 diagrams the principles of steam turbine generation of electrical power using fossil and nuclear fuels.

Radioactive fission products accumulate in the fuel elements as the fuel itself is consumed. Most of the 200 or so fission products are short-lived and decay in the reactor core during tenure of the fuel element. This decay energy is mostly absorbed within the reactor vessel and converted to useful heat. However, some of the fission products are longer-lived and have a high probability of decaying sometime after the fuel element is spent and is removed from the reactor. Some of the fission products produced in a nuclear reactor that are of potential biological importance are listed in Table 8. The biological importance of the fission products is determined by a combination of

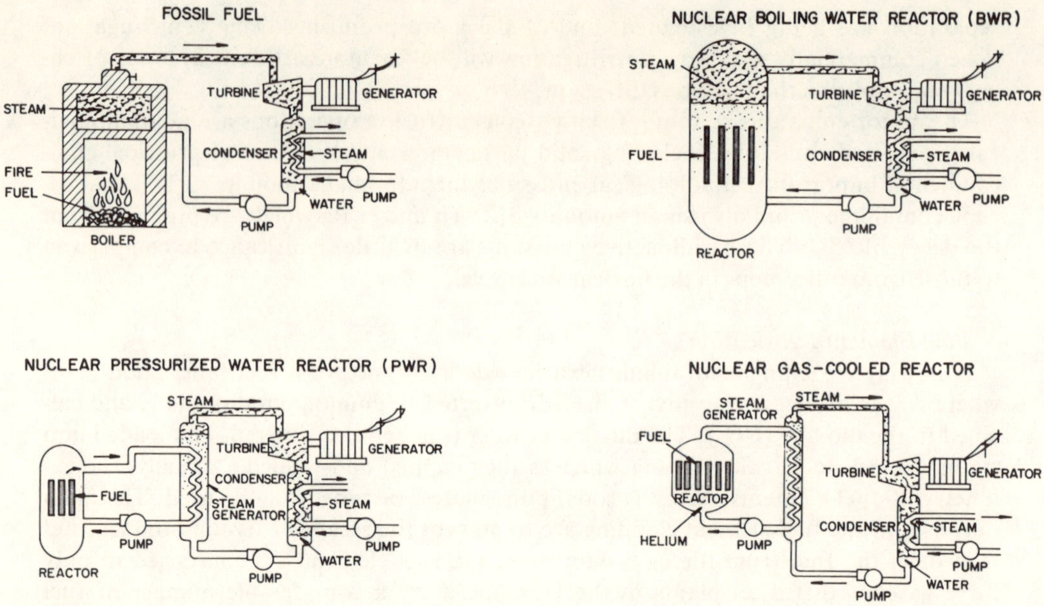

FIGURE 9. Types of steam-electric generating plants. (Photograph courtesy of U.S. Atomic Energy Commission, Washington, D.C.)

Table 8
SOME FISSION PRODUCT RADIONUCLIDES OF POTENTIAL BIOLOGICAL IMPORTANCE

Radionuclide	Fission yield (%)[a]	Radiation	Half-life	Important element analogs
^{3}H	0.01	β	12 year	H
^{85}Kr	0.29	β,γ	10 year	—
^{90}Sr	5.77	β	28 year	Ca
^{89}Sr	4.79	β	51 day	Ca
^{137}Cs	6.15	β,γ	27 year	K
^{131}I	3.1	β,γ	8.1 day	I
^{129}I	0.9	β,γ	1.7×10^{7} year	I
^{144}Ce[b]	6.0	β,γ	285 day	—
^{103}Ru[b]	3.0	β,γ	40 day	—
^{106}Ru[b]	0.38	β,γ	1.0 year	—
^{95}Zr[b]	6.2	β,γ	65 day	—
^{140}Ba[b]	6.32	β,γ	12.8 day	Ca
^{91}Y	5.4	β,γ	58 day	—
^{143}Ce[b]	5.7	β,γ	33 hr	—
^{147}Nd[b]	2.7	β,γ	11 day	—

[a] Based upon thermal neutron fission of ^{235}U.
[b] Decay to radioactive daughters.

Table 9
SOME RADIONUCLIDES PRODUCED BY NEUTRON ACTIVATION IN NUCLEAR REACTORS WHICH ARE OF POTENTIAL BIOLOGICAL SIGNIFICANCE

Radionuclide	Radiation	Half-life	Important element analogs
^{3}H	β	12.3 year	H
^{14}C	β	5568 year	C
^{24}Na	β,γ	15 hr	Na
^{32}P	β	14 day	P
^{35}S	β	87 day	S
^{41}A	β,γ	110 min	—
^{45}Ca	β	164 day	Ca
^{54}Mn	γ	291 day	Mn
^{55}Fe	χ (EC)[a]	2.6 year	Fe
^{59}Fe	β,γ	45 day	Fe
^{57}Co	γ	270 day	Co
^{58}Co	β^{+},γ	71 day	Co
^{60}Co	β,γ	5.2 year	Co
^{65}Zn	β^{+},γ	245 day	Zn
^{239}Pu	α,γ	24,360 year	—
^{239}Np	β,γ	2.3 day	—
^{241}Am	α,γ	470 year	—
^{242}Cm	α,γ	163 day	—

[a] Electron capture.

factors such as fission yield, half-life, chemical properties, solubility, etc., as will be discussed in Chapter 5.

Neutrons, in addition to maintaining the chain reaction, react with nonfissile nuclei to produce a large array of radioactive elements, called ''activation products'' (Table 9). The quantity and type of activation products produced is governed by the neutron flux and the kinds and amounts of elements within the range of the neutrons. Impurities and functional elements within the fuel, the coolant, and the structural parts of the reactor are nearly all candidates for neutron activation.

Fission and neutron activation products, then, are the major classes of radionuclides produced in nuclear reactors. The fuel element cladding and closed-loop cooling designs of modern reactors (Figure 9) almost totally preclude the escape of radionuclides to the environment. Yet, small amounts do escape and these can be measured in low concentrations in gaseous and liquid reactor effluents.[7] Perfect integrity of the myriad of fuel elements and pipes has not and is never likely to be attained. Pinholes and hairline fractures occasionally occur in fuel cladding and other structures under the high temperatures of a reactor core. In addition, some elements, and ^{3}H in particular, can slowly migrate through solid barriers. Eisenbud[7] presents a detailed discussion of the mechanisms of radionuclide escape from reactor systems.

Despite the lack of perfect containment of reactor systems, which allows a very small fraction of the core inventory to escape, reactor discharges have been remarkably low and well within permissible release limits which are based upon dose to the public surrounding each power plant (Tables 10 and 11). Radionuclide release limits are established by state and federal agencies for each reactor. Site-specific limits are determined on the basis of meteorology, area ecology, human population, reactor design, and other criteria.

Table 10
RELEASES OF RADIOACTIVITY IN GASEOUS EFFLUENTS FROM U.S. POWER REACTORS FROM 1967 THROUGH 1970

Facility	Noble and activation gases			Halogens and particulates		
	Released (Ci)	Permissible (Ci)	% of permissible	Released (Ci)	Permissible (Ci)	% of permissible
Oyster Creek	110,000	9.4×10^6	1.2	0.32	130	0.25
Dresden I	910,000	1.8×10^7	5.2	3.3	75	4.3
Nine Mile Point	9,500	2.7×10^7	0.037	<0.06	63	<0.1
Dresden III	250,000	2.2×10^7	1.1	1.6	110	1.4
Humboldt Bay	540,000	1.6×10^6	34	0.35	5.6	6.2
San Onofre	4,200	5.7×10^5	0.75	<0.0001	0.8	<0.001
Ginna	10,000	3.6×10^5	2.8	0.05	1.7	3
Big Rock Point	280,000	3.0×10^7	0.88	0.13	37	0.35
Connecticut Yankee	700	2.8×10^5	0.24	0.0015	0.2	0.7
Saxton	2,200	3.7×10^3	59	0.15	10	1.5
Indian Point I	1,800	5.3×10^6	0.03	0.075	7.6	1
Peach Bottom	5.7	1.9×10^5	0.003	<0.0006	0.1	<0.6
Yankee Rowe	17	6.6×10^3	0.26	Not measured		—
La Crosse	950	3.2×10^5	0.3	<0.063	1.6	<4

Data from Eisenbud, M., *Environmental Radioactivity,* 2nd ed., Academic Press, New York, 1973. With permission.

Table 11
RELEASES OF RADIOACTIVITY IN LIQUID EFFLUENTS FROM U.S. POWER REACTORS DURING 1970

Facility	Mixed fission and corrosion products			Tritium	
	Released (Ci)	Concentration limit (10^{-7} μCi/mℓ)	% of limit	Released (Ci)	% of limit
Oyster Creek	18.5	1	71	22	0.0028
Dresden I	8.2	1	25	5	0.0005
Nine Mile Point	28	1	21	50	0.0015
Dresden II	13	1	15	31	0.0012
Humboldt Bay	2.4	1	15	<7	<0.0014
San Onofre	7.6	1	12	4800	0.26
Ginna	10	3	5	110	0.005
Big Rock Point	4.7	15	3.1	54	0.018
Connecticut Yankee	6.7	3.6	2.8	7400	0.38
Saxton	0.012	1	0.33	10	0.01
Indian Point I	7.8	70	0.28	410	0.03
Peach Bottom	0.006	1	0.15	<50	<0.05
Yankee Rowe	0.034	1	0.14	1500	0.21
La Crosse	6.4	300	0.067	20	0.002

Data from Eisenbud, M., *Environmental Radioactivity,* 2nd ed., Academic Press, New York, 1973. With permission.

7. Fuel Reprocessing

Two major options exist for spent fuel elements that have "cooled" for a period of time (>3 months) to permit decay of shorter-lived radionuclides.[35] In one option, the elements can be chemically reprocessed to recover residual ^{235}U and ^{239}Pu formed from ^{238}U or ^{233}U formed from ^{232}Th. In the other, the spent fuel can be disposed of directly. Reprocessing conserves uranium ore and may ultimately have economic advantages, but it has the disadvantage of requiring another major facility which utilizes land and releases some chemical and radioactive material to the environment. The "no recycle" option wastes natural fissile material, and alters the technical problems of waste disposal, but may have the advantage of making it more difficult for terrorists to steal purified fissile material and divert it for military purposes. The latter option has received considerable discussion.[32,36]

Government-owned reprocessing facilities have operated at Hanford, Wash., Oak Ridge, Tenn., Savannah River, Plant, S.C., and Idaho Falls. In the past, the major purpose of these facilities was to recover plutonium for national defense. Three commercial reprocessing facilities have been constructed in the U.S. They are located at West Valley, N.Y., Morris, Ill., and Barnwell, S.C. Development of commercial reprocessing capability has not kept pace with the needs of the nuclear industry, in part because of the expensive, elaborate processes required, and in part because of indecision about the reprocessing option itself. Commercial reprocessing is somewhat more difficult than reprocessing for weapons plutonium because the commercial fuel elements reside in the reactor 5 to 10 times longer, which boosts the radionuclide inventory substantially. For economic and technical reasons, none of the commercial reprocessing plants are operating at present (1979).

Fuel reprocessing involves shearing the elements into small pieces and dissolution in nitric acid. A solvent extraction procedure, termed the "Purex process", employs tributyl phosphate to separate uranium and plutonium from the fission products. Uranium is chemically converted to UF_6 for shipment to the isotopic enrichment facility, while plutonium is converted to the dioxide form. It is desirable to convert the fission products to solid forms for disposal. In some cases, long-lived ^{90}Sr and ^{137}Cs are chemically separated and encapsulated for storage or disposal.

The intense radioactivity of the spent fuel elements complicates the reprocessing scheme. The plant must be designed for remote operation to protect workers and the process design must account for radiolytic decomposition of chemicals. The large fission-product inventory requires elaborate systems to prevent dangerous quantities of radioactivity from escaping to the environment.

Eisenbud[7] cites data from Magno et al. and Schleien on environmental releases and resulting dose commitments from the Nuclear Fuel Services reprocessing plant at West Valley, N.Y. During a 6-month period in 1969, the plant released 1700 Ci ^{3}H, 52 Ci ^{106}Ru, 8 Ci each of ^{90}Sr and ^{137}Cs, and lesser quantities of other fission and activation products. Detectable levels of these nuclides were found in deer, fish, and dietary items. Although it appears that fuel reprocessing tends to release substantially more radioactivity to the environment than power reactors, the amounts released under normal conditions do not appear to constitute a biological hazard, but they do warrant careful monitoring.

8. Waste Disposal

"Waste" is considered to be material for which there is no need, or which is in such a form that it is too expensive to reclaim. Virtually every component of the nuclear industry, and other industries too, produce radioactive wastes. In the eyes of the public, safe disposal of radioactive wastes is perceived as probably the greatest technical problem of the nuclear industry.[37] This concern is understandable because extremely

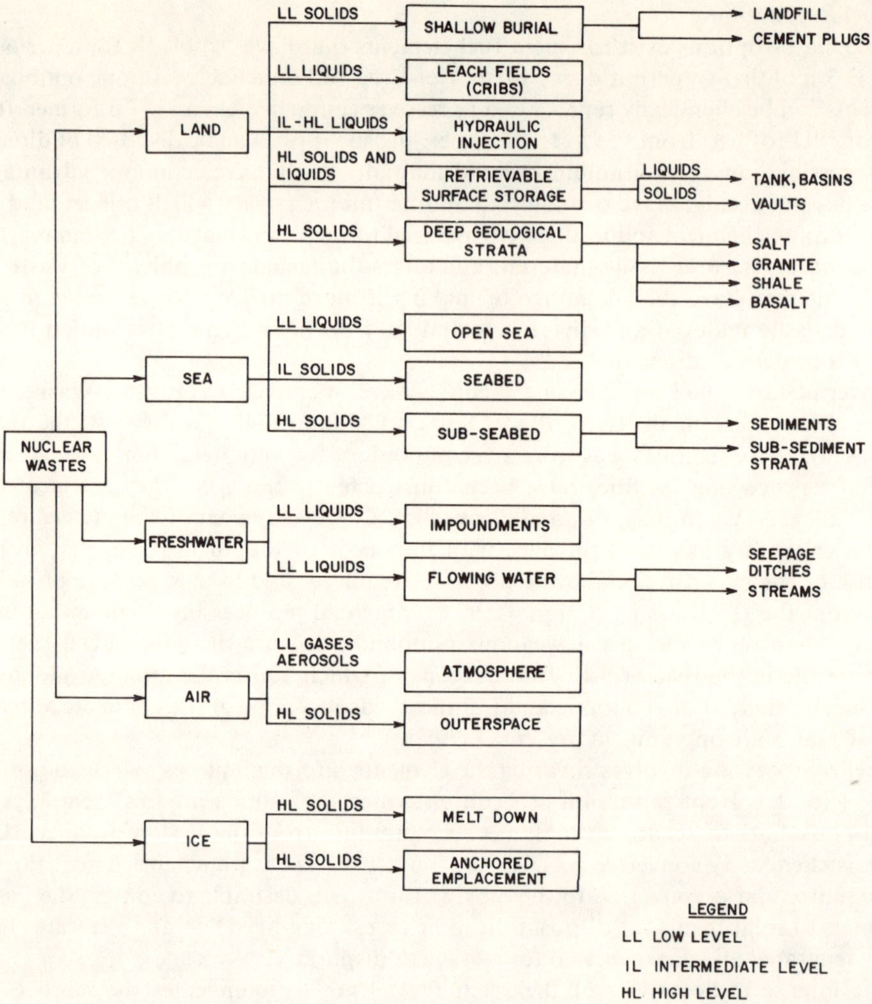

FIGURE 10. Potentially feasible disposal sites and methods for nuclear waste materials.

large quantities of radioactivity are involved, absolutely "safe" storage has not been demonstrated, and present storage systems have confirmed public suspicion that man-made barriers are not perfectly reliable for the lifetime of the wastes.

On the other hand, those professionally involved with nuclear waste management are confident that there are numerous satisfactory options for waste disposal, and that it is mostly a matter of choosing the most feasible of these (Figure 10). The choice between options will be governed by technical, economic, moral, and political criteria.[38]

The existing inventory of nuclear wastes in the U.S. is composed largely of high-level defense or military wastes and commercial wastes. Some 75 million gallons of high-level defense wastes are currently stored in roughly 200 underground tanks at Hanford, Savannah River, and Idaho Falls U.S. Department of Energy (DOE) installations. Hundreds of megacuries of fission products and hundreds of kilograms of plutonium and other actinides are included in these wastes. In addition, some 2500 metric tons of spent fuel has been discharged from the 64 commercial reactors now (1975) operating. Most of this material is in temporary storage at reactor sites, awaiting

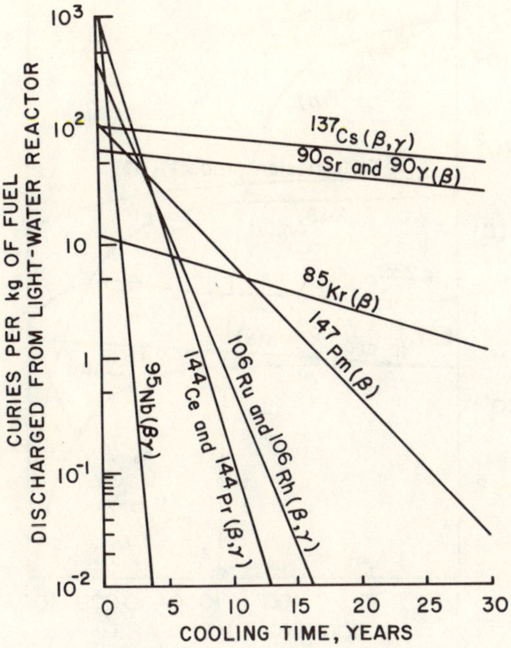

FIGURE 11. Isotopic activity (Ci/kg of U) of fission products in waste resulting from a kilogram of uranium charged to a light water reactor with a burn-up of 33,000 MWD/metric ton. (Reproduced from Committee on Radioactive Waste Management, *Interim Storage of Solidified High-Level Radioactive Wastes,* 1975, With permission of the National Academy of Sciences, Washington, D.C.)

decisions about recycling and disposal options. Representative kinds and quantities of commercial wastes are illustrated in Figures 11 and 12. In addition, over 1 million cubic meters of solid low-level wastes, containing over 8 MCi of fission products and 900 kg of transuranium nuclides were stored and buried at DOE sites, as of 1974.[40]

An important characteristic of nuclear wastes is the large range of activities and half-lives of the constituent radionuclides. This results in a total hazard which varies with time, according to the kinds and amounts of specific radionuclides present. For example, Figure 13 illustrates the quantities of water needed to dilute a given volume of solidified high-level waste to allowable concentration guidelines, as a function of time. This scale may be termed a "hazard index". Note that the fission products (^{137}Cs and ^{90}Sr, primarily) present the greatest potential hazard for some 500 to 1000 years. Beyond 1000 years, the long-lived actinides clearly present a greater hazard than the fission products. The hazard from ^{99}Tc remains stable for some 30,000 years because of its 2×10^5 year half-life. Natural uranium ore presents an essentially constant hazard for millions of years because ^{238}U has a half-life of 4.5 billion years. While a number of radionuclides in nuclear wastes have extremely long half-lives, let us not forget that chemical toxins such as lead, arsenic, and mercury have lifetimes that are infinite.

There are two basic philosophical options for the handling of radioactive wastes (or any other toxic substance, for that matter). One option is to dilute the material in the environment (air, soil, water) to the point that the concentrations become innocuous. This, in effect, was the fate of huge quantities of radioactivity generated during the atmospheric testing of nuclear weapons. In practice, this option is routinely used for

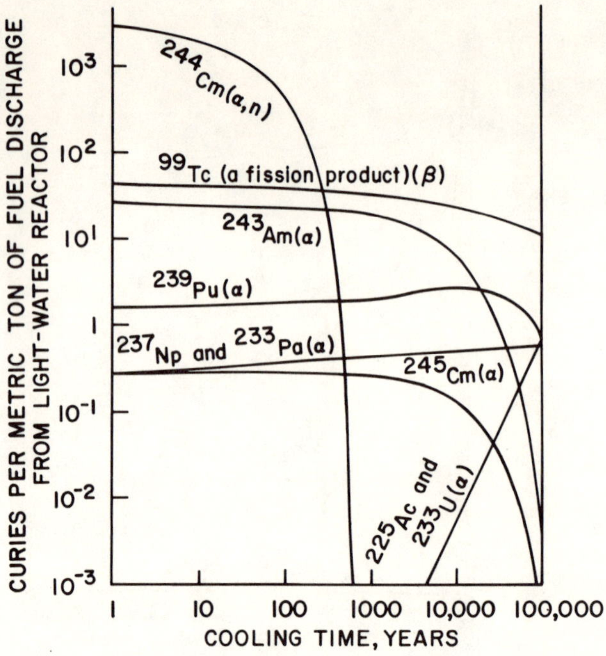

FIGURE 12. Isotopic activity (Ci/t of U) of actinides and their daughters in waste resulting from a metric ton of uranium charged to a light water reactor, with a burn-up of 33,000 MWD/metric ton. (Reproduced from Committee on Radioactive Waste Management, *Interim Storage of Solidified High-Level Radioactive Wastes,* 1975, with permission of the National Academy of Sciences, Washington, D.C.)

disposal of very low-level radioactive wastes generated by the nuclear as well as other industries. The second option is that of containment and isolation. Containment can be achieved through the use of barriers, both engineered and natural. Isolation may be achieved through placement strategy so that the material is far removed from man and the biosphere. The containment/isolation option, though far more expensive than dilution, is necessary for high-level wastes.

Referring to Figure 10, it is clear that wastes can be put in several places. A major factor governing the disposal medium and method is the physical form and activity level of the waste. Low-level solids for instance, are normally destined for shallow burial at an approved site (Figure 14).[40] The waste material may be unpackaged, or rather elaborate packaging in metal drums, concrete vaults, etc., may precede shallow burial, depending upon the degree of protection needed. Low-level liquids have been handled in various ways, including underground discharge using cribs, or discharges to the ocean or a body of freshwater. Underground liquid discharges rely upon the adsorptive capacity of geologic material to prevent the spread of radioactivity, while discharges to open bodies of water depend upon dilution and/or adsorption by sediments to reduce the radiation hazard. Low-level radioactivity in gaseous or aerosol form may be discharged to the atmosphere for dispersion and dilution.

Wastes in the more or less intermediate level category have been disposed of by hydraulic injection into deep geological strata and by dumping into the ocean. The hydraulic injection technique is similar to that commonly used by the natural gas industry to fracture gas-bearing formations which increases gas permeability and, hence, gas availability. Oceanic dumping has generally involved solid wastes that are carefully

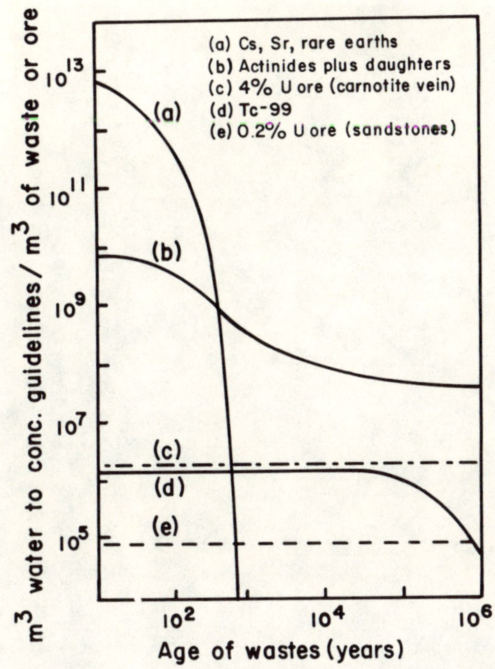

FIGURE 13. The ingestion hazard index (volume of water needed to dilute a volume of waste or ore to allowable concentration guidelines for solidified HLW from the reprocessing of enriched-uranium light water reactor fuel. It is assumed that 99.5% of the uranium and plutonium in the spent fuel and all of the krypton and iodine are removed before solidification. For comparison, the hazard indexes for a fairly rich Colorado carnotite vein ore and for more typical 0.2% concentration sandstone ores are also indicated (Adapted from Rochlin G.I., *Science,* 195(4273), 23, 1977. Copyright 1977 by the American Association for the Advancement of Science.)

packaged, in concrete-filled drums, for instance, prior to dumping. These drums sink to the seabed where they reside intact for some period of years. The nature of wastes disposed of in this way has been such that by the time the drums deteriorate, the activity has been reduced by decay and the sea water concentrations are reduced with distance by dilution and adsorption to the sediments.

It is largely the question of what to do with high-level wastes that has attracted so much public as well as scientific attention in recent years.[41] The principal options for high-level wastes include retrievable surface storage (which is where the existing high-level wastes are now located); entombment in deep geological strata, either below land or sea; injection into outer space; and placement in thick ice sheets. The latter two options have been all but ruled out. Space injection would be risky and very expensive. Placement in ice is also risky because the long-term behavior of ice cannot be well-predicted. Present surface storage systems, while adequate as an interim measure, probably do not offer long-term (hundreds to thousands of years) security. While surface storage systems can be vastly improved,[39] the question of long-term integrity still cannot be answered with certainty. These concerns leave deep geological storage as the most attractive option for high-level wastes.

FIGURE 14. A shallow burial trench for low-level solid wastes.

The seabed option for high-level wastes has been considered in substantial detail.[42] The attractiveness of this option lies in the predictability and stability of certain lithospheric plates, the comparative tranquility of such deep ocean floors, the international location, and other factors. High-level wastes can be placed in high-integrity forms (such as glass) and covered with a canister. Entombment of such canisters in deep rock or sediments would provide multiple barriers to radionuclide migration. If small quantities of long-lived radionuclides were to eventually migrate to the ocean floor, the vastness of the ocean would provide tremendous dilution capacity.

A major advantage of deep geologic disposal under land is that the risks associated with transportation would be diminished, as compared to seabed disposal. A number of potential geologic disposal sites in the U.S., including salt mines and hard rock formations, have been considered. At Hanford, where the majority of high-level defense wastes now reside, a major program is underway to examine the feasibility of storage in deep basalt underneath the existing waste storage area. Major questions being asked include the integrity and stability of the rock, and the hydrologic scheme. The presence of ground water and its movement, now and in the future, would present a credible radionuclide transport mechanism and thus is of great concern.

The question of radionuclide migration in geological strata is similar, whether the strata are under land or sea. This was discussed recently by de Marsily et al.[43] Certainly, the ultimate choice of disposal locations will take into account the rates at which the long-lived radionuclides can migrate through a multiple barrier system. It will need to be demonstrated that the waste material will have decayed to nonhazardous levels by the time it can possibly reach the biosphere.

What are the present radioecological concerns in regard to nuclear waste management? The existing high-level wastes are reasonably well isolated from the biosphere and, discounting a highly improbable catastrophic accident or event, they are likely to remain isolated. Although some large high-level waste tanks have developed leaks and some radioactivity has migrated into geological strata and aquifers, hazardous quantities have not reached the earth's surface. Of greater immediate ecological concern are the low-level wastes in shallow land-fills, seepage ditches, and retention ponds. While there is no evidence that ionizing radiation in such areas is causing measurable ecological perturbations, biological processes can cause dispersal of radioactivity from such areas to the general environment. Such processes include, for example, radionuclide uptake by deep-rooted plants which later blow away, burrowing into wastes by mammals, emergence of contaminated aquatic insects, and utilization of contaminated ponds by migrating waterfowl and other birds. Such processes are currently under investigation by radioecologists at Hanford, Idaho Falls, Oak Ridge, Savannah River, and other major nuclear facilities.

9. Radiation Dose from the Nuclear Fuel Cycle

Radiation doses to the public from each component of the nuclear fuel cycle have been estimated on a generic basis for terrestrial and aquatic exposure pathways.[44,45] Such doses have been calculated on the basis of expected routine releases of radionuclides for humans living at nuclear site boundaries and deriving sustenance from food produced at the boundaries of such sites. Internal doses from inhalation and ingestion are considered, as well as external exposure from radionuclides in the environment. This approach, while intended to provide potential doses to man, usually gives a rough (order of magnitude) idea of doses to plants and other animals at the same location, and thus has some ecological significance and utility. Doses to man are compared with applicable standards, generally accepted as conservative by one or two orders of magnitude, but there are no established "standards" for doses to nonhuman receptors.

Table 12
DOSE RATES VIA ALL TERRESTRIAL PATHWAYS TO INDIVIDUALS AT THE SITE BOUNDARY OF MODEL LIGHT WATER REACTOR NUCLEAR FUEL-CYCLE FACILITIES

		Dose equivalent[a] (rem/year of operation)						
Installation	Receptor	Total body	Bone	Lung	Thyroid	Kidneys	Liver	Lower large intestine
Uranium mill	Adult	0.18[b]	0.35[c]	0.22[c]	0.18[c]	0.047[c]	0.017[d]	0.049[c]
	Child	0.45[b]	0.75[b]	0.47[b]	0.45[b]	0.021[d]	0.0078[d]	0.0031
Fluorination (UF_6)	Adult	0.019[d]	0.095[c]	0.034[c]	0.019[d]	0.017[d]	0.0046	0.0028
	Child	0.04[c]	0.087[c]	0.053[c]	0.040[c]	0.0047	0.0025	0.0010
Uranium enrichment	Adult	0.0013	0.010[d]	0.0072[d]	0.0014	0.0026	0.0011	0.0015
(gaseous diffusion)	Child	0.0015	0.014[d]	0.010[d]	0.0016	0.0012	0.0013	0.00060
Fuel fabrication	Adult	0.00057	0.0047	0.0033	0.00060	0.0012	0.00046	0.00046
	Child	0.00068	0.0063[d]	0.0046	0.00071	0.00050	0.00058	0.00021
Pressurized water reactor	Adult	0.0021	0.0036	0.00097	0.0015	0.0012	0.0013	0.0016
	Child	0.0021	0.0036	0.00095	0.0073	0.0012	0.0013	0.0016
Boiling water reactor	Adult	0.01[d]	0.014[d]	0.0092[d]	0.0096	0.0076[d]	0.0084[d]	0.014[d]
	Child	0.01[d]	0.014[d]	0.0087[d]	0.020[d]	0.0076[d]	0.0084[d]	0.0091[d]
Fuel reprocessing plant	Adult	0.01[d]	0.020[d]	0.0092[d]	0.46[c]	0.0098[d]	0.0098[d]	0.033[c]
	Child	0.013[d]	0.023[d]	0.010[d]	2.1[e]	0.012[d]	0.016[d]	0.0098[d]

[a] These values are hypothetical dose equivalents calculated to identify potentially critical pathways of radionuclide transport from model fuel-cycle facilities.
[b] Potentially approaches (>25%) International Commission on Radiological Protection (ICRP) recommended dose limits.
[c] Potentially exceeds U.S. Environmental Protection Agency (EPA) proposed standards.
[d] Potentially approaches (>25%) EPA proposed standards.
[e] Potentially exceeds ICRP recommended dose limits.

From Hoffman, F. O. and Kaye, S. V., Terrestrial exposure pathways: potential exposures of man from the environmental transport of waste nuclides, in Proc. Int. Symp. Management of Wastes from the LWR Fuel Cycle, U.S. ERDA Rep. CONF-76-0701, U.S. Energy Research and Development Administration, Washington, D.C., 1976, 524.

Table 13
POTENTIAL INDIVIDUAL HUMAN DOSE RATES FROM LIQUID DISCHARGES

Installation	Individual dose equivalent (mrem/year)		
	Thyroid	Total body	Bone
Open pit mine			Nil
Uranium mill			13
Fluorination (UF_6)			2
Uranium enrichment			0.7
Fuel fabrication			0.6
Light-water reactor[a]	0.7	6	5
Fuel reprocessing plant	10		

[a] Includes crop irrigation pathway.

Data adapted from Soldat, J. K., Aquatic exposure pathways—potential exposure of man from environmental transport of waste radionuclides, in Proc. Int. Symp. Management of Wastes from the LWR Fuel Cycle, U.S. ERDA Rep. CONF-76-0701, U.S. Energy Research and Development Administration, Washington, D.C., 1976, 539.

Representative radiation doses calculated on the basis of these criteria are shown in Tables 12 and 13. It seems clear, that under normal operating conditions, the maximum credible* individual doses from the nuclear fuel cycle are within applicable standards. These doses apply to very limited populations and generally represent a small fraction of typical background radiation exposures. While these and similarly derived numbers are reassuring, they are not grounds for complacency, and site-specific cases should continue to be evaluated in the future.

B. Nuclear Explosions

Among human activities which have led to a ubiquitous distribution of artificial radioactivity in the biosphere, fallout from atmospheric nuclear explosions is perhaps the most noteworthy. The initial discovery that neutrons could cause uranium nuclei to fission, and that such fission released enormous amounts of energy, eventually led to the development of explosive devices as well as nuclear reactors. An unavoidable result of the fission reaction is the formation of radioactive fragments which are released to the surroundings in a nuclear explosion. If released to the atmosphere, radioactive products of nuclear explosions ultimately "fallout" or settle to earth. The ubiquitous distribution of fallout is illustrated by the likelihood that every organism within the biosphere contains some of the radionuclides from nuclear weapons testing.

Nuclear explosive devices have been employed as weapons of war in the cases of Hiroshima and Nagasaki, Japan. However, it has been the testing of such devices for both peaceful and military applications that has released the preponderance of artificial radioactivity that is now dispersed globally.

1. Historical Perspective

The first nuclear explosion in history was from a 19 KT (TNT-equivalent) device at the Trinity Site near Alamogordo, N.M., on July 16, 1945. On August 5, 1945, a nominal yield bomb (~20 KT) was dropped on Hiroshima by the U.S. and 4 days

* Maximum credible dose implies that set of conditions, however unusual, that would result in the maximum possible dose. The "probable" dose will likely be considerably lower than the maximum credible dose.

Table 14
APPROXIMATE FISSION AND TOTAL YIELDS (IN MEGATONS) OF NUCLEAR WEAPONS TESTS CONDUCTED IN THE ATMOSPHERE BY ALL NATIONS THROUGH 1962

Inclusive years	Fission yield (MT)	Total yield (MT)
1945—1951	0.76	0.76
1952—1954	38	60
1955—1956	13	28
1957—1958	40	85
1959—1960	Test moratorium	
1961—1962	155	442
Totals	247	616

Data from Federal Radiation Council, Estimates and Evaluation of Fallout in the United States from Nuclear Testing Conducted Through 1962, Rep. No. 4, Federal Radiation Council, Washington, D.C., 1963.

later, a similar device was dropped on Nagasaki. A long history of nuclear weapons testing began in 1946 with U.S. activities in the Pacific. On August 29, 1949, the U.S.S.R. detonated its first nuclear device and in 1952 the U.K. also demonstrated nuclear weapons capability with an explosion in Australia. Since 1960, France, the People's Republic of China, and India have also detonated nuclear explosives, which clearly points out that nuclear capability is within the technological reach of many nations.

Thermonuclear capability was demonstrated in 1952 by the U.S. and in 1957 by the U.S.S.R. It was demonstrated in subsequent years that thermonuclear devices could release energy in the megaton range. The first thermonuclear device was detonated by the U.S. in 1954 at Bikini Atoll. The radioactive debris from this device fell out on Japanese fishermen on the ship Lucky Dragon and on Marshall Islanders. As a result of these unintentional exposures, and the ubiquitous presence of radioactive fallout, worldwide concern about testing began to surface.

In 1963, the countries considered as major nuclear powers at that time (U.S., U.K., and U.S.S.R.) signed the Limited Test Ban Treaty. This treaty forbade nuclear testing in the atmosphere, underwater, or in outer space, but permitted underground detonations. The U.S. and U.S.S.R. have maintained active underground testing programs since 1963, and France and the People's Republic of China, neither of whom signed the 1963 Test Ban Treaty, have conducted atmospheric tests since that time. In March 1976, the U.S. and U.S.S.R. agreed without a formal treaty to limit the size of underground detonations to 150 KT or less.

In spite of testing activities since 1963, the vast majority of radioactive debris released to the biosphere over the history of nuclear testing was done so by the U.S. and U.S.S.R. prior to that time. Table 14 lists fission and total yields generated by all nations through 1962. Total yield includes explosive yields from fission plus fusion. Clearly, the peak periods of biospheric contamination from weapons testing occurred in the early and late 1950s and early 1960s.

Table 15
NUMBER OF NUCLEAR TESTS CONDUCTED BY MAJOR NUCLEAR POWERS FOR THE 10-YEAR PERIOD FOLLOWING THE LIMITED NUCLEAR TEST BAN TREATY OF 1963, AND FOR THE PERIOD 1945—1973

Period	Environment	U.S.A.	U.S.S.R.	U.K.	France	China	Total
1963—1973	Atmosphere	0	0	0	29	14	43
	Subsurface	261	121	2	9	1	394
	Total	261	121	2	38	15	437
1945—1973	Atmosphere	193	161	21	33	14	422
	Subsurface	372	124	4	13	1	514
	Total	565	285	25	46	15	963

Adapted from Barnaby, F., *I.A.E.A., Bull.*, 15(4), 13, 1973.

Table 15 indicates the number of nuclear tests conducted by major nuclear powers between 1945 and 1973. While the number of tests is not necessarily a good indication of the amounts of radioactivity released, it does clarify the important periods of testing. As of June 30, 1976, the U.S. has conducted a grand total of 588 nuclear tests (including peaceful applications), with 466 of these being at the Nevada Test Site.[48]

Geographic locations of nuclear testing sites are of historical interest and have influenced the global distributions of fallout radionuclides. Table 16 lists major test sites by countries. Most testing has been conducted in the northern hemisphere, the U.K. and France being the only countries to test in the southern. The geographical list is not entirely complete, since specific nuclear experiments have been conducted in other locales. For instance, the U.S. has conducted underground tests on Amchitka Island in Alaska, Mississippi, New Mexico, and Colorado. The latter two locations were used to test the effectiveness of buried nuclear devices to release natural gas from impermeable geologic formations.

The amounts of radioactivity that have been produced by nuclear testing can be estimated from the total fission yields and the radionuclide activities per megaton of yield. Table 17 lists the approximate yields of several fission and activation products. Using these data, it is estimated that some 40 MCi ^{137}Cs and 25 MCi ^{90}Sr were produced by testing through 1962. As of 1972, some 12 MCi of ^{90}Sr could be accounted for in the biosphere through the worldwide sampling network of the U.S. Atomic Energy Commission's (USAEC) Health and Safety Laboratory.[52] Thermonuclear devices also produce large quantities of activation products because of the abundance of neutrons released. In addition, transuranics, such as ^{239}Pu, have been produced by nuclear testing in sufficient abundance to make possible its detection in the biosphere.

The future of nuclear testing seems to be uncertain because of the advancing technical capabilities of most nations and the unpredictable political climate. Underground testing by the major nuclear powers is likely to continue because of potential peaceful as well as military applications. The prospect of nuclear warfare is not pleasant, but the risk of it will persist into the foreseeable future.

2. Formation of Fallout

Radioactive fallout results from nuclear explosions in which all or part of the debris cloud is dissipated in the atmosphere. The type and composition of the nuclear device markedly affects the kinds of radioactivity produced, while the size and location of detonation determine the quantity of radioactivity which is released to the biosphere. Nuclear devices may operate from the fission process alone or from a combination of

Table 16
MAJOR NUCLEAR EXPLOSIVE TEST SITES

U.S.
- Nevada Test Site
- Central Nevada Test Site
- Amchitka Island, Alaska
- Pacific Proving Ground
  - Bikini Atoll
  - Eniwetok Atoll
  - Johnston Island
  - Christmas Island
- Plus other underground sites in the contiguous U.S.

U.K.
- Monte Bello Island, Australia
- Woomera, Australia
- Maralinga, Australia
- Christmas Island

U.S.S.R.
- Novaya Zemlya
- Semipalatinsk
- Ural Test Site plus other underground sites in the contiguous U.S.S.R.

Republic of France
- Reggan, Algeria
- In Ekker, Algeria
- Tuamotu Islands

People's Republic of China
- Lop Nor

India
- Rajasthan Desert

From Hardy, E. P., Jr., Rivera, J., and Collins, W. R., Jr., Eds./Comps., Fallout Program Quarterly Summary Rep. (September 1, 1963 through December 1, 1963), U.S. AEC Rep. HASL-142, Health and Safety Laboratory, U.S. Atomic Energy Commission, Washington, D.C., 1964.

fission and fusion processes. The former device, called the "atom" or "nuclear" bomb, produces large quantities of fission products and is therefore referred to as a "dirty" bomb. The fission-fusion device, known as the "thermonuclear" bomb, derives most of its energy from the fusion process and produces less radioactivity. Therefore, for equivalent energy produced, the thermonuclear bomb is "clean" relative to the pure fission bomb.

In nuclear fission, a free neutron which enters the nucleus of ^{235}U or ^{239}Pu will cause a fissioning of the nucleus into smaller fragments and the simultaneous release of energy and two or three neutrons. The fission-produced neutrons are free to interact with other fissionable nuclei and thus the chain reaction is initiated. Providing there are enough ^{235}U or ^{239}Pu nuclei packed within a sufficiently small space to give a "critical mass", then the number of fissions will increase geometrically within a fraction of a second and the explosive release of energy promptly follows. The basic principle of a fission device is the rapid joining of subcritical masses of highly purified fissile material to form a supercritical mass. This may be accomplished with a conventional explosive. A nuclear explosion equivalent to 1000 ton of TNT would only require about 56 g of fissile uranium or plutonium, assuming complete fission.[7]

Table 17
APPROXIMATE YIELDS OF SEVERAL IMPORTANT RADIONUCLIDES PER MEGATON OF FISSION[7,50,51]

Radionuclide	Half-life	MCi
Fission products		
^{89}Sr	53 day	20
^{90}Sr	28 year	0.1
^{95}Zr	65 day	25
^{103}Ru	40 day	19
^{106}Ru	1 year	0.29
^{131}I	8 day	125
^{137}Cs	30 year	0.16
^{144}Ce	290 day	3.7
Activation products (air)		
^{3}H	12 year	<1
^{14}C	5600 year	3.4×10^4
^{39}A	260 year	59
Activation products (soil)[a]		
^{24}Na	15 hr	2.8×10^{11}
^{32}P	14 day	1.9×10^8
^{42}K	12 hr	3×10^{10}
^{45}Ca	152 day	4.7×10^7
^{56}Mo	2.6 hr	3.4×10^{11}
^{55}Fe	2.9 year	1.7×10^7
^{59}Fe	46 day	2.2×10^6

[a] Surface detonation assumed.

In the thermonuclear device, the fusion reactions liberate enormous quantities of energy, but very high temperature are required for their initiation. A fission device can trigger the reaction, since it produces ample heat. Therefore, in the thermonuclear bomb, the fission reaction used to initiate fusion results in radioactive fragments and the large number of neutrons produced by both fission and fusion reactions produce activation radionuclides. Unreacted ^{3}H and ^{235}U or ^{239}Pu also contribute to the total inventory of radioactive products.

Depending upon the way in which the fissionable nuclei split, some 80 or 90 different primary radioactive fragments may be formed, most having masses in the range of 80 to 108 and 126 to 154.[53,54] These decay to form daughter nuclides, many of which are in turn radioactive. The total fission product mixture may contain up to about 200 radioactive species immediately following the explosion, but only a small fraction of these persist more than a few hours as many are very short-lived. Neutron activation of various elements in soil, air, water, and other materials surrounding the device, is another source of radioactivity from nuclear devices. In large fission-fusion devices, activation products may provide substantial quantities of radioactivity, especially if detonated in or near the earth.[55] Neutron-activated radionuclides generally differ from the principal fission products (Tables 8 and 9). Tritium, however, arises from several mechanisms, including neutron activation, deuterium-deuterium reactions, and ternary fission.

In a nuclear explosion, tremendous quantities of heat are produced within a small fraction of a second. The nuclear fuel, fission fragments, structural parts of the device, and the immediate surroundings are raised to a temperature of several million degrees, vaporized, and form a bright fiery ball of gases descriptively called the "fireball". If

FIGURE 15. Photograph of the mushroom cloud from a nuclear explosion showing the rising fireball and the column of debris being drawn into it. (Photograph courtesy of the U.S. Department of Energy, Washington, D.C.)

formed at or above the earth's surface, the fireball expands and begins to rise into the atmosphere (Figure 15). As it rises, gradual cooling occurs, allowing the vaporized materials to condense into small droplets which solidify to form small particles that eventually fall back to earth.

Nuclear devices may be detonated at various elevations in the atmosphere, at the surface of the earth and at various depths below ground or underwater, depending upon the particular application of the explosion. The type of explosion which would result in the greatest atmospheric release of radioactivity would be one conducted at or slightly above the earth's surface. In such an explosion essentially all the fission products would be released to the atmosphere and additional radioactivity would result from the activation of soil particles that would be drawn up into the rising fireball. By contrast, a deeply buried device may do little more than create a large subterranean cavern, with little or no escape of radioactivity to the atmosphere. Devices buried at depths of a few hundred feet will rupture the earth's surface and eject large quantities of earth from the site, leaving a crater (Figure 16).[56]

Crater-forming explosions allow a variable fraction of the total radioactivity to escape into the environs.[57] The remainder is trapped by soil and rock falling back into the crater. More deeply buried devices may fail to rupture the surface but small air passages may form between the fireball cavern and the surface. This situation is called "venting", and it allows the preferential escape of the more volatile, gaseous radionuclides such as ^{131}I, ^{133}Xe, ^{85}Kr, and ^{37}Ar to the atmosphere. A nuclear detonation conducted under water would vent into the atmosphere and in addition, would contaminate the aquatic environment. Hydrogen and various elements dissolved in the water would be activated with neutrons to form a large array of nonfission product radionuclides. Klement[51] has prepared an extensive review of potential radionuclides produced in nuclear detonations under various conditions.

3. Properties of Fallout

The fate and potential effects of fallout debris depend to a large extent upon the properties of the debris itself. The properties with which we must concern ourselves include radiological characteristics, such as the species of radionuclides present and their respective radiations and half-lives, and physical characteristics of the debris, such as particle size, density, and solubility.

A list of some of the biologically important fission products was given in Table 8. Of the large array of fission products, a few are of particular biological importance. Factors which tend to make certain of the fission products particularly important include high yield, a reasonably long half-life, energetic radiations, and chemical characteristics which permit their transport through food chains and deposition in biological tissues. Strontium-89 and -90, for instance, are particularly important because when ingested by animals, they tend to lodge in tissues such as bone because of their chemical similarity to calcium. Strontium-90, because of its long half-life and slow release from bone, constitutes a potential long-term hazard. Cesium-137 is very important because it follows the course of potassium in ecosystems and it persists in the environment for many years as a result of its long half-life. Cesium-137 is a potential genetic hazard because it accumulates in many types of tissues and its penetrating gamma rays reach all cells of the body. Iodine-131 concentrates in the thyroid gland with remarkable efficiency because it behaves like ordinary iodine. It delivers a relatively large amount of radiation within a few days. All three nuclides, ^{90}Sr, ^{137}Cs, and ^{131}I can form relatively soluble compounds, which enhances their biological and ecological mobility.

Some of the fission products listed in Table 8 do not have nutrient element analogs, but still are of potential biological importance. Some, although comparatively insolu-

FIGURE 16. Photograph of the crater produced in July 1962, by project Sedan at the Nevada Test Site. This crater, measuring 1200 ft in diameter and 323 ft deep, was produced by a 100 KT device detonated at a depth of 635 ft in alluvium. The crater volume corresponds to the removal of about 5 million tons of earth and rock. The crater lip height varies from 18 to 95 ft. (Photograph courtesy of the U.S. Department of Energy, Washington, D.C.)

ble, such as ^{95}Zr, contribute significant external radiation exposures to plants and animals. Others, including ^{144}Ce and $^{103,106}Ru$, accumulate to a measurable extent in certain biological tissues.[58]

Some radionuclides of possible biological significance that are produced by the neutrons released in a nuclear explosion are listed in Table 9. Many other possible activation-produced radionuclides are not listed because their production is very low or because they have very short half-lives. With a few exceptions, the neutron-induced radionuclides are isotopes of either major or minor nutrient elements and can be efficiently transferred in food chains. Plutonium-239, although not related to a nutrient element and quite immobile, is of interest because of its slow release from biological tissues and high radiotoxicity. Tritium is of interest because of its hydrogen analog, high mobility, and 12 year half-life.

The total radioactivity and radionuclide composition of fallout varies greatly with time. The composition of the radioactivity is determined mainly by initial composition and half-lives of the individual radionuclides.[59,60] Shortly after a nuclear detonation, for example, the total radioactivity is dominated by short-lived materials. The decay rate of a mixture of fission products was investigated in 1948 by Way and Wigner[61] and it was shown that the total radioactivity, R_t, as a function of time following fission, can be estimated by

$$R_t = R_1 \, t^{-1.2} \tag{2}$$

where R_1 = radioactivity at an initial reference time and t = time (in multiples of the reference time). This relationship is reasonably valid from an hour to 6 months after fission. During this period, an approximate tenfold decrease in radioactivity is achieved for every sevenfold increase in time after fission. Figure 17 illustrates the quantities of fission and activation product activities after a 1 MT explosion buried in typical crustal material.

Radionuclides fractionate to varying degrees following formation in a nuclear explosion from causes other than differential decay rates. Some fractionation may be caused by differences in volatility of the various fission products, and fractionation may also occur in relation to the size of fallout particles.[55,62] Once fallout debris settles out of the atmosphere, the fate of the individual nuclides is governed largely by solubility of the particles and the chemical properties of each nuclide. Relatively soluble nuclides, which are isotopes or analogs of essential nutrient elements, will cycle through ecological systems much like their nutrient element counterparts. Insoluble constituents of fallout usually find their way to soil or sediment relatively soon after deposition and are not significantly recycled into the food web of the system.

Many of the behavioral characteristics of fallout, from the time of formation to the time of entry into biological systems, can be explained on the basis of the physical characteristics of the material. Airborne fallout is usually associated with small particles, which are variously shaped and which may range from less than 10^{-3} to over 100 μm in diameter.[62] Some of these particles are formed as the gases within the fireball cool, condense, and solidify. Such particles may trap or occlude considerable radioactivity within the particle. Other fallout particles may consist of ground surface dust that is drawn into the mushroom cloud of the explosion but not vaporized, and dust normally present in the atmosphere. Radionuclides in the atomic state may attach to the surfaces of these particles and their transport is subsequently governed by that of the particle. Some of the more volatile radionuclides, such as ^{131}I, may remain partially in the gaseous state.[63]

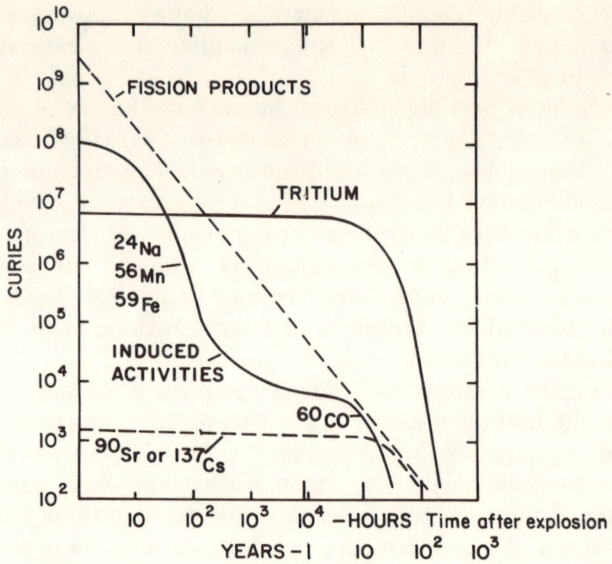

FIGURE 17. Activities of radionuclides at various times after a nominal 1 MT explosion (10 KT fission) in average crustal material. (From Scientific Committee on the Effects of Atomic Radiation, Ionizing Radiation: Levels and Effects, Vol. 1, United Nations, New York, 1972.)

The average size of radioactive fallout particles is dependent upon the height of the burst above the ground, the magnitude of the explosion, and the type of support and containment apparatus of the device.[64] As a rule, the higher the burst above the ground, the smaller is the amount of debris which is sucked into the fireball, resulting in smaller fallout particles. Particles less than 20 μm generally behave like gases and undergo normal gaseous diffusion processes. On the other hand, particles more than 20 μm have measurable gravitational settling velocities and therefore deposit earlier.

The fraction of the radioactivity of fallout particles which is water soluble markedly affects the degree to which the critical radionuclides such as ^{90}Sr and ^{137}Cs will enter biological systems. The solubility of radionuclides associated with fallout particles is quite variable and dependent upon conditions of the explosion. If large quantities of earth are incorporated into the fireball, the resulting fallout particles may contain large amounts of silicates which are relatively insoluble. On the other hand, detonations high in the atmosphere usually produce very small particles from which much of the radioactivity can be leached with water.[63] Neel and Larson[65] found that the solubility and potential biological availability of ^{90}Sr in fallout increased with distance from ground zero. The fact that fallout particle size also decreased with distance from ground zero suggests that particle size and solubility may be inversely correlated.

4. Transport of Fallout

Fallout from nuclear explosions may be classified either as "local" or "worldwide." That which is made up of relatively large (>35 μm diameter) particles normally falls to earth by gravity within 24 hr after the explosion and is termed "local" fallout. That consisting of smaller particles may become widely dispersed over the earth and is therefore called "worldwide" or "delayed" fallout. Worldwide fallout may be further subdivided into "tropospheric" and "stratospheric" fallout, depending upon the altitude band in which the debris cloud is dispersed.

Local fallout, consisting initially of many shorter-lived radionuclides, may lead to extremely high radiation doses in the vicinity of the explosion. For example, a 1 MT fission bomb could lead to ground-level exposures exceeding 1000 R within the first day over an area 45-mi long and 5- to 10-mi wide.[55] Under conditions of uniform wind speed and direction and uniform topography, ellipsoid dose patterns usually result from early local fallout. Variable wind speed and direction following an explosion would lead to a fallout field of irregular shape. Topographic and climatic variables can drastically alter the deposition of local fallout.

Fallout which originates and is transported solely in the troposphere, but which remains airborne longer than 24 hr, is called "tropospheric worldwide fallout". Tropospheric fallout usually results from atmospheric explosions of relatively low yield[66] detonated near the earth's surface. This type of fallout may also be formed by underground or underwater explosions which vent or produce craters. Tropospheric fallout usually reaches the earth's surface within a month from its date of production. The troposphere is that portion of the earth's atmosphere extending from the surface to the stratosphere, some 30 to 60 thousand feet above sea level. In the troposphere, clouds and weather patterns develop and temperature decreases rapidly with altitude. The primary agents which cause deposition of tropospheric debris include gravity, wind currents, and precipitation. Deposition usually occurs along a band near the same latitude as the detonation. Weather patterns affect deposition along the band and therefore deposition is somewhat unpredictable. Deposition "hot spots" may occur hundreds or even thousands of miles from the detonation site. For example, a significant fraction of the radioactivity in a debris cloud could be washed from the troposphere by a rainstorm and deposited within a distant local area. A notable example of this occurred in Troy, N.Y., in 1953, 36 hr after a nuclear explosion at the Nevada Test Site, some 2300 miles away.[67] A violent thunderstorm deposited fallout to the extent that Geiger-Mueller (G-M) counter readings of 5 mR/hr were recorded in the downtown area and hot spots as high as 120 mR/hr were found. Normally, fallout has not been found to increase significantly the radiation background, but the normal background of 0.015 mR/hr was exceeded by orders of magnitude for a short period in Troy. The most important constituents of tropospheric fallout, from the standpoint of total external dose delivered, would include some of the radionuclides of intermediate half-lives, such as ^{131}I, ^{140}Ba, ^{147}Nd, ^{133}Xe, ^{103}Ru, ^{95}Zr, ^{141}Ce, and ^{89}Sr.

The majority of worldwide fallout produced to date has been stratospheric. Stratospheric fallout arises from relatively high yield devices which have sufficient energy to propel the radioactive mushroom cloud into the stratosphere. The stratosphere is that portion of the earth's atmosphere which extends from about 30,000 to 60,000 ft to about 160,000 ft above sea level. Air in the stratosphere is relatively stable, free of clouds, and its temperature gradually rises with altitude because of the heat absorption qualities of ozone which is produced in the upper atmosphere by cosmic rays. Stratospheric fallout remains airborne considerably longer than tropospheric debris because stratospheric air mixes quite slowly with the lower atmosphere and there is no precipitation to scavenge the particles. Mean residence time of debris in the stratosphere may range from less than a year to over 5 years, depending upon the height and latitude of the material. In general, the higher into the stratosphere the radioactivity is injected, the longer it will take to reach the troposphere. Fallout injected into the stratosphere at polar latitudes will reach the earth sooner than material injected at tropical latitudes. Material injected in winter will fallout more rapidly than material injected in summer, other factors being equal. The time during which fallout is in the stratosphere allows many of the short- and intermediate-lived radionuclides to decay and the longer-lived materials, such as ^{90}Sr, ^{137}Cs, ^{144}Ce, and ^{14}C, become relatively more important.

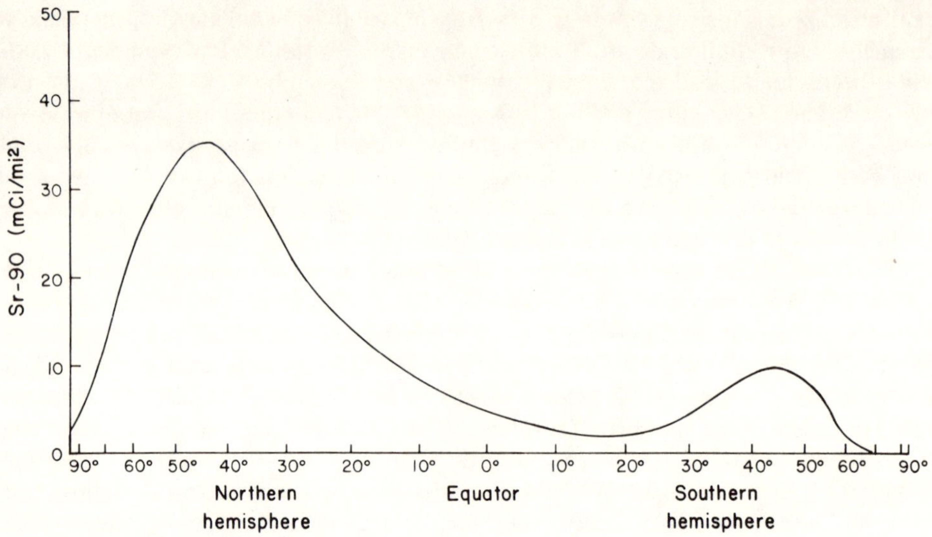

FIGURE 18. Deposition of Sr on the earth's surface as a function of latitude. Estimated from world-wide sampling data. (From Machta, L. and List, R. L., in *Fallout: A Study of Superbombs, Strontium 90, and Survival,* Fowler, J. M., Ed., Basic Books, New York, 1960, 26.)

The phenomena of atmospheric fallout transport can be partially explained by the structure of the tropopause and by the general circulation patterns in the atmosphere.[68] The tropopause is the boundary between the troposphere and the stratosphere. The boundary is a temperature discontinuity which normally prevents significant vertical mixing of air between the troposphere and stratosphere. However, British scientists Brewer and Dobson have postulated that warm air rises from the troposphere and enters the stratosphere at equatorial latitudes. Once in the stratosphere and constrained by the tropopause, air gradually moves toward the poles. The cold stratospheric air then sinks back into the troposphere at temperate and polar latitudes. This process is aided considerably by gaps which develop in the tropopause, allowing accelerated mixing. This mixing process is greatest at temperate latitudes and it is also greatest during the late winter and spring months when the air is comparatively cold and dense.

Observed spatial and temporal patterns of fallout seem compatible with the Brewer-Dobson model. For example, a plot of ^{90}Sr deposition vs. latitude indicates definite peaks at 40 to 50° in both hemispheres (Figure 18), which agrees with the concept of stratospheric air entering the troposphere at midtemperate latitudes. The reason for the peak in the northern hemisphere being larger than in the southern hemisphere is that most nuclear testing has been north of the equator and mixing of air between hemispheres is minimal.

The observation that sinking of stratospheric air is accelerated in late winter and spring, would lead one to expect that the troposphere would be enriched in fallout, and therefore, deposition would increase during those months. Such observations have indeed been made.[68-70] In many temperate regions, precipitation is also highest in the spring, which maximizes the potential for fallout deposition. However, there are cases in which fallout deposition has been higher in fall than in spring because of a dominating influence of fall precipitation.[71]

Along given latitudinal belts, considerable variation in fallout deposition is apparent. Much of this variation can be explained by differences in precipitation. There is little doubt that increased precipitation leads to increased fallout deposition, assuming other factors are constant. One of the better correlations between rainfall and fallout deposition is provided by Hardy and Alexander.[72] In mountainous regions where much of the annual precipitation is in the form of snow, snowfall may be a dominant mechanism of fallout deposition.[73] Other phenomena which have been postulated as having potential influence on fallout within latitudes are the position of the jet stream, the effect of mountains in causing vertical air motions, and the timing, severity, and altitude of rainstorms.[74]

In arid regions, a large fraction of the total fallout is deposited as dry particles, without the aid of precipitation.[63] The larger particles in "dry" fallout may settle out of the air primarily by gravity, whereas smaller particles may be deposited principally by impaction when air becomes deflected by a solid object. The latter mechanism is particularly important in the situation of turbulent winds and irregular land surfaces.

Thus far, we have considered processes which distribute fallout on a primarily global or continental scale. Fallout deposition and subsequent redistribution on a local scale is also important. In fact, variations in local fallout distribution may exceed considerably the continental and worldwide variations. In addition to macrometeorological effects, fallout in local sites is subjected to complex microclimatic and ecosystem processes.

Wind, coupled with topographic and physiognomic variation is a particularly dominant force in the deposition and redistribution of fallout.[73,75] Turbulence and eddys are caused when air passes over or around solid objects ranging in size from rocks to mountain ranges. This process creates added opportunity for aerosol particles to impact and lodge on surfaces. Also, there are situations in which the overall rate of air motion is reduced in the lee of obstructions and in valleys or depressions. This allows particles suspended in the air to settle out by gravitational forces. Particles deposited in wind-protected sites are apt to accumulate since resuspension may be lower than in exposed terrain.

Wind is also a major force in the distribution of soil particles, pollens, dust, and snow, and it is known that these materials scavenge considerable radioactivity from the air. Therefore, the fate of these materials is that of the fallout, at least for a time. These materials are picked up from exposed areas by wind and they gradually accumulate in protected sites. The alpine regions in the Rocky Mountains are a particularly good place to observe the distribution of materials by wind. Protected depressions and exposed ridges on a micro- as well as macroscale cause highly uneven distribution patterns of snow and solid debris, and therefore, of fallout (Figure 19). Since these forces also affect the development of soil and plant communities in the alpine tundra, certain organisms receive much higher doses of radiation from fallout than others.

Fallout that settles directly on surface water, such as a lake or a stream, is transported, but seldom for great distances, by the water. Most fallout radionuclides quickly adsorb to sediment, plankton, and algae upon entry into an aquatic system. The movement of these organic and inorganic materials determines the fate of the radioactivity. Water in the upper reaches of a watershed may contain considerable amounts of dissolved radionuclides, but the activity of the water normally diminishes rapidly as it flows to lower elevations. Fallout ^{3}H would not be expected to diminish in concentration through a watershed as rapidly as other radionuclides, since only dilution and not absorption would operate.

Water flow, however, can and does act as an important agent of radionuclide concentration and redistribution. Surface runoff, particularly if sufficient to cause erosion, may significantly alter the distribution of fallout attached to small soil and litter particles. This process tends to cause redeposition of fallout in sites of silt accumulation such as microdepressions, ponds, reservoirs, and meandering sections of streams. In mountains, large quantities of snow melt trickle through mats of sedges, mosses, and other vegetation which effectively filter particles and absorb dissolved ions. This

FIGURE 19. Photograph of a mountain ridge showing the uneven patterns of snow distribution caused by wind and topographic variations.

process purifies the water and sometimes leads to a large accumulation of radioactivity in the vegetation-soil complex.

Significant movement and nonuniform distribution of fallout radioactivity within ecosystems is accomplished by the functions of plants and animals as well as the physical processes just mentioned. The importance of these physical and biological processes is emphasized at this point because of their relevance to the entire subject of environmental radioactivity and its potential consequences. The accumulation and movement of radionuclides in ecological systems will be treated in considerable detail in Chapter 5.

5. Peaceful Applications of Nuclear Explosives

Numerous peaceful applications of nuclear explosives have been proposed. These proposals include the use of nuclear detonations to control runaway oil or gas wells, to stimulate natural gas and oil production, to create geothermal cavities, and to move large quantities of geological materials to reach minerals or to form harbors or canals.[76] From a technical and economic standpoint, many of these proposed applications appear feasible. However, the peaceful application of nuclear explosives has not, at least in the U.S., lived up to its potential. Despite exhaustive safety measures,[77] a major deterrent to such applications appears to be public concern about radioactive contamination. Although the U.S. program in this area, called "Plowshare", is now relatively inactive, the Soviet Union has a vigorous program at present, and both France and India intend to pursue this technology.

Most applications proposed for peaceful nuclear explosive technology involve buried devices. A deeply buried device will create a subterranean cavity that can effectively contain the radioactive debris (Figure 20). In some such experiments, however, fissures to the surface have occurred, allowing some escape of gaseous radionuclides such as ^{85}Kr, ^{133}Xe, and ^{131}I. The refractory radionuclides are effectively trapped in the cavity. Earth-moving applications, on the other hand, involve shallow burial so that the energy of the explosion can eject large quantities of earth to form a crater (Figure 16). The sequential development of a nuclear crater is illustrated in Figure 21. A string of such devices could for instance, create a canal. Earth-moving applications, however, could be expected to release significant quantities of radioactive debris to the biosphere, and, largely for this reason, their application has met with public disfavor. Other environmental concerns about buried nuclear devices include damage from ground motion, as well as conventional physical damage from heavy equipment and people.

Factors such as cost and compact size make the idea of nuclear explosives attractive. For instance, Glasstone[78] states, "...the price quoted officially in the United States for a nuclear explosive with an energy yield equivalent to 2 megatons... is $600,000 or 30 cents per ton TNT equivalent. Conventional explosives, however, would cost about $100 for the equivalent of 1 ton of TNT." The small size of a powerful nuclear device would simplify the problems of transportation and emplacement. For instance, a 10 MT nuclear explosive can be contained in a canister 65 in. in diameter and 30-ft long, whereas a conventional device of similar power would involve a volume of 10 ft by 10 ft by 400 mi.[7]

Table 18 provides a list of proposed and actual nuclear explosion engineering events sponsored by the U.S., U.S.S.R., and France. It is clear that the majority of explosion events have been experimental and for the purpose of observing phenomena associated with underground detonations. The first major experiment conducted by the U.S. to study the feasibility of large-scale earth moving was project "Sedan", conducted at the Nevada Test Site in July 1962 (Figure 16). A 100 KT device, detonated 635 ft beneath the surface in weakly consolidated alluvium, created a 323-ft deep, 1200-ft diameter crater. Approximately 5 million tons of earthen material was ejected from the crater. In addition to engineering data, the experiment provided the opportunity for numerous ecological and radioecological investigations.[57,80-82] Such studies demonstrate measurable accumulation of radionuclides in wildlife around the Sedan crater, however, blast and dirtfall appeared to cause far more ecological damage than radioactivity. The senior author documented a steep rise in the ^{131}I content of mule deer *(Odocoileus hemionus)* thyroids in northcentral Colorado a few days following the Sedan test,[83] which confirmed that radionuclides from large cratering events could be continentally dispersed.

In the mid- to late 1960s, several other cratering experiments were conducted in

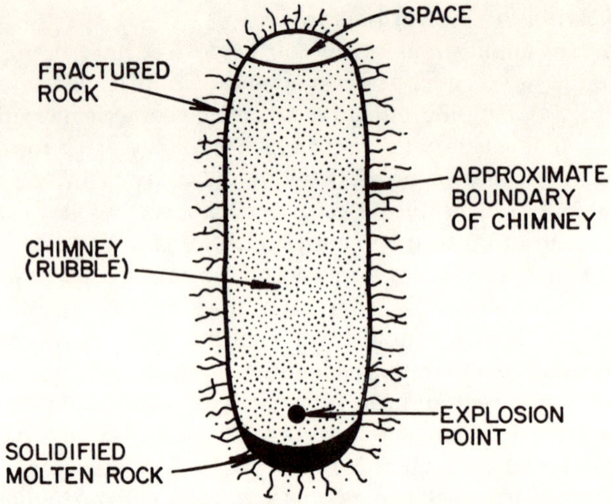

FIGURE 20. Schematic diagram of the final geological configuration resulting from a nuclear detonation placed sufficiently deep to be contained. The initial spherical cavity is usually filled by the collapse of material above the detonation point which forms a "chimney" of broken rock. A void space usually forms at the top of the chimney and condensation and cooling of material vaporized or melted from the heat of the detonation forms a solidified puddle at the bottom of the chimney. (Adapted from Glasstone, S., Public Safety & Underground Nuclear Detonations, U.S. AEC Rep. TID-25708, U.S. Atomic Energy Agency, Washington, D.C., 1971.)

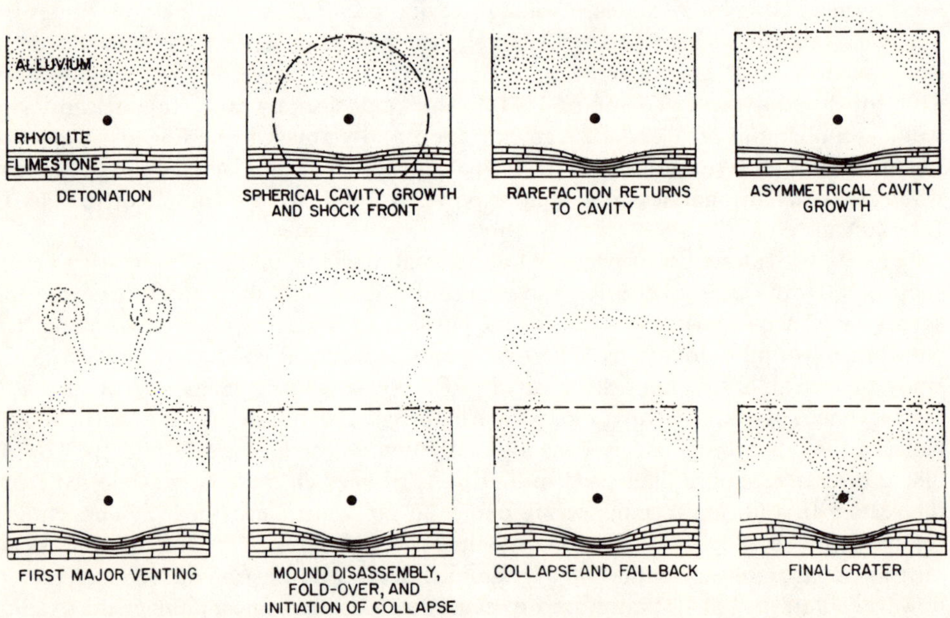

FIGURE 21. Stages in the formation of an excavation crater following detonation of a buried nuclear device. Rarefaction refers to the tension wave which is reflected from the surface back into the ground. (Adapted from Glasstone, S., Public Safety & Undergroun Nuclear Detonations, U.S. AEC Rep. TID-25708, U.S. Atomic Energy Commission, Washington, D.C., 1977.)

Table 18
SUMMARY OF PEACEFUL NUCLEAR EXPLOSION ENGINEERING EVENTS AS OF 1975

	U.S.		U.S.S.R.		France	
Application	Actual	Proposed	Actual	Proposed	Actual	Proposed
Excavation						
Phenomenology Experiments	7		5	?		
Highways		1[a]				
Harbors		2[a]				
Canals		1[a]		1		
Water resources			1	1		
Overburden removal				1		
Contained						
Phenomenology Experiments	Many	?	Many	?	13	?
Gas stimulation	3	1	1	1		
Gas-fire blowout			2			
Oil stimulation			2	1		
Oil shale		1				
Storage		1	4	1		1
Geothermal energy		1		1		
Mining		2	1	1		

[a] Inactive.

From Kruger, P., Peaceful Nuclear Explosives, U.S. ERDA Rep. ERDA-44, U.S. Energy Research and Development Administration, Washington, D.C., 1975.

Nevada. These involved varied configurations in varied types of geological strata. Individual events carried names like Palanquin, Cabriolet, Schooner, and Buggy. Project Buggy involved simultaneous detonation of a row of five devices, each with a yield of about 1 KT. The ditch resulting from this experiment was 65-ft deep, 255-ft wide, and 855-ft long. Projects Palanquin and Cabriolet were of particular radioecological interest because convincing evidence was found that sagebrush *(Artemisia)* shrubs downwind of the craters were damaged or killed by beta radiation.[84]

Two major nuclear excavation projects that were proposed by the U.S. are worthy of mention. In early 1958, the USAEC authorized planning for Project Chariot, an excavation project near Cape Thompson, Alaska. From 1959 to 1961, some 40 separate environmental investigations were carried out to obtain baseline information for the proposed project.[85] These studies, initiated prior to the National Environmental Policy Act of 1969, are a tribute to the environmental sensitivity and farsightedness of the late Dr. Wolfe (then with the USAEC) and members of the environmental studies advisory committee for Project Chariot. However, in 1962, the USAEC decided to table the project. Then, in 1965, President Lyndon B. Johnson formed the Atlantic-Pacific Interoceanic Canal Study Commission and authorized them to study the feasibility and ramifications of constructing an interoceanic canal with conventional and nuclear excavation techniques.[86] After several years of extensive study and safety analysis, it was concluded that the use of nuclear explosives for canal construction was feasible.[87,88] Again however, consideration of this project was tabled, probably because of political ramifications.

The increasingly serious shortage of natural gas in the U.S. has led to a search by industry and government for new methods of obtaining marketable gas. One such

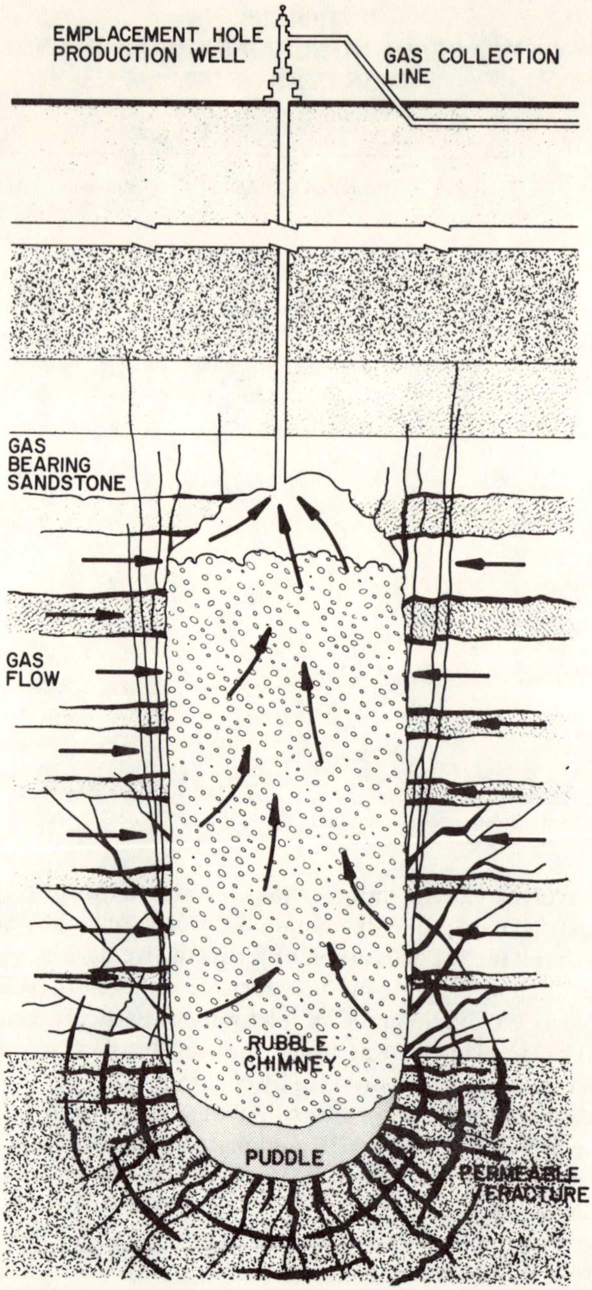

FIGURE 22. Concept of nuclear gas stimulation.

method involves nuclear stimulation of gas-bearing formations which are not sufficiently permeable to allow adequate gas flow into a conventional well. Nuclear stimulation of such gas-bearing formations involves detonation of a nuclear device within the formation, which creates an underground chimney of broken rock and fissures outside the chimney. Natural gas, diffusing into the large area of increased permeability, flows into the chimney and ultimately through a reentry well (Figure 22).

Three nuclear gas stimulation experiments have been conducted in the U.S. The first, Project Gasbuggy, was conducted at a depth of 4240 ft near Farmington, N.M., in 1967, using a 29 KT nuclear explosive. Project Rulison, conducted near Rifle, Colo., in 1969, involved a 43 KT device at a depth of 8426 ft. The most recent experiment of this kind, Project Rio Blanco, was conducted in 1973 near Rifle, Colo. This experiment involved three separate 30 KT devices, detonated simultaneously at depths ranging from 5838 to 6690 ft in the same bore hole.

The nuclear gas stimulation experiments thus far conducted have produced increased gas flow, but the overall effectiveness of the method now seems in doubt. Nuclear gas stimulation experiments have released only innocuous amounts of radioactivity to the biosphere, but ground motion produced by the detonations has produced quantifiable ecological damage.[89] Project Wagon Wheel, a nuclear gas stimulation experiment proposed for an area in west-central Wyoming, suffered the same fate as Project Chariot and the interoceanic canal. Despite an extensive environmental impact analysis,[90] the Wagon Wheel Project has been tabled on the basis of disappointment in the success of Project Rio Blanco and public opposition.

C. Other Sources of Environmental Radioactivity

Although the nuclear fuel cycle and nuclear explosions have been and will likely continue to be the major sources of man-generated radioactivity in the environment, many other significant sources of radioactive contamination exist. These mostly fall in the category of radioisotope applications. There are literally thousands of actual and potential uses of radionuclides in the areas of medicine, industry, agriculture, scientific research, and education. In addition to the actual uses of radionuclides in specific fields, millions of curies of radioactivity are transported from producer to consumer each year. Production, transportation, use, and disposal of radioisotopes all present opportunity for radioactivity to find its way, either routinely or by accident, to the environment.

The quantities of radioisotopes sold for miscellaneous applications are large and sales have shown dramatic and continuous increase. For example, in 1968 over 3 million curies were distributed by Oak Ridge National Laboratory (ORNL) alone,[91] which amounted to a 9% increase over the previous year and a sevenfold increase over 1961. Far greater quantities are no doubt being sold by other producers. The types of radionuclides produced for specific applications include naturally occurring materials such as radium, uranium, and thorium, and "byproduct materials" which include fission products, and isotopes induced from neutrons and particle accelerators.

Routine use of many different radionuclides is growing in the fields of medicine, industry, and agriculture. Although individual applications seldom involve large quantities of radioactivity, the collective inventory of radionuclides being used throughout the world is enormous. The variety of routine radioisotope uses is so great and likewise so well known that there is little reason for detailed enumeration here. The main point we wish to make is that radionuclides are now an integral part of modern society. Eventually, a portion of these radionuclides ends up somewhere in the environment and are, therefore, a concern.

Radionuclides may be used as tracers or as radiation sources. As tracers, they are used to determine the movement and behavior of many things, including atoms, molecules, cells, organisms, fluids, particles, gases, etc. As radiation sources, radionuclides can be used to measure the quality and quantity of materials and to produce various changes in organic or inorganic material. Sealed radiation sources are used in such diverse applications as insect eradication, medical therapy, food sterilization, smoke detectors, and material density and thickness gauging. In general, sealed sources do not provide much potential for environmental contamination. However, there are circumstances which could lead to rupture of source containers and loss or misplacement. Large sealed gamma sources are normally shielded with several inches or more of lead or other material to reduce radiation exposures in the vicinity of the source.

Tracer applications are inherently more likely to lead to environmental contamination. For example, in medical applications, such as measurement of blood flow or thyroid function with ^{131}I, some of the radioactivity ends up in the sewer in excretory products. This is also true for many other diagnostic tests involving radionuclides. In industry, tracing of leaks, measurement of gas or liquid flow, and determination of machine wear involves radioisotopes, part of which winds up in a sewer or industrial effluent. Many agricultural applications involve direct application of radionuclides into the environment. For example, ^{32}P is often mixed with fertilizers to test their utilization by crops. Some of the ^{32}P may end up in runoff water, blowing dust, and in plants and animals. Such procedures are entirely safe when used in compliance with regulatory guidelines, but research is usually necessary to establish these guidelines and to confirm them for different situations.

The diversity of radionuclide uses in research is probably even greater than routine applications. Radionuclides constitute one of the most powerful tools in modern research and applications are found in nearly all disciplines in the natural sciences, from fundamental physics to natural resource management. Radioactive tracers provide the only feasible means in many cases of determining complex chemical transformations and functional relationships. The use of ^{14}C to elucidate the path of carbon in photosynthesis is an outstanding example. In nearly every modern scientific laboratory, one finds an array of G-M counters, scintillation detectors, and other instruments used in radionuclide measurement and handling. Likewise, liquid, solid, and gaseous effluents from such laboratories usually contain at least traces of radioactivity.

The need to understand the behavior and effects of radionuclides in the environment has led to many studies in which radioactive materials or radiation are purposefully introduced into local study environments. Such studies are seldom undertaken without prior laboratory testing and preevaluation of the results. However, nature can seldom if ever be duplicated or simulated with precision in the laboratory. Therefore, *in situ* tests are often necessary to obtain credible data. With care, a great deal can be learned from such experiments, without causing stress to the system or to the workers involved. In some cases, small portions of ecological systems are purposely irradiated with sealed sources of sufficient strength to produce dramatic effects.[92] Such studies provide the only means of accurately determining the effects of varying doses of radiation on natural ecosystems.

The training of doctors, scientists, and engineers also involves the use of radioisotopes. Increasing numbers of professional personnel need the ability to work with radioisotopes in order to perform their normal duties. Also, persons specializing in the safety aspects of radioisotopes, and nuclear energy in general, must be trained. The profession dedicated to the safe handling, application and disposal of radioactive materials is called "health physics". Health physicists are needed by universities, hospitals, private and government laboratories, reactor facilities, and various types of nuclear energy installations. Training programs in radiation science are available in many universities and several institutes and governmental agencies. Some radiation science is even being taught in high schools. Such training programs utilize radioisotopes and thus provide added potential for environmental contamination with radioactivity. However, most training applications can be conducted with very small quantities of radioactivity and this source is generally insignificant in relation to others.

The vast majority of miscellaneous radionuclide applications involve comparatively small amounts of material, distributed diffusely over the earth's surface. Because of this, individual disposals or accidents have very limited potential for contamination of large areas. However, the space applications of radionuclide-powered systems is worthy of brief discussion because of the potential for global contamination. Radiation interaction with matter produces heat and heat can be converted into electrical power.

Conversion of heat to electrical power is accomplished by thermoelectric phenomena with conversion efficiencies approaching 10%.[7] These facts have led to the development of electrical generators which are powered from the decay of radionuclides. Several radionuclides are practical as sources of heat. Remote weather stations, satellites, and spacecraft are examples for which nuclear generators are practical. Such generators are attractive because the energy of decay is totally reliable and predictable, and sufficient power can be produced with a very compact device.

Of interest is the case history of a navigational satellite known as SNAP 9-A, which was launched in April 1964. This satellite, equipped with a nuclear generator containing 17,000 Ci ^{238}Pu, failed to reach orbital velocity and reentered the earth's atmosphere at about 150,000 ft over the Indian Ocean.[7] Reentry caused burnup of the satellite and the release of ^{238}Pu to the atmosphere, which was subsequently detected in worldwide fallout.[3] Following dispersal in the earth's atmosphere, the 17,000 Ci of ^{238}Pu was diluted to innocuous, but nevertheless, detectable levels. This is but one of many examples which illustrate the finite capacity of the atmosphere to dilute foreign substances.

REFERENCES

1. **Lapp, R. E. and Andrews, H. L.,** *Nuclear Radiation Physics,* 4th ed., Prentice-Hall, Englewood Cliffs, N.J., 1972.
2. **Wolfendale, A. W.,** *Cosmic Rays,* Philosophical Library, New York, 1963.
3. **Compton, A. H.,** A geographic study of cosmic rays, *Physiol. Rev. 2nd Ser.,* 43(6), 387, 1933.
4. **Libby, W. F.,** Dosages from natural radioactivity and cosmic rays, *Science,* 122(3158), 57, 1955.
5. **Van Allen, J. A.,** Corpuscular radiations in space, *Radiat. Res.,* 14(5), 540, 1961.
6. **Perkins, R. W. and Nielsen, J. M.,** Cosmic-ray produced radionuclides in the environment, *Health Phys.,* 11(12), 1297, 1965.
7. **Eisenbud, M.,** *Environmental Radioactivity,* 2nd ed., Academic Press, New York, 1973.
8. **Blatz, H., Ed.,** *Radiation Hygiene Handbook,* McGraw-Hill, New York, 1959.
9. **Cowan, G. A.,** A natural fission reactor, *Sci. Am.,* 235(1), 36, 1976.
10. **Adams, J. A. S.,** The geological origins of radioactive anomalies, in *Int. Symp. Areas High Natural Radioactivity.* Cullen, T. L. and Penna Franca, E., Eds., Academia Brasileira de Ciencias, Rio de Janeiro, 1977, 5; *Energy Res. Abstr.,* 2(16), 40196.
11. **Osburn, W. S.,** Primordial radionuclides: their distribution, movement, and possible effect within terrestrial ecosystems, *Health Phys.,* 11(12), 1275, 1965.
12. **Adams, J. A. S. and Lowder, W. M., Eds.,** *The Natural Radiation Environment,* University of Chicago Press, Ill., 1964.
13. **National Council on Radiation Protection and Measurements,** Environmental radiation measurements, NCRP Rep. 50, National Council on Radiation Protection, Washington, D.C., 1976.
14. **Beck, H. L.,** The physics of environmental radiation fields, in The Natural Radiation Environment. II, Adams, J. A. S., Lowder, W. M., and Gesell, T., Eds., U.S. ERDA Rep. CONF-720805-P1, U.S. Energy Research and Development Administration, Washington, D.C., 1972, 101.
15. **Lowder, W. M., Condon, W. J., and Beck, H. L.,** Field spectrometric investigations of environmental radiation in the U.S.A., in *The Natural Radiation Environment,* Adams, J. A. S. and Lowder, W. M., Eds., University of Chicago Press, Ill., 1964, 597.
16. **Gruneberg, H., Bains, G. S., Berry, R. J., Rites, L., Smith, C. A. B., Weiss, R. A.,** A search for genetic effects of high natural radioactivity in South India, *Med. Res. Counc. (London) Spec. Rep. Ser.,* SRS 307, 1966.
17. **Herbst, W.,** Investigations of environmental radiation and its variability, in *The Natural Radiation Environment,* Adams, J. A. S. and Lowder, W. M., Eds., University of Chicago Press, Ill., 1964, 781.
18. **Mericle, L. W. and Mericle, R. P.,** Reassessing the role of background terrestrial radiation as a constituent of the natural environment, *Health Phys.,* 11(12), 1607, 1965.
19. **Roser, F. X. and Cullen, T. L.,** External radiation levels in high-background regions of Brazil, in *The Natural Radiation Environment,* Adams, J. A. S. and Lowder, W. M., Eds., University of Chicago Press, Ill., 1964, 835.

20. **Maslov, V. I., Maslova, K. I., and Verkhovskaya, I. N.,** Characteristics of the radioecological groups of mammals and birds of biogeocoenoses with high natural radiation, in *Radioecological Concentration Processes,* Åberg, B. and Hungate, F. P., Eds., Pergamon Press, New York, 1967, 561.
21. **Phair, G. and Gottfried, D.,** The Colorado Front Range, Colorado, U.S.A., as a uranium and thorium province, in *The Natural Radiation Environment,* Adams, J. A. S. and Lowder, W. M., Eds., University of Chicago Press, Ill., 1964, 7.
22. **Gopal-Ayengar, A. R., Sundaram, K., Mistry, K. B., and George, K. P.,** Current status of investigations on biological effects of high background radioactivity in the monazite bearing areas of Kerala Coast in south-west India, in *Int. Symp. Areas High Natural Radioactivity,* Cullen, T. L. and Penna Franca, E., Eds., Academia Brasileira de Ciencias, Rio de Janeiro, 1977, 19; *Energy Res. Abstr.,* 2(16), 40196.
23. **Penna Franca, E.,** Review of Brazilian investigations in areas of high natural radioactivity. Part II: Internal exposure and cytogenetic survey, in *Int. Symp. Areas High Natural Radioactivity,* Cullen, T. L. and Penna Franca, E., Eds., Academia Brasileira de Ciencias, Rio de Janeiro, 1977, 29; *Energy Res. Abstr.,* 2(16), 40196.
24. **Blanchard, R. L.,** Relationship between ^{210}Po and ^{210}Pb in man and his environment, in *Radioecological Concentration Processes,* Åberg, B. and Hungate, F. P., Eds., Pergamon Press, New York, 1967, 281.
25. **Hill, C. R.,** Routes of uptake of ^{210}Po into human tissues, in *Radioecological Concentration Processes,* Åberg, B. and Hungate, F. P., Eds., Pergamon Press, New York, 1967, 297.
26. **Kauranen, P., and Miettinen, J. K.,** ^{210}Po and ^{210}Pb in environmental samples in Finland, in *Radioecological Concentration Processes,* Åberg, B. and Hungate, F. P., Eds., Pergamon Press, New York, 1967, 275.
27. **Holtzman, R. B. and Ilcewicz, F. H.** Lead-210 and polonium-210 in tissues of cigarette smokers, *Science,* 153(3741), 1259, 1966.
28. **Anon.,** Statistical Data of the Uranium Industry, January 1, 1977, U.S. ERDA Rep. GJO-100 (77), Grand Junction Office, U.S. Energy Research and Development Administration, Grand Junction, Colo., 1977.
29. **Scientific Committee on the Effects of Atomic Radiation,** Ionizing Radiation: Levels and Effects, Vol. 1, United Nations, New York, 1972; *Nucl. Sci. Abstr.,* 27(7), 15039.
30. **Ashby, W. C., Beggs, J. N., Kastner, J., Oltman, B. G., and Moses, H.,** Ecological dosimetry: radiation levels influenced by plant growth, *Science,* 155(3768), 1430, 1967.
31. **IAEA,** Effects of Ionizing Radiation on Aquatic Organisms and Ecosystems, Tech. Rep. Ser. No. 172, International Atomic Energy Agency, Vienna, 1976.
32. **Beckmann, P.,** *The Health Hazards of Not Going Nuclear,* Golem Press, Boulder, Colo., 1976.
33. **Directorate of Licensing,** Fuels and Materials, Environmental Survey of the Nuclear Fuel Cycle, U.S. AEC Rep., U.S. Atomic Energy Commission, Washington, D.C., 1972.
34. **Anon.,** Environmental Analysis of the Uranium Fuel Cycle, Part 1, U.S. EPA Rep. EPA 520/9-73-003-13, U.S. Environmental Protection Agency, Washington, D.C., 1973.
35. **Metz, W. D.,** Reprocessing: how necessary is it for the near term?, *Science,* 196(4258), 43, 1977.
36. **Union of Concerned Scientists,** *The Nuclear Fuel Cycle: A Survey of the Public Health, Environmental and National Security Effects of Nuclear Power,* rev. ed., MIT Press, Cambridge, 1975.
37. **Brooks, H.,** The public concern in radioactive waste management, in Proc. Int. Symp. Management of Wastes from the LWR Fuel Cycle, U.S. ERDA Rep. CONF-76-0701, U.S. Energy Research and Development Administration, Washington, D.C., 1976, 52.
38. **Rochlin, G. I.,** Nuclear-waste disposal: two social criteria, *Science,* 195(4273), 23, 1977.
39. **Committee on Radioactive Waste Management,** *Interim Storage of Solidified High-Level Radioactive Wastes,* National Research Council, National Academy of Sciences, Washington, D.C., 1975.
40. **Committee on Radioactive Waste Management,** *The Shallow Land Burial of Low-Level Radioactivity Contaminated Solid Waste,* National Research Council, National Academy of Sciences, Washington, D.C., 1976.
41. **Frosch, R. A.,** Disposing of high-level radioactive waste, *Oceanus,* 20(1), 5, 1977.
42. **Hollister, C. D.,** The seabed option, *Oceanus,* 20(1), 18, 1977.
43. **de Marsily, G., Ledoux, E., Barbreau, A., and Margat, J.,** Nuclear waste disposal: can the geologist guarantee isolation?, *Science,* 197(4303), 519, 1977.
44. **Hoffman, F. O. and Kaye, S. V.,** Terrestrial exposure pathways: potential exposures of man from the environmental transport of waste nuclides, in Proc. Int. Symp. Management of Wastes from the LWR Fuel Cycle, U.S. ERDA Rep. CONF-76-0701, U.S. Energy Research and Development Administration, Washington, D.C., 1976, 524.
45. **Soldat, J. K.,** Aquatic exposure pathways — potential exposure of man from environmental transport of waste radionuclides, in Proc. Int. Symp. Management of Wastes from the LWR Fuel Cycle, U.S. ERDA Rep. CONF-76-0701, U.S. Energy Research and Development Administration, Washington, D.C., 1976, 539.

46. Federal Radiation Council, Estimates and Evaluation of Fallout in the United States from Nuclear Testing Conducted Through 1962, Rep. No. 4, Federal Radiation Council, Washington, D.C., 1963.
47. **Barnaby, F.**, TBT, *I.A.E.A., Bull.*, 15(4), 13, 1973.
48. Office of Public Affairs, Announced United States Nuclear Test Statistics. Through December 31, 1977, Nevada Operations Office, U.S. Department of Energy, Las Vegas, 1978.
49. **Hardy, E. P., Jr., Rivera, J., and Collins, W. R., Jr., Eds./Comps.**, Fallout Program Quarterly Summary Rep. (September 1, 1963 through December 1, 1963), U.S. AEC Rep. HASL-142, Health and Safety Laboratory, U.S. Atomic Energy Commission, Washington, D.C., 1964.
50. **Klement, A. W., Jr.**, Radioactive fallout phenomena and mechanisms, *Health Phys.*, 11(12), 1265, 1965.
51. **Klement, A. W., Jr.**, A Review of Potential Radionuclides Produced in Weapons Detonations, U.S. AEC Rep. WASH-1024, U.S. Atomic Energy Commission, Washington, D.C., 1959.
52. **Volchok, H. L. and Toonkel, L.**, Worldwide deposition of ^{90}Sr through 1973, in Health and Safety Laboratory Fallout Program Quarterly Summary Report (June 1, 1974 Through September 1, 1974), U.S. AEC Rep. HASL-286, Hardy, E. P., Jr., Ed., Health and Safety Laboratory, U.S. Atomic Energy Commission, Washington, D.C., 1974, 117.
53. **Comar, C. L.**, Fallout from nuclear tests, Understanding the Atom Series, U.S. Atomic Energy Commission, Washington, D.C., 1963.
54. **Steinberg, E. P. and Glendenin, L. E.**, Survey of radiochemical studies of the fission process, *Proc. Int. Conf. Peaceful Uses of Atomic Energy*, 7, 3, 1956.
55. **Glasstone, S. and Dolan, P. I., Eds./Comps.**, The Effects of Nuclear Weapons, 3rd ed., U.S. Department of Defense, U.S. Department of Energy, Washington, D.C., 1977.
56. **Gerber, C. R., Hamburger, R., and Seabrook Hull, E. W.**, Plowshare, Understanding the Atom Series, U.S. Atomic Energy Commission, Washington, D.C., 1966.
57. **Turner, F. B., Rowland, R. H., and Wood, R. A.**, Nuclear engineering and wildlife: radioactivity in jackrabbits after the Sedan test, *J. Wildl. Manage.*, 30(2), 433, 1966.
58. **Whicker, F. W., Walters, R. A., and Dahl, A. H.**, Fall-out radionuclides in Colorado deer liver, *Nature (London)*, 214(5087), 511, 1967.
59. **Katcoff, S.**, Fission-product yields from U, Th and Pu, *Nucleonics*, 16(4), 78, 1958.
60. **Hunter, H. F. and Ballou, N. E.**, Fission-product decay rates, *Nucleonics*, 9(5), C2, 1951.
61. **Way, K. and Wigner, E. P.**, The rate of decay of fission products, *Physiol. Rev. 2nd Ser.*, 73(11), 1318, 1948.
62. **Klement, A. W., Jr., Ed.**, Radioactive Fallout from Nuclear Weapons Tests, Proc. 2nd Conf., Germantown, Maryland, November 3 to 6, 1964, (AEC Symp. Ser. 5), U.S. AEC Rep. CONF-765, U.S. Atomic Energy Commission, Washington, D.C., 1965.
63. **Perkins, R. W., Thomas, C. W., and Nielsen, J. M.**, Measurements of airborne radionuclides and determination of their physical characteristics, in Radioactive Fallout from Nuclear Weapons Tests (AEC Symp. Ser. 5), U.S. AEC Rep. CONF-765, Klement, A. W., Jr., Ed., U.S. Atomic Energy Commission, Washington, D.C., 1965, 198.
64. **Larson, K. H.**, Continental close-in fallout: its history, measurement and characteristics, in *Radioecology*, Schultz, V. and Klement, A. W., Jr., Eds., Reinhold, New York, 1963, 19.
65. **Neel, J. W. and Larson, K. H.**, Biological availability of strontium-90 to small native animals in fallout patterns from the Nevada Test Site, in *Radioecology*, Schultz, V. and Klement, A. W., Jr., Eds., Reinhold, New York, 1963, 45.
66. **Ferber, G. J.**, Distribution of radioactivity with height in nuclear clouds, in Radioactive Fallout from Nuclear Weapons Tests (AEC Symp. Ser. 5), U.S. AEC Rep. CONF-765, Klement, A. W., Jr., Ed., U.S. Atomic Energy Commission, Washington, D.C., 1965, 629.
67. **Clark, H. M.**, The occurrence of an unusually high-level radioactive rainout in the area of Troy, N.Y., *Science*, 119(3097), 619, 1954.
68. **Machta, L. and List, R. L.**, The global pattern of fallout, in *Fallout: a Study of Superbombs, Strontium 90, and Survival*, Fowler, J. M., Ed., Basic Books, New York, 1960, 26.
69. **Perkins, R. W., Nielsen, J. M., and Thomas, C. W.**, Air concentrations of twelve radionuclides from 1962 through mid-1964, *Science*, 146(3645), 762, 1964.
70. **Whicker, F. W., Farris, G. C., Remmenga, E. E., and Dahl, A. H.**, Factors influencing the accumulation of fallout ^{137}Cs in Colorado mule deer, *Health Phys.*, 11(12), 1407, 1965.
71. **Hvinden, T., Hveding, D., Lillegraven, A., and Small, S. H.**, Caesium-137 fall-out over Norway since September 1956, *Nature (London)*, 188(4757), 1177, 1960.
72. **Hardy, E. and Alexander, L. T.**, Rainfall and deposition of strontium-90 in Clallam County, Washington, *Science*, 136(3519), 881, 1962.
73. **Osburn, W. S.**, Ecological concentration of nuclear fallout in a Colorado mountain watershed, in *Radioecological Concentration Processes*, Åberg, B. and Hungate, F. P., Eds., Pergamon Press, New York, 1967, 675.

74. **List, R. J., Machta, L., Alexander, L. T., Allen, J. S., Meyer, M. W., Valassis, V. T., and Hardy, E. P., Jr.,** Strontium-90 on the earth's surface. III, in Radioactive Fallout from Nuclear Weapons Tests (AEC Symp. Ser. 5), U.S. AEC Rep. CONF-765, Klement, A. W., Jr., Ed., U.S. Atomic Energy Commission, Washington, D.C., 1965, 359.
75. **Whicker, F. W. and Loveless, C. M.,** Relationships of physiography and microclimate to fallout deposition, *Ecology,* 49(2), 363, 1968.
76. **West, R. G. and Kelley, R. C.,** A Selected, Annotated Bibliography of the Civil, Industrial and Scientific Uses for Nuclear Explosives, U.S. AEC Rep. TID-3522, U.S. Atomic Energy Commission, Washington, D.C., 1971, Rev. 9.
77. **Glasstone, S.,** Public Safety & Underground Nuclear Detonations, U.S. AEC Rep. TID-25708, U.S Atomic Energy Commission, Washington, D.C., 1971.
78. **Glasstone, S.,** *Sourcebook on Atomic Energy,* 3rd ed., D Van Nostrand, New York, 1967.
79. **Kruger, P.,** Peaceful Nuclear Explosives, U.S. ERDA Rep. ERDA-44, U.S. Energy Research and Development Administration, Washington, D.C., 1975.
80. **Koranda, J. J. and Martin, J. R.,** Persistence of radionuclides at sites of nuclear detonations, in Biological Implications of the Nuclear Age (AEC Symp. Ser. 16), U.S. AEC Rep. CONF-690303, Goulden, A. M., Ed., U. S. Atomic Energy Commission, Washington, D.C., 1969, 159.
81. **Martin, W. E.,** Early food-chain kinetics of radionuclides following close-in fallout from a single nuclear detonation, in Radioactive Fallout from Nuclear Weapons Tests (AEC Symp. Ser. 5), U.S. AEC Rep. CONF-765. Klement, A. W., Jr.,, Ed., U.S. Atomic Energy Commission, Washington, D.C., 1965, 758.
82. **Turner, F. B. and Gist, C. S.,** Influence of a thermonuclear cratering test on close-in populations of lizards, *Ecology,* 46(6), 845, 1965.
83. **Hanson, W. C., Dahl, A. H., Whicker, F. W., Longhurst, W. M., Flyger, V., Davey, S. P., and Greer, K. R.,** Thyroidal radioiodine concentrations in North American deer following 1961-1963 nuclear weapons tests, *Health Phys.,* 9(12), 1235, 1963.
84. **Rhoads, W. A. and Platt, R. B.,** Beta radiation damage from close-in fallout from two nuclear detonations, *BioScience,* 21(22), 1121, 1971.
85. **Wilimovsky, N. J. and Wolfe, J. N., Eds.,** Environment of the Cape Thompson Region, Alaska, U.S. AEC Rep. PNE-481, U.S. Atomic Energy Commission, Washington, D.C., 1966.
86. **Martin, W. E.,** Radioecology and the feasibility of nuclear canal excavation, in Symposium on Radioecology, Nelson, D. J. and Evans, F. C., Eds., U.S. AEC Rep. CONF-670503, U.S. Atomic Energy Commission, Washington, D.C., 1969, 9.
87. **Atlantic-Pacific Interoceanic Canal Study Commission,** Interoceanic Canal Studies; Annex 1 to 5; Vol. 1 to 6, Washington D.C., 1970.
88. **Klement, A. W., Jr.,** Radiological safety research for nuclear excavation projects-interoceanic canal studies, in Symp. Public Aspects of Peaceful Uses of Nuclear Explosives, Rep. SWRHL-82, Southwestern Radiological Health Laboratory, U.S. Public Health Service, Las Vegas, Nev., 1969, 652.
89. **Alldredge, A. W., Whicker, F. W., and Hanson, W. C.,** Some environmental impacts associated with Project Rio Blanco, in *Radioecology and Energy Resources,* Cushing, C. E. et al., Eds., Dowden, Hutchinson & Ross, Stroudsburg, Pa., 1976, 65.
90. **Whicker, F. W., Alldredge, A. W., Fisser, H. G., Hanson, W. C., Johnson, J. E., and Post, G.,** Ecological Considerations for Project Wagon Wheel, U.S. ERDA Rep. PNE-WW-35, U.S. Energy Research and Development Administration, Washington, D.C., 1974.
91. **U.S. Atomic Energy Commission,** Annu. Rep. to Congress of the Atomic Energy Commission for 1968, Washington, D.C., 1969.
92. **Woodwell, G. M.,** The ecological effects of radiation, *Sci. Am.,* 280(6), 2, 1963.
93. **Krey, P. W.,** Atmospheric burnup of a plutonium-238 generator, *Science,* 158(3802), 769, 1967.

ADDITIONAL READINGS

Bartels, O. G., Estimate of volcanic contributions to the atmosphere and volcanic gases and sublimates as the source of the radioisotopes ^{10}Be, ^{35}S, ^{32}P, and ^{22}Na, *Health Phys.*, 22(4), 387, 1972.

Baxter, M. S. and Walton, A., Fluctuations of atmospheric ^{14}C concentrations during the past century, *Proc. R. Soc. London, Ser. A.*, 321, 105, 1971.

Bernat, M. and Goldberg, E. D., Thorium isotopes in the marine environment, *Earth Planet. Sci. Lett.*, 5, 308, 1969.

Broecker, W. S. and Walton, A. Radiocarbon from nuclear tests, *Science*, 130(3371), 309, 1959.

Broecker, W. S., Gerard, R., Ewing, M., and Heezen, B. C., Natural radio-carbon in the Atlantic Ocean, *J. Geophys. Res.*, 65(9), 2903, 1960.

Burton, J. D., Radioactive nuclides in sea water, marine sediments and marine organisms, in *Chemical Oceanography*, Vol. 2, Riley, J. P. and Skirrow, G., Eds., Academic Press, New York, 1965, 425.

Dean, M. H., A survey of the uranium content of vegetation in Great Britain, *J. Ecol.*, 54(3), 589, 1966.

Duursma, E. K., Geochemical aspects and applications of radionuclides in the sea, *Oceanogr. Mar. Biol. Ann. Rev.*, 10, 137, 1972.

Fedorov, A. F., Natural radioactivity in some ocean regions, *Radioact. Sea (Monaco)*, (15), 1, 1967; *Nucl. Sci. Abstr.*, 21(17), 30775.

Folsom, T. R. and Beasley, T. M., Contributions from the alpha emitter, polonium-210, to the natural radiation environment of the marine organisms, in Radioactive Contamination of the Marine Environment, Publ. STI/PUB/313, International Atomic Energy Agency, Vienna, 1973, 625.

Joseph, A. B., Gustafson, P. F., Russell, I. R., Schuert, E. A., Volchok, H. L., and Tamplin, A., Sources of radioactivity and their characteristics, in *Radioactivity in the Marine Environment*, Panel on Radioactivity in the Marine Environment, National Academy of Sciences, Washington, D.C., 1971, 6.

Kastner, J., The natural radiation environment, Understanding the Atom Series, U.S. Atomic Energy Commission, Washington, D.C., 1968.

Kenny, A. W., Radioactive contamination arising from the peaceful uses of atomic energy: past, present and foreseeable future, in Environmental Contamination by Radioactive Materials, Publ. STI/PUB/226, International Atomic Energy Agency, Vienna, 1969, 15.

Klement, A. W., Jr., Natural radionuclides in foods and food source materials, in *Radioactive Fallout, Soils, Plants, Foods, Man*, Fowler, E. B., Ed., American Elsevier, New York, 1965, 113.

Kohman, T. P. and Saito, N., Radioactivity in geology and cosmology, *Annu. Rev. Nucl. Sci.*, 4, 401, 1954.

Langham, W. H., Considerations of biospheric contamination by radioactive fallout, in *Radioactive Fallout, Soils, Plants, Foods, Man*, Fowler, E. B., Ed., American Elsevier, New York, 1965, 3.

Lough, S. A. and Solon, I. R., The natural radiation environment, in *Radiation Biology and Medicine*, Claus, W. D., Ed., Addison-Wesley, Reading, Mass., 1958, 405.

Mauchline, J. and Templeton, W. L., Artificial and natural radioisotopes in the marine environment, *Oceanogr. Mar. Biol. Ann. Rev.*, 2, 229, 1964.

Miyake, Y., Artificial radioactivity in the sea, in *The Sea*, Vol. 2, Hill, M. N., Ed., Interscience, New York, 1963, 78.

Miyake, Y., The sources of the artificial radioactive substances in the oceans, in *Impingement of Man on the Oceans*, Hood, D. W., Ed., John Wiley & Sons, New York, 1971, 565.

Morozova, N. G. Natural radioelements in contemporary bottom sediments in freshwater lakes of Estonia, in *Radioecology of Water Organisms. II. Distribution and Migration of Radionuclides in Freshwater and Seawater Biocenoses*, Andrushaitis, G. P., Ed., Zinatne, Riga, 1973, 116; AEC-tr-7606, U.S. Atomic Energy Commission, Washington, D.C., 1975, 70; (English transl.)

Nikiforov, V. S. and Ovchenkov, V. Ya., Determination of radon concentrations in the burrows and refuges of some burrowing animals, in *Radioecological Investigations in Natural Biogeocenoses*, Verkhovskaya, I. N., Ed., Nauka, Moscow, 1972, 230; *Nucl. Sci. Abstr.*, 26(23), 56136.

Parfenov, Yu. D., Polonium-210 in the environment and in the human organism, *At. Energy Rev.*, 12(1), 75, 1974.

Preston, A., Artificial radioactivity in the sea, in *The Sea*, Vol. 5, Goldberg, E. D., Ed., John Wiley & Sons, New York, 1974, 817.

Reiter, E. R., Atmospheric Transport Processes, Part 4 (DOE Crit. Rev. Ser.), U.S. DOE Rep. TID-27114, U.S. Department of Energy, Washington, D.C., 1978.

Rice, T. R. and Duke, T. W., Radioactivity in the sea, in *The Encyclopedia of Marine Resources*, Firth, F. E., Ed., Van Nostrand Reinhold, New York, 1969, 566.

Sackett, V. M., Mo, T., Spaulding, R. F., and Exner, M. E., A revaluation of the marine geochemistry of uranium, in Radioactive Contamination of the Marine Environment, Publ. STI/PUB/313, International Atomic Energy Agency, Vienna, 1973, 757.

Schell, W. R., Jokela, T., and Eagle, R., Natural ^{210}Pb and ^{210}Po in a marine environment, in Radioactive Contamination of the Marine Environment, Publ. STI/PUB/313, International Atomic Energy Agency, Vienna, 1973, 701.

Schreiber, B., Radionuclides in marine plankton and in coastal sediments, in *Radioecological Concentration Processes,* Åberg, B. and Hungate, F. P., Eds., Pergamon Press, New York, 1967, 753.

Schultz, V. and Longhurst, W. M., Accumulation of strontium-90 in yearling Columbian black-tailed deer, 1950-1960, in *Radioecology,* Schultz, V. and Klement, A. W., Jr., Eds., Reinhold, New York, 1963, 73.

Suess, H. E., Natural radiocarbon, *Endeavour,* 32(115), 34, 1973.

Thomson, J. and Walton, A., Natural radioactive decay series elements in the oceans and sediments, *Proc. R. Soc. Edinburgh, Sect. B,* 72(15), 167, 1972.

Verkhovskaya, I. N., Vavilov, P. P., and Maslov, V. I., Distribution and redistribution of uranium, radium, and thorium in natural biogeocenoses, in *Radioecological Investigations in Natural Biogeocenoses,* Verkhovskaya, I. N., Ed., Nauka, Moscow, 1972, 243; *Nucl. Sci. Abstr.,* 26(23), 56137; *Nucl. Sci. Abstr.,* 30(8), 21080; (English transl. RFP-trans-162).

Volchok, H. L., Bowen, V. T., Folsom, T. R., Broecker, W. S., Schuert, E. A., and Bien, G. S., Oceanic distributions of radionuclides from nuclear explosions, in *Radioactivity in the Marine Environment,* Panel on Radioactivity in the Marine Environment, National Academy of Sciences, Washington, D.C., 1971, 42.

Yliruokanen, I., A chemical study on the occurrence of rare earths in plants, *Ann. Acad. Sci. Fenn., Ser. A,* II(176), 1, 1975.

Chapter 5

RADIONUCLIDE BEHAVIOR IN ECOSYSTEMS

I. INTRODUCTION

When introduced into air or water, radionuclides are dispersed and diluted, spatially redistributed, and they ultimately accumulate in specific components of the environment. The fate of radionuclides in the environment is governed by a highly complex mosaic of physical, chemical, and biological factors. The fact that nearly all components of the environment contain at least traces of natural and man-produced radioactivity has stimulated considerable interest in radionuclide behavior in most major ecosystems important to man. Of primary interest are rates of movement of radionuclides within and between ecosystem components, mechanisms responsible for such movements, and degree of concentration of radionuclides within biotic and abiotic components of ecological systems.

Interest in the environmental behavior of radionuclides is usually motivated by concern for the biological effects which they may elicit or by the desire to understand geochemical and ecological processes through observations of radionuclide transport. In order to answer the question of biological effects, one must know or be able to predict (1) the movement and concentration of the material within the system of concern and (2) the radiation or chemical toxicity of the expected concentrations to the biotic components of the system. Experimental research is necessary in order to formulate predictive models for both the behavior and the effects of specific radioactive materials. In order to investigate the behavior of a given radionuclide in a given system, three approaches can be used: (1) relative concentrations of the radionuclide can be measured in the system's components following a contaminating event such as fallout, (2) the radionuclide can be introduced in a controlled manner into the system (or a small portion of it) by the investigator who subsequently measures the behavior of the material through time, and (3) the system can be theoretically modeled and applicable transport rate and other data can be used in the model to predict concentrations. Radiation ecologists have used all of these approaches, sometimes singly and sometimes in combination. To achieve maximum confidence in any predictive model and maximum insight into the ecological mechanisms involved, a combination of theoretical and experimental approaches is required.

Inasmuch as it is largely the nature and functions of an ecological system which control the environmental behavior of radionuclides, the observed behavior of radionuclides can elucidate a great deal of basic information about ecosystems. For example, a considerable portion of our knowledge on atmospheric motions, ocean currents, and mineral transport, has been developed through the measurement of radioactive contaminants. Contamination that has resulted from radioactive fallout has provided an environmental tracer which has been used on a global scale to infer the movement of environmental materials. Data from radioactive fallout provides a useful basis for the prediction of the movement of other airborne contaminants.

This chapter will explore some general concepts and apparent regularities regarding behavior of specific radionuclides or groups of radionuclides, and it will review actual observations on radionuclide behavior in some particularly interesting biomes and ecosystems. Using this approach, the authors hope to highlight some apparent commonalities in the ways that ecosystems cycle specific groups of radionuclides. They also hope to portray differences and uncertainties in their knowledge of radionuclide behavior in the environment.

II. GENERAL CONCEPTS

Radionuclide behavior is most readily understood within the framework of general concepts regarding transport processes and factors which modify such processes. While many transport processes are common to a large number of radionuclides, the quantitative importance of such processes varies with the properties of the radionuclides in question. In addition, characteristics of organisms and ecosystems are also of major importance in determining transport and concentration processes and their quantification. This section will qualitatively review transport processes as they apply to radionuclides and ecosystems in general, and then discuss how properties of the radionuclide, the organism, and the ecosystem can modify such transport processes.

A. General Transport Processes

When radionuclides are released to the environment through a set of conditions and circumstances called a "source term", the material becomes spatially dispersed in abiotic media, mainly air and water (Figure 1). Following such dispersal, various processes operate to funnel quantities of the material through components or "compartments" of the ecosystem. Compartments receiving the material may be living or nonliving, and the quantities that accumulate in each compartment are dependent upon the quantitative importance of competing transport processes.

The source term (Figure 1) is normally a description of the release, and to be most useful should describe the type of radionuclide, its physical and chemical form, the quantity released over time, the geometrical configuration of the release, and other pertinent circumstances. In the nuclear industry, most environmental releases of radioactive material are either in the form of suspended or dissolved materials in liquid effluents, or gases or particulates in airborne effluents. Radionuclides which enter a body of water or the airstream are immediately subject to processes such as turbulent and molecular diffusion which cause dispersion of the material. The general effect of dispersion is a reduction in concentration of the radionuclide in air or water with distance from the effluent source.[1,2] The extent and rapidity with which material is dispersed varies tremendously according to the degree of turbulence of the air or water medium.

Radionuclides in air or water are immediately subject to depositional phenomena.[3] Such phenomena include gravitational settling, precipitation scavenging, impaction, and chemical adsorption or exchange. Gravitational settling of suspended particles is increasingly important as the particles exceed 20 μm in diameter. However, in turbulent air or water, diffusion may temporarily override the effects of gravitational settling even for particles substantially larger than 20 μm. Precipitation scavenging refers to processes by which aerosol particles are washed from the atmosphere by ice crystals and water droplets.[4] Impaction is a process in which particles suspended in a stream of air or water impinge upon a solid object when the stream is diverted around such an object. Like gravitational settling, the importance of this process also increases with particle size. Chemical sorption or exchange is somewhat more complex in that it is dependent upon physical and chemical properties of the radionuclide, as well as of the surface upon which sorption or exchange occurs.

When a parcel of air or water medium containing radioactive particles comes into close proximity with solid materials such as rock, suspended sediment, vegetation, plankton, bacterial cells, and such, the particles generally have a reasonable probability of attaching to the surface of the materials. The attachment may result from various kinds of physical or chemical forces and it will vary in strength and duration, depending upon the properties of the radionuclide and the surface in question. The radionuclide ^{137}Cs tends to attach very strongly to many materials, especially clay minerals

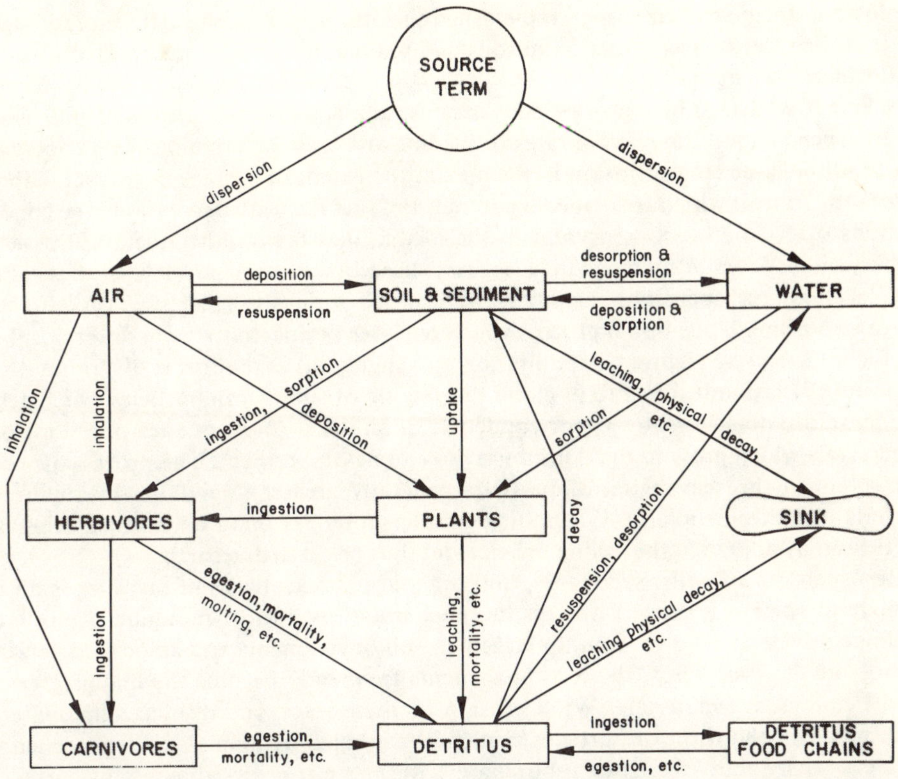

FIGURE 1. Major processes which effect radionuclide transport in ecosystems. Boxes represent ecosystem components while arrows represent flow of materials through functional processes.

and organic detritus, as do many other radionuclides. Radionuclides of the noble gases such as ^{85}Kr are noteworthy exceptions in that they generally do not attach to solid objects or surfaces. Noble gases tend to remain in the atmosphere until physical decay occurs. Tritium and ^{14}C may exhibit behavior similar to that of the noble gases if they happen to be in the gaseous state.

In general, solid materials accumulate most radionuclides to equilibrium concentrations which are considerably higher than in the surrounding air or water medium. Frequently the extent of adsorption and ultimate concentration increases as the surface area per unit mass or volume increases. For example, a kilogram of fine sediment will accumulate more radioactivity from a given aqueous medium than a kilogram of coarse gravel, assuming similar surface properties. Similarly, finely branched, hairy vegetation generally accumulates airborne radioactivity more efficiently than smooth surfaced vegetation.

Adsorption of radionuclides to surfaces from air or water generally proceeds rather rapidly and in a confined, static situation, such as a small aquarium with a fine-grained sediment, an acute introduction of chemically active radionuclide into the water will likely lead to equilibration between water and sediment within hours or perhaps a few days. However, if the air or water media is being continually replenished, and if the media contains a significant concentration of radionuclide over a long period of time, then the surface in question will accumulate the radionuclide and the radionuclide will increase in concentration over a much longer period of time. Examples of this are soils and sediments which accumulated radioactive fallout over a period of several years, particularly in the late 1950s and through the early and mid-1960s. Periodic nuclear explosions during this time span replenished the atmosphere with sufficient radioactivity that the rates of accumulation in soils and sediments exceeded the rates of loss over a considerable length of time.

Whether a surface increases or decreases in radioactivity over a span of time essentially depends upon the relative rates of income and loss. The rate of income increases with radionuclide concentration in the media, the extent of surface in contact with the media, the rate at which fresh media contacts the surface, with decreases in the concentrations of competing ions present in the media, and many other factors. Losses of radionuclides from adsorbing surfaces are caused by phenomena such as physical decay, leaching, resuspension, or spatial reallocations of the absorbing surfaces.

An important depositional process which removes radioactivity from the atmosphere is precipitation scavenging.[4] Precipitation scavenging takes the forms of "rainout" or "washout". Rainout refers to in-cloud scavenging of submicron particles which act as condensation nuclei. Water vapor condenses upon the surfaces of such nuclei to form ice crystals which grow in mass until gravity can act upon them. The process of washout refers to the scavenging of particles generally greater than 1 μm by falling ice crystals or water droplets. As a falling droplet impinges upon aerosol particles, the particles may adhere to the falling droplet and thus be carried to earth.

Resuspension of soils, sediments, and organic particles by wind or water is an extremely important transport process in many situations.[3] Most radionuclides in most ecological systems tend to wind up largely in soil or sediments and associated deposits of organic detritus. Thus, these compartments frequently become the major reservoir of the radioactive materials. Wind erosion of these reservoirs displaces the material into the air in the form of particles that can be inhaled by animals or redeposited on surfaces where they can enter food chains. Resuspension of sediments by turbulent water can move such sediments and sorbed radioactivity to distant locales. Sediments suspended in water can likewise enter food chains of aquatic biota.

Radionuclides in abiotic compartments of the environment can enter plants, the base of the grazing food chain, by uptake from soil, deposition from air, or sorption from water in the case of aquatic plants. Uptake from soil by plants normally refers to the passage of atoms (usually ions) or molecules from the soil solution through the root membrane and into plant tissues where translocation can take place. Uptake, therefore, usually implies internal incorporation within tissues of the organism in question. Frequently radioactive materials in soil adhere to the root or shoot surfaces of plants, but are not actually incorporated into plant tissues. Radionuclides in air or water can be deposited on or otherwise attached to plant surfaces directly, without passage through the soil. Such material may be held only temporarily on the surface, and later be washed or blown off, or it may attach quite strongly. Some surface-deposited material may even be assimilated into plant tissues through stomates or epidermal tissues. In many studies of radionuclide contamination of plants, it is technically difficult to distinguish between true uptake and surface deposition. Yet this remains an important question, both from a practical and theoretical viewpoint.

Passage of radioactive material from plants to herbivores is accomplished mainly by ingestion. If entire leaves are consumed, then both internally incorporated and externally attached material will enter the consumer. In some cases, the radioactive material incorporated within the plant may be in a more soluble form than surface material and thus a higher fraction may be assimilated by the consumer. Herbivores also ingest radionuclides associated with soils or sediments. In some cases, soil ingestion by animals is deliberate, but measurable quantities can also be inadvertently ingested in the process of foraging on low-growing plants, and licking or preening of fur, feathers, or young. Inhalation of aerosol particles provides an additional pathway of radionuclide intake by animals. The inhalation pathway may be more important than ingestion, in some cases, such as comparatively insoluble radionuclides in arid, dusty environments.

Radionuclides in tissues of herbiovores can be ingested by carnivores through predation. Each step in the food chain provides some form of discrimination, in that radionuclides are channeled physiologically into certain tissues, depending on their physical/chemical properties. Others may be poorly assimilated and simply pass through the digestive tract. Since some predators consume only a select portion of their prey, only those radionuclides in the ingested tissues can be assimilated. Frequently, however, the prey is consumed whole, and in such cases, radioactivity within the tissues as well as on the surface of the skin or in the gastrointestinal tract has opportunity for assimilation by the predator.

Death of plants and animals, or their parts, as well as secretions and excretions provide radionuclide input to the organic detritus reservoir. Although not indicated in Figure 1, organic detritus also receives radionuclides directly from air or water through deposition and sorption phenomena. Detritus serves as a significant reservoir of radioactive material which can recycle through biotic compartments by means of detritus food chains. Mineralization (decay) of detritus, accomplished largely by microorganisms, releases radionuclides to soil and sediment, where they can enter the grazing food chain, or be inhaled if the material is suspended into the air. Small detritus particles can be suspended into the atmosphere as well. Radionuclides in aquatic organic detritus can be solubilized or resuspended to reenter aquatic food chains, or be transported to a distant locale.

Environmental compartments which receive radioactive material but do not release it to other compartments may be termed "sinks". Sinks frequently consist of deeper strata of soils and sediments which tightly bind most radionuclides, which have negligible contact with biological processes, and which are protected from wind or water erosion. Radionuclides can migrate downward from surface soils, sediments and organic detritus by leaching and percolation, and become tightly bound in the deeper strata by sorption phenomena. Biological processes in soil, such as animal burrowing, tunneling by worms and other invertebrates, and plant root penetration can also effect downward transport of radionuclides. Soil can also be brought from lower depths to the surface by physical processes such as frost-heaving and needle ice, and by biological activities. This can effectively bury some surface contamination beneath less contaminated soil, which can make it less subject to transport. On the other hand, surface-casting of contaminated soil might enhance the probability of its erosion and dispersal. Physical decay of radioactive material essentially prevents transport to other compartments and thus enhances the effectiveness of sinks. The "environment" could serve as a sink in cases where the environmental media air or water removes radioactivity from a particular locale, with little or no opportunity for feedback.

The matrix of rates at which radionuclides move between compartments of ecosystems essentially determines how the material will ultimately be distributed and how rapidly the ultimate or steady-state distribution will be reached. However, rates of intercompartmental transport vary with the radionuclide, nature and activities of the biota, and properties of the ecosystem. The authors shall next look at some of the factors which affect transport rates, and hence, compartmental inventories.

B. Factors Affecting Radionuclide Transport and Accumulation

Before a discussion of how specific factors affect transport rates of radionuclides and their accumulation in specific ecosystem compartments, it is well to show how transport rates and compartment inventories are interrelated. To do this, the authors shall introduce a few fundamental concepts of tracer kinetics. A compartment shall be considered as a recognizable structural entity within the ecosystem which is reasonably homogeneous throughout, and which receives, loses, and maintains an inventory of chemical substances. For our purposes, the boxes in Figure 1 represent examples of such compartments. However, recognize that the definition of compartments is sometimes arbitrary and based solely upon convenience. One ordinarily defines q as the inventory or total content of a radionuclide within a compartment and it can be symbolized as

$$\boxed{q}$$

where the box represents the compartment of interest and q the amount of radionuclide contained within it. The average concentration (C) of the radionuclide in a compartment is simply

$$C = \frac{q}{M} \tag{1}$$

where M is the mass of the compartment.

The principal factors which determine q or C are the rates of income to and loss from the compartment. Symbolically

$$\text{Input rate} \longrightarrow \boxed{q} \longrightarrow \text{loss rate}$$

The input rate R_{in} is usually expressed as amount entering the compartment per unit time (example: μCi/day), and the loss rate R_{out} is usually expressed as the amount leaving the compartment per unit time (e.g.: μCi/day). Therefore, the derivative of q with respect to time (t) is

$$\frac{dq}{dt} = R_{in} - R_{out} \tag{2}$$

This equation says that if $R_{in} > R_{out}$, dq/dt is positive and q will increase with time. Likewise, if $R_{in} < R_{out}$, q will decrease and if $R_{in} = R_{out}$, q remains constant. In the latter case, the compartment is said to be in a steady-state, since $dq/dt = 0$.

Radionuclides are commonly lost from compartments according to first-order processes. In such cases, the loss rate is proportional to the compartmental inventory

$$R_{out} = kq \tag{3}$$

where k is called a "rate constant" and has the units of time $^{-1}$. Thus, if a compartment obeys first-order kinetics,

$$\frac{dq}{dt} = R_{in} - kq \tag{4}$$

As will be shown in Volume II, Chapter 1, k is related to a term called the "effective half-time" (T_{eff})

$$k = \frac{\ln 2}{T_{eff}} \tag{5}$$

The effective half-time is frequently used to express the rapidity with which a radionuclide is lost from a compartment. This term takes into account all mechanisms which cause compartment loss, such as physical decay, excretion, transport, etc.

Equation 4 can easily be integrated so that q can be expressed as a function of R_{in}, k, and t. Doing this with the assumptions that R_{in} is constant in time, and that q = 0 at t = 0

$$q = \frac{R_{in}}{k}(1 - e^{-kt}) \tag{6}$$

This equation shows that q is proportional to R_{in}, inversely proportional to k, and builds up with time until the steady-state condition is realized (when $1\text{-}e^{-kt} \rightarrow 1.0$).

The preceding equations show that the amount or concentration of a radionuclide in a compartment is governed by the relative rates of income and loss. One may be able to define and measure m different rates of income and n different rates of loss acting on a single compartment. In this more general case:

$$\frac{dq}{dt} = \sum_{i=1}^{m} R(in)_i - q\sum_{j=1}^{n} k_j \tag{7}$$

The important concept to gain at this point is that compartments increase or decrease their radionuclide burdens at a rate governed by all rates of income and loss. Individual rates of income and loss represent various mechanisms, the quantitative importance of each being described by associated values of R_{in} and k. Now the authors are ready to proceed with their discussion of important factors which affect rates of income to and loss from ecological compartments.

1. Properties of Radionuclides

If all possible radionuclides were simultaneously released at one point in the biosphere, subsequent distribution in space and time would be unique for each individual nuclide. The general movement and concentration of radionuclides in ecosystems is governed by many factors which depend in part upon the physical and chemical nature of the nuclides themselves. Elements have chemical properties unique to each, and different radioisotopes of the same element generally have differing physical half-lives and decay schemes. These variations more or less assure different environmental behavior through time.

The abundance of a radionuclide is obviously important but it may not be obvious that inherent properties can affect abundance. For example, the process of nuclear fission preferentially forms fission fragments having atomic masses between 80 and 105 and between 125 and 155. Neutron activation selectively forms radionuclides from those elements having a high neutron capture probability or "cross section". Radon is abundant in atmospheres of mines, basements, and other subterranean cavities because of its inert gas properties. In general, readily solubilized radionuclides having mineral nutrient counterparts are extensively and rapidly transferred through ecosystems. Persistence of radionuclides in ecological systems is directly related to physical half-life, biological half-life, and persistence of the system's components per se. Individual ecosystem components which have maximum income and minimum loss rates will have maximum radionuclide concentrations over long periods of time. A consumer organism within a food web will ingest radionuclides at a rate which is proportional to its food consumption rate, and to the average radionuclide concentration in the food. The latter becomes a matter of the extent of contamination, the distribution of the contamination, and the feeding habits of the consumer being considered. The consumer organism will lose radionuclides at a rate dependent upon the physical and biological half-lives of the radionuclides. While physical half-lives are fixed, biological half-lives are extremely variable, depending upon physical and chemical properties of the radionuclides and physiology of the organism.

Many radionuclides behave, at least qualitatively, like essential nutrient analogs. However, an important difference needs pointing out. Most essential nutrients are relatively abundant and organisms have generally evolved rather specific needs for these nutrients. Inasmuch as most of these nutrients are continually taken up or ingested, organisms have developed mechanisms for rejecting whatever quantities are not needed. In this way, most nutrient elements are homeostatically controlled to maintain nutrient concentrations within certain limits. The kidney is an organ which has the osmoregulation of body fluids as one of its most important functions. Most radionuclides, however, generally exist in chemical concentrations so minute, in comparison to nutrients, that they are not osmotically regulated. Even in concentrations sufficient to cause extreme biological damage, most radionuclides behave according to their individual atomic chemistry and not to mass action or colligative properties. For example, if ^{137}Cs were concentrated in the human body sufficiently to give a radiation dose of 500 rad/day (more than enough to cause lethality within a month), there would be less than 0.2 μg ^{137}Cs/g tissue. This is less than 10^{-4} the concentration of potassium, cesium's analog, and for every atom of cesium, there would be over 30,000 atoms of potassium. Therefore, most radionuclides, particularly those with half-lives of less than 100 years, generally behave as tracers in ecological systems and their ability to concentrate in biological tissues is not likely to be dampened by the mass of the radionuclide present. This cannot necessarily be said, however, for extremely long-lived radionuclides such as ^{40}K, ^{232}Th, and ^{238}U, and all of which have very low specific activities in terms of curies per gram.

Radionuclides which are in a soluble form, and which are chemically analogous to essential nutrient elements, will tend to follow ecological pathways in similar fashion to the nutrient analogs. For example, radionuclides such as ^{90}Sr, ^{89}Sr, ^{140}Ba, ^{226}Ra, and ^{45}Ca will behave like calcium; ^{137}Cs, ^{86}Rb, and ^{40}K will generally follow the movement of potassium; ^{131}I and ^{129}I will mimic the movement of stable iodine; and tritium will qualitatively behave like hydrogen. One of the most remarkable concentration processes is the accumulation of ^{131}I in the thyroids of herbivores following contamination of foliage. The thyroid may concentrate ^{131}I up to 10^4-fold higher than the foliage. This apparently occurs primarily for the reason that ^{131}I behaves chemically analogously to stable iodine which is needed by the thyroid for the synthesis of thyroxine and related compounds. Strontium-90 is an important radionuclide which may reach significant concentrations in bony tissues because of its chemical similarity to calcium. In general, ^{90}Sr is replaced or is otherwise removed from bone at a slow rate, thus, if it is chronically ingested over a long period of time, it may accumulate to concentrations that are very high relative to prevailing environmental concentrations. On the other hand, some important radionuclides possess physical and chemical characteristics which are quite unique and whose behavior cannot be inferred from the common nutrient elements. Some examples of these radionuclides include ^{144}Ce, ^{106}Ru, ^{95}Zr, ^{85}Kr, and ^{239}Pu. The fact that these radionuclides do not have nutrient analogs does not preclude them as a potential biological hazard. Cerium-144 will accumulate in the liver, ^{239}Pu is a very toxic bone seeker, and the other nuclides mentioned can pose an external or gastrointestinal tract dose problem.

The physical half-life of a radionuclide has an important bearing on the extent to which it will accumulate in plants or animals. If extremely short-lived, as many are (i.e., half-lives of less than an hour or so), they are likely to decay before significant biological concentration is possible. If extremely long-lived (i.e., half-lives greater than 10^5 years or so), the specific activity of the radionuclide may be so low that the rate of energy release from a significant concentration in terms of mass may be very small.

In general, radionuclides with half-lives in the range of a few days to a thousand years or so are most likely to achieve high activity concentrations in biological systems.

Physical form of radionuclides also affects the degree of biological concentration. In general, chemically active, soluble forms are most readily taken up and concentrated by organisms. Soluble ^{137}Cs is readily assimilated by animals but if the nuclide is encapsulated in insoluble or inert particles, it cannot be assimilated and concentrated in living tissues. This does not preclude a nuclide from having a biological impact since insoluble particles can lodge in hair, lungs, and on the surfaces of vegetation, potentially creating an external radiation problem. Also, radionuclides as insoluble particles can be selectively accumulated by filter feeding aquatic organisms. The size and shape of insoluble particles to which radioactivity may be attached affects behavior and distribution. As a rule, the smaller particles can travel farther and lodge more tightly on surfaces.

In summary, intrinsic properties affect the behavior of radionuclides. Input rates to ecological compartments are generally enhanced by abundance, high solubility, chemical and radiological activity, existence of nutrient analogs, small particle size, and intermediate half-life. Loss rates are enhanced by high solubility, existence of abundant nutrient analogs, and short half-lives. Later in this chapter the authors shall consider groups of radionuclides which, owing to their inherent properties, behave in reasonably common fashion and contrast them with groups which exhibit totally different behavior.

2. Characteristics and Behavior of Organisms

Innate characteristics and behavioral attributes of organisms markedly affect the degree to which they accumulate radionuclides. Frequently, studies have shown that different species belonging to a common biotic community vary tremendously in their content of radionuclides, even though they have been exposed to a common source. Such interspecific differences hold true over a large spectrum of radionuclides and environments. Some of the important physical characteristics and behavioral patterns which lead to such differences require discussion.

External morphology of organisms has an important bearing on this propensity to accumulate radioactive materials from the environment, particularly if surface adsorption is an important vector. Organisms with a large surface area relative to mass generally accumulate comparatively large quantities of radionuclides from air, water, or soil. Thus, small organisms generally adsorb more contamination from their environment per unit mass than large organisms, simply because they present greater surface area per unit mass. Morphological configuration is also important. Spherical shapes present less surface area than other morphological forms having the same mass. Flattened, branched, or convoluted configurations can present large surface areas, and, in fact, some organisms such as lichens have evolved into such forms, which enhance absorptive surface area (Figure 2).

The nature of biological surfaces is also extremely important in the adsorption of particles from air, water, or soil. For instance, many plants have extremely "hairy" leaves (Figure 3). Such a carpet of fine epidermal projections can efficiently trap and hold contaminated particles. Epidermal hairs may have absorptive capabilities for soluble materials and they sometimes excrete oily or waxy substances which further serve to hold particles. Animal hairs also act to trap dust particles, and the grooming or licking of fur provides a vector of ingestion to many animals. The stomates of plant leaves allow epidermal penetration of gases and small particulates. The lower epidermis of apple leaves may have nearly 30,000 stomata per square centimeter.[5]

Metabolism and physiological characteristics have great influence upon radionuclide intake, assimilation, tissue distribution, and retention by plants and animals. Active

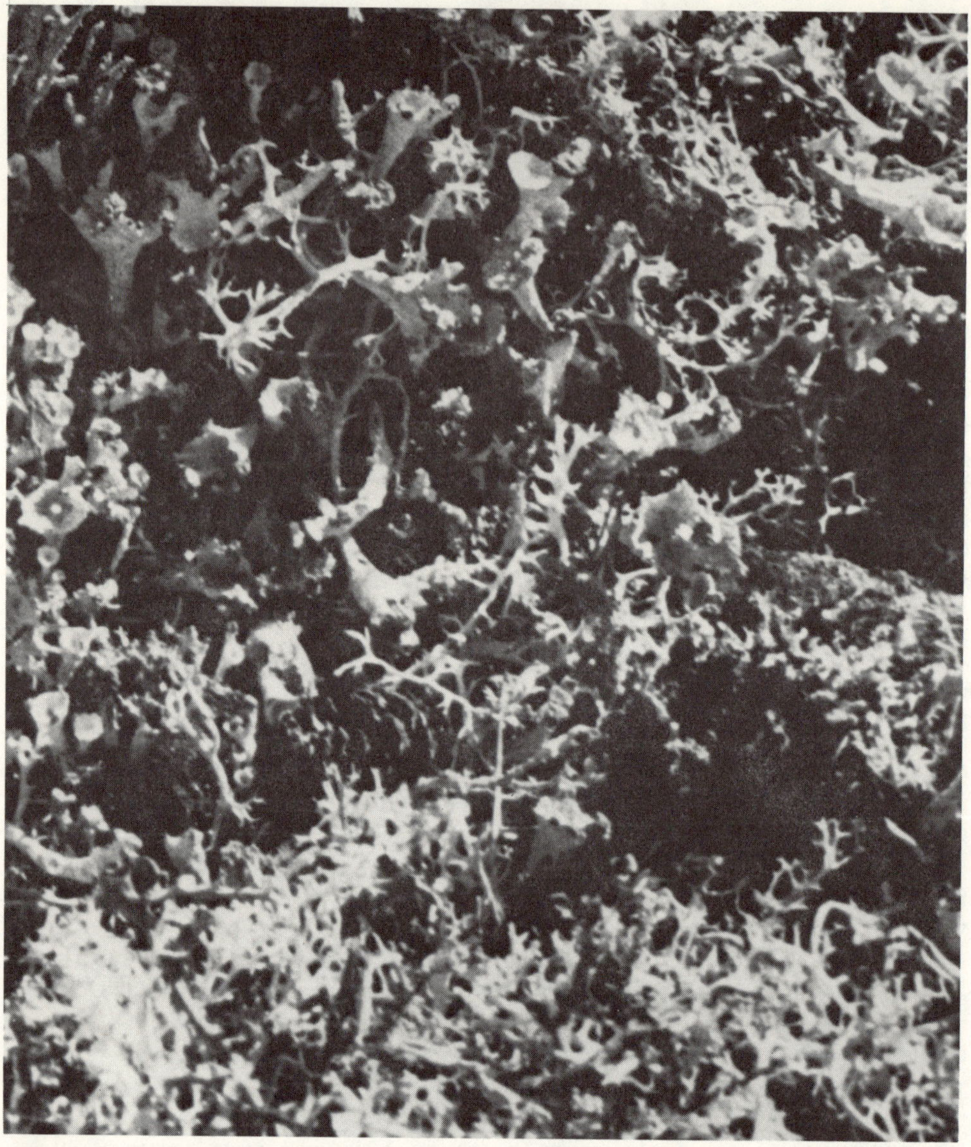

FIGURE 2. Photograph of *Cladonia* sp. lichens. The morphological configuration of these organisms provides a large surface-to-mass ratio which enhances adsorption of nutrients, water, and radionuclides from the atmosphere.

metabolism requires intake of nutrients and energy and such intake carries radionuclides along with the stable elements. Radionuclides which are soluble and chemically similar to metabolically active elements cross biological membranes and become distributed within biological tissues. In general, the greater the metabolic rate, the greater the flux of energy and minerals through the organism. This is reflected by increased food consumption, which results in an increased radionuclide ingestion rate in animals. Increased metabolic activity also increases the excretion rate of most minerals because they must be homeostatically maintained in the body within certain limits. Radionuclides may or may not follow the same pattern. Since they are usually present in tracer concentrations, they are not homeostatically maintained like major

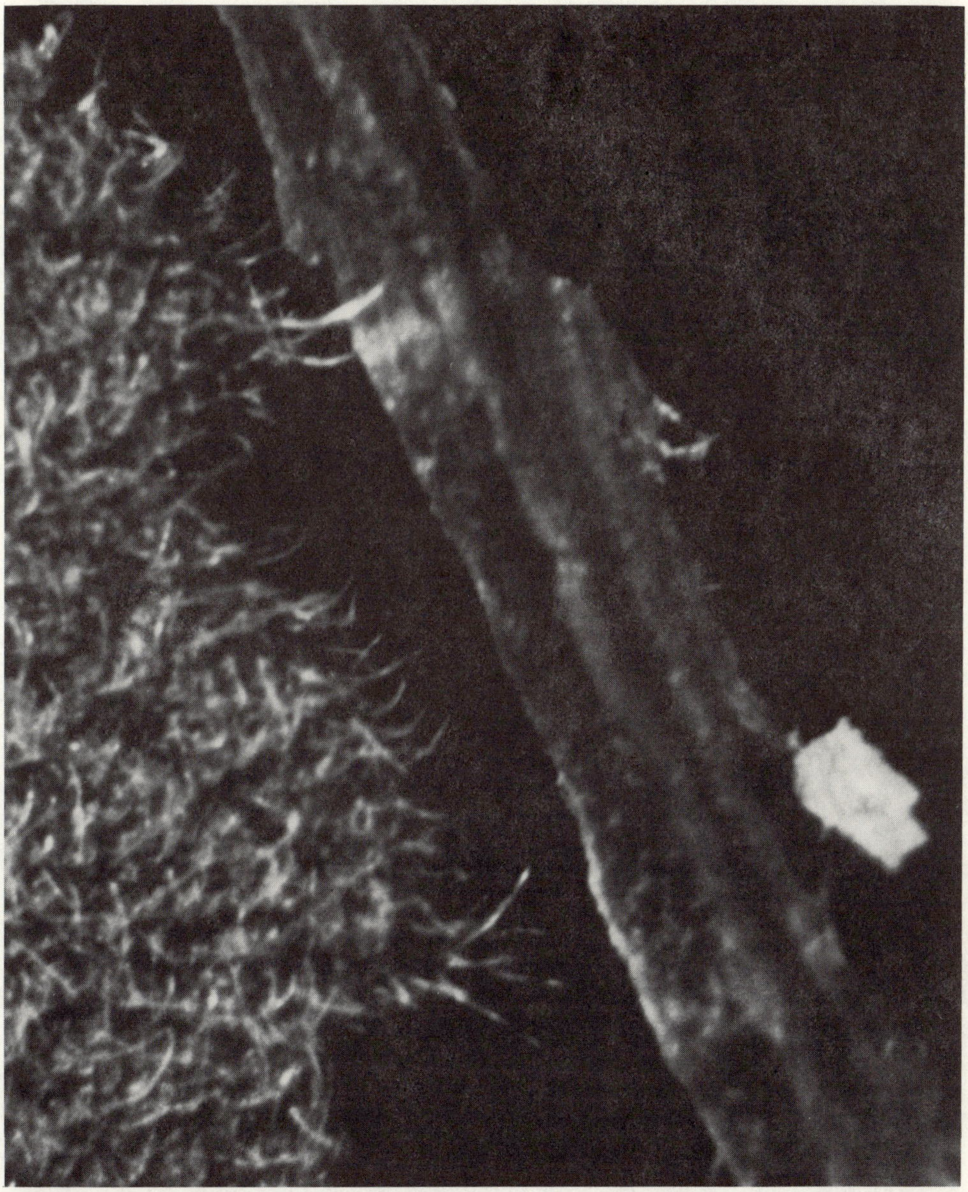

FIGURE 3. Photograph of the leaf surface of mountain mahogany *(Cerococarpus montanus)* showing epidermal hairs which efficiently trap aerosol particles.

nutrients. Therefore, it is possible that increased metabolic activity will result in increased intake of a radionuclide, but if it becomes tightly bound in the body, it may not be excreted at a proportionately high rate and thus will show net accumulation in tissues.

Metabolism and physiology vary with age, sex, season, health, diet, and other factors. For example, young, growing animals usually have a comparatively high food intake rate, assimilation of elements is frequently enhanced over that of adults, and a fraction of the energy and nutrients ingested is stored as new tissue. The net result of this is frequently higher radionuclide burdens in young, especially for certain radionuclides. On the other hand, some radionuclides, especially long-lived bone seekers, may continually accumulate with age and show higher concentrations in older animals. Physiological variations between males and females, usually mitigated by hormonal differences, sometimes lead to sex-related variations in radionuclide burdens. Variables associated with season include temperature, light, and hormonal factors. Poikilothermic (cold-blooded) animals respond strongly to temperature changes and plants respond dramatically to light, temperature, and moisture. Such seasonal changes lead to marked temporal fluctuations in the intake and loss rates of radionuclides. Fluctuations in intake of mineral nutrients can also affect the assimilation and retention of analogous radionuclides. For instance, a high flux of calcium will depress strontium assimilation and also increase the rate of strontium excretion.

Longevity is a factor of particular significance for longer-lived radionuclides. In general, organisms which themselves are long-lived or which have long-lived tissues have opportunity for chronic accumulation of persistent radionuclides. If inputs are continual and if retention is high (i.e., the loss rate is low), nuclide burdens can continue to build up over long periods of time. For instance, several long-lived bone-seeking radionuclides when chronically ingested by man do not reach equilibrium, even after 50 years.[6] Many lichen species are long-lived, slow-growing, and retain radionuclides with effective half-times exceeding a decade. This is a major factor contributing to the comparatively high concentrations of fallout ^{137}Cs and other radionuclides in lichens. Plant leaves are efficient accumulators of many radionuclides, but in deciduous species the leaves are precluded from achieving the theoretical maximum burdens because of annual leaf fall. Some evergreens retain their leaves two or more years, which provides additional time for radionuclide accumulation. Life spans within the plant and animal kingdoms range from hours or days to several centuries. The importance of organism longevity diminishes as one proceeds from a chronic to an acute intake situation and from long- to short-lived radionuclides. The cycling of radionuclides in ecosystems depends heavily in many cases upon longevity; thus one needs to understand life-spans and their relation to the kinetics of radionuclides in the environment.

In terms of behavioral attributes that affect radionuclide intake and loss by animals, one needs to mention movements, habitat selection, food habits, and other factors. Because the sources of environmental radioactivity and species abundance are not uniformly distributed, the proximity of plants and animals to contamination varies geographically. Geographic distribution of organisms is governed by physical tolerances and interactions with biotic and abiotic features of the environment, while the distribution of radionuclides is dictated by man, as well as geology, atmospheric motions, latitude, and elevation. Some natural populations, owing to their proximity to radioactively contaminated areas, accumulate radionuclides in measurable quantities. Migratory waterfowl periodically contact radionuclides when they utilize areas containing nuclear activities.[7] Seasonally migrating mule deer ingest greater amounts of fallout radionuclides in summer at the higher elevation in the Colorado mountains.[8] Food habits of caribou, which change as the animals traverse over large areas in Alaska, dramatically affect radionuclide intake and body burdens.[9]

Because radionuclides usually contaminate environments in heterogeneous patterns, the kinds of habitat occupied and utilized by plants and animals can markedly affect contact with radioactive material. Vertical zonation in biotic communities is a good example. In most ecosystems, radioactive contamination usually tends to accumulate near the soil or sediment surface. Thus, those species in most intimate contact with the soil surface have good opportunity to adsorb or ingest radionuclides. Those species which largely occupy the forest canopy, for example, have less opportunity for radionuclide ingestion. Topography, which affects the distributions of radionuclides as well as biota, also provides varied opportunity for radionuclide accumulation.[10] For instance, protected depressions collect moisture, develop deep soils, and nurture certain populations which are quite distinct from populations using adjacent ridges of exposed terrain. Radionuclides tend to accumulate in protected depressions, but disperse from exposed terrain. This may lead to a significant contrast in radionuclide burdens of plants or animals which utilize these systems.

Food habits of animals are modified by the species supported within habitat types, as well as by innate selectivity. A large matrix of factors operates to cause substantial variability in radionuclide concentrations between different organisms or different tissues within given organisms. Two plant species, growing side by side, may vary in radionuclide concentration by an order of magnitude.[8] Likewise, the seeds, leaves, and stems of a plant may also vary by an order of magnitude or more. Possible contrasts in radioactivity between animal groups, compared on the basis of types of food eaten, have been clearly demonstrated by Osburn.[10] Similarly, Hanson[9] has shown the dramatic variations in radionuclide burdens of a single species through time in response to changes in food habits. Variability in radionuclide burdens is a function of food preference diversity in animals. Therefore, radionuclide burdens of animals having a broad range of choices of energy sources, are less predictable than those of animals having limited food choices.

Other aspects of animal behavior which affect radionuclide accumulation include ingestion of soil or sediments, burrowing, and grooming. It is well established that both aquatic and terrestrial animals ingest soils and sediments.[11,12] In some cases, soil ingestion may be deliberate and possibly for the purpose of obtaining mineral nutrients or grit. In other cases, soil ingestion is inadvertent. Inadvertent soil ingestion may arise from dust on the surfaces of plants, or consumption of food near the soil or sediment surface. Soil ingestion also occurs in animals which groom or lick their fur, which contains dust picked up from the air or soil. Since soil is frequently a major reservoir of radionuclides, this pathway should receive due consideration in radiological assessment. Burrowing animals are in very close contact with the soil by virtue of this habit. Because the atmospheres of subterranean burrows generally have poor circulation and exchange, significant concentrations of natural radon and progeny may accumulate and pose a measurable radiation dose to burrow occupants through inhalation.

3. Ecosystem Attributes

Numerous attributes of the environment best described at the ecosystem level also play significant roles in determining the behavior of radionuclides. Such attributes may be physical, chemical, or biological in nature, or of some combination. The tremendous diversity of ecosystems of the biosphere offers added dimensions of complexity to the problem of radionuclides in the environment. However, such diversity also offers some explanations for many experimental observations, which can lead to a better overall understanding of the subject.

Pertinent ecosystem attributes, which may be classified as physical, include proximity, climate, topography, and volume and shape. Proximity of ecosystems to sources of radionuclides is an obvious feature that needs little elaboration beyond saying that it is a major variable affecting the quantities of radionuclides entering ecological systems. Climate, on the other hand, requires a bit more discussion. The climate of geographical areas strongly influences the development of ecosystems by helping to determine which species can persist and how dominant each might be. In return, climate, especially near the ground, is influenced by the development of ecosystems. Climate at the same time affects the deposition and transport of radionuclides. For instance, precipitation enhances the deposition of airborne radionuclides and plays a role in the redistribution of deposited radionuclides through leaching and erosional processes. Wind disperses radionuclides, while the lack of it leads to deposition of particulate matter. Temperature affects the physical state of water, which may have a profound effect on the behavior of radionuclides.

Topography interacts with climatic variables in ways which can determine whether ecosystems will tend to accumulate or lose radionuclides through time. While some ecosystems tend to concentrate radionuclides, other systems are "flushed" continually by turbulent air or water motions and this not only helps to prevent radionuclide buildup, but it also causes radionuclide dispersion away from that location. Examples of such ecosystems are mountain tops or high ridges which are exposed to strong wind action (Figure 4), river channels, which are periodically scoured by turbulent water and suspended debris (Figure 5), and seacoasts exposed to violent wind and water action. On the other hand, ecosystems which have evolved in quiet, protected areas tend to accumulate radioactive debris. The lee sides of mountain peaks and ridges are usually protected to some extent from winds which allows suspended airborne debris and snowflakes to settle to the ground. A large fraction of atmospheric radioactivity is frequently attached to suspended particles and thus finds its way to the ground through the gravitational mechanism. High mountain lakes, which are located in protected glacial cirques and valleys, provide an excellent example of a "debris accumulating" system (Figure 6). Flushing rates in such lakes are generally small and large snowbanks which accumulate during winter and spring months contain substantial amounts of stable and radioactive debris. Such debris is released into the lake during spring thaw. Terrestrial ecosystems located in similar topographic depressions also accumulate relatively large quantities of air- and waterborne debris. Standing water tends to hold radionuclides beneath the surface. In addition to preventing losses by wind erosion, standing water is a reasonably efficient trap for dust and other atmospheric aerosols. Temporary pools of standing water also tend to increase radionuclide movement from the abiotic substrate to biotic components of the system. Quiet pools, bays, and sounds provide examples of aquatic systems which undergo long-term accumulation of suspended debris. For instance, silt deposition at the mouth of the Columbia River contains radionuclides which entered the river several hundred miles upstream at the Hanford reactors.

In aquatic ecosystems, volume is of considerable importance in terms of inherent capacity to dilute radioactivity. For aerially deposited material, initial water concentration is inversely proportional to mean mixing depth per unit surface area. The mixing depth usually relates approximately to the depth of the epilimnion. Within this water layer, wind primarily effects rapid mixing. Deeper layers are usually comparatively stagnant, although they do eventually play a role in dilution. In the oceans, the mixing depth averages around 75 m, while in freshwater lakes, it averages 10 m or less. The difference in mixing depth is a major (but not the only) reason why marine biota concentrate less fallout ^{137}Cs and ^{90}Sr than freshwater counterparts. In regard to flowing streams, a large, turbulent river would dilute a given effluent volume far more effectively than a small stream.

Turning to chemical factors, ecosystems which contain abundant quantities of available mineral nutrients usually harbor organisms with reduced levels of radionuclides. This is particularly true for radionuclides which are chemically analogous to essential nutrients. Conversely, nutrient-poor or nutrient-limited systems lead to comparatively high concentrations of such radionuclides in organisms. A basic reason for this phenomenon is that available or soluble mineral nutrients chemically dilute analogous radionuclides. This leads to reduced assimilation and enhanced excretion of such nuclides, and hence, reduced tissue accumulation. Furthermore, biota in nutrient-limited environments may evolve nutrient-conserving mechanisms, usually at the physiological level, which also leads to conservation of some radionuclides.

FIGURE 4. Strong wind and water action on topographically exposed sites do not usually tend to accumulate radioactive material. Instead, any deposited debris becomes dispersed from such sites by wind and water.

Table 1
TYPICAL CONCENTRATION FACTORS FOR CESIUM AND STRONTIUM IN MARINE AND FRESHWATER ORGANISMS

		Concentration factors		
Element	**Ecosystem**	**Mollusks**	**Crustaceans**	**Fish muscle**
Cesium	Freshwater	600	4000	3000
	Marine	8	23	15
Strontium	Freshwater	600	200	200
	Marine	1	3	0.1

Data adapted from Polikarpov, G. G., *Radioecology of Aquatic Organisms,* Schultz, V. and Klement, A. W., Jr., Eds., North-Holland, Amsterdam, 1966.

The effect of nutrient availability has been amply demonstrated in a variety of ecosystems, especially for fallout ^{137}Cs, ^{90}Sr, and ^{131}I. For instance, marine and freshwater organisms show dramatically different concentration factors for cesium and strontium (Table 1). The concentration factor, which is loosely defined at this point as the amount of radionuclide per gram organism ÷ amount of radionuclide per milliliter water, will be discussed in detail later. The competing nutrient elements, potassium and calcium, average around 400 ppm in seawater, but usually less than 10 ppm in freshwaters. Trout bone from specimens collected from 18 mountain lakes in Colorado contained ^{90}Sr levels that were highly correlated with calcium concentrations in water (Figure 7).[14] Similar relationships between ^{137}Cs and K have also been reported for

FIGURE 5. Stream channels that are periodically scoured and eroded by high, rapidly moving water do not tend to accumulate radionuclides over long time periods.

biota in freshwater lakes.[15] As will be shown later, such relationships occur in terrestrial as well as aquatic ecosystems.

In addition to abundance of mineral nutrients, other chemical factors may exert significant influence on radionuclide behavior. Such factors are frequently associated with the chemistry of soils, sediments, and water. Acidity or alkalinity of soils can affect mobility and availability of many radionuclides. Mineral composition of soil is also important, as it affects particle size, surface area, and surface chemistry. Some soils particularly those high in montmorillonite and illite clays, tend to act as sinks for cesium and other elements. Such soils may bind ions so tightly that radionuclides be-

FIGURE 6. High mountain lakes are good examples of debris-accumulating systems because they are usually located in protected glacial cirques where snow and dust can accumulate over long time periods.

come essentially unavailable for uptake or leaching. Other substrates, such as the sandy, low cation exchange capacity soils of the lower coastal plain of the southeastern U.S. do not readily bind most radionuclides. This leads to high availability of many elements and a large fraction of the radionuclide inventory of such ecosystems can reside in biotic components. Another chemical factor that is often considered is the presence of organic acids, such as the humic and folvic acids. Such substances can act as natural chelating agents which can effectively increase the mobility of many radionuclides, even some of the transuranium nuclides which normally exhibit very low mobility. The chemistry of natural chelating agents is very complex, and especially difficult to study at normal environment concentrations; it appears that there is much to be learned in this area.

In terms of biological attributes of ecosystems, physiognomic features frequently operate which enhance radionuclide accumulation. For example, vegetation acts as an effective filter to remove suspended materials from air and both dissolved and suspended materials from water. Thus, well-vegetated ecosystems tend to trap debris to which the system is exposed; in many cases, the debris will enter functional food chains and mineral cycles within the system. In this manner, the debris is ''held'' within the system. Vegetation also acts to hold radionuclides within a system by modifying microclimatic variations and preventing erosion of soil by wind and water and in some cases by recycling of nuclides from the soil which might otherwise be lost to subterranean sinks through leaching.

Communities which have evolved in rigorous climates, such as arctic and alpine systems, frequently tend to accumulate unusually high concentrations of radionuclides

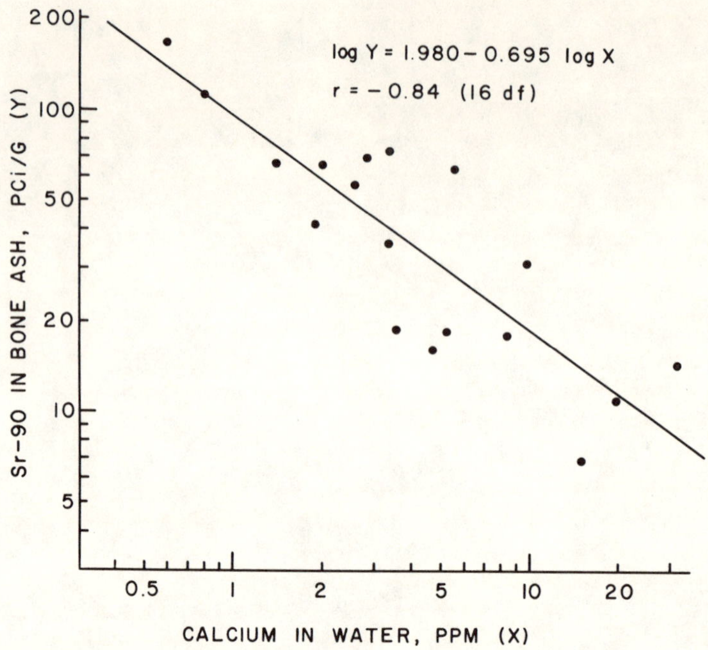

FIGURE 7. Strontium-90 concentrations in bone samples of trout from 18 mountain lakes in Colorado vs. dissolved calcium concentrations in water samples. (Reproduced from *Health Physics,* Volume 23, Issue 4, page 519, 1972. By permission of the Health Physics Society.)

for the reason that a given quantity of contamination deposited over a unit area is contained within a comparatively small standing crop of biomass. Prostrate, low-growing plants, characteristic of arctic and alpine regions, are also in intimate contact with melting snow which encourages direct transfer of radionuclides to the vegetation. Tree canopies usually do not have this type of association with melting snow, particularly if wind occurs prior to snowmelt. Community biomass may also affect the partitioning of radionuclides between biotic and abiotic compartments of ecosystems. Systems which support lush communities of high biomass are more likely to contain a large fraction of the total radionuclide inventory than sparsely populated ecosystems. However, other factors could override this tendency.

Ecosystem diversity relates to the movement and accumulation of radionuclides within the system. A diverse array of plant and animal species is likely to constitute a relatively complex matrix of food web relationships with many alternative pathways. The flow of minerals, energy, and radionuclides through such a system will be subdivided and partitioned to an extent commensurate with the diversity of the system. In systems of relatively low diversity, a larger fraction of the total energy, nutrient and radionuclide flux will tend to flow through certain species, since the total number of species is less and since a few species or groups of species may exercise a high degree of functional dominance. This concept is strengthened by numerous observations on radionuclide behavior, particularly the observations on radionuclide passage through arctic food webs, one of the better examples of systems which are comparatively low in biotic diversity.

III. BEHAVIOR OF SPECIFIC GROUPS OF RADIONUCLIDES

Individual radionuclides exhibit unique sets of characteristics, and therefore, somewhat unique behavior. However, certain groups of radionuclides, owing to their chemical properties, behave similarly in the environment. In this section, the authors discuss the ecological behavior of some of the more important radionuclides, and at the same time, point out some commonalities and differences in behavior that result from chemical attributes. They have chosen to treat this topic by element groups, including nonmetals, light metals, noble gases, heavy metals, rare earths, and the actinides. Example radionuclides that fall within each element group are discussed, with emphasis upon those which are particularly abundant in the environment and which have received substantial study. In Volume II, Appendix B lists several major reviews on the behavior of specific radionuclides in the environment.

A. Nonmetals (H, C, P, I)

Nonmetals are chemically distinguished from metals on the basis that nonmetals tend to gain valence electrons when they chemically combine with other elements, while metals tend to lose electrons. The nonmetals include some, but not all of the elements within groups IIIA through VIIA of the periodic table. Hydrogen can behave either as a metal or nonmetal, and is included in this group largely as a matter of convenience. Within the nonmetal group considerable diversity in chemical and environmental behavior is evident. Of the 15 or so elements which might be placed in this group, the authors shall consider four, each of which is biologically essential and has one or more significant radioisotopes.

The lightest and one of the most ubiquitously distributed radionuclides is tritium (^{3}H). Tritium is produced continually in the atmosphere by cosmic rays and also by nuclear reactors and nuclear detonations.[16] Tritium decays with a half-life of about 12 years to ^{3}He by the emission of beta particles of 18 keV or less. Tritium closely follows the reactions of ordinary hydrogen, but because of the large relative mass discrepancy between ^{3}H and ^{1}H, chemical reaction and diffusion rates show measurable differences. The vast majority of environmental tritium is in the form of tritiated water (HTO), but small amounts in the form of HT or tritiated carbon compounds also exist. Upon release to the environment, tritium generally follows the hydrologic cycle, but its initial behavior is determined by the source term. If released as a gas or vapor to the atmosphere, substantial dispersion can be expected and the rapidity of deposition is largely dependent upon climatic factors. Precipitation is the principal mechanism of deposition.[17] If released in liquid form, the HTO is diluted in surface waters and it is subject to physical dispersion, percolation, and evaporation.

Tritium exhibits ecological behavior unlike most other radionuclides. In aqueous bodies, it does not have the tendency to adsorb to sediments or biotic surfaces, which is so common for most other radionuclides. For this reason, HTO makes an excellent tracer for H_2O. Tritium can undergo exchange reactions with surfaces, but at normal environmental concentrations there are roughly 10^{18} atoms of ^{1}H for every few atoms of ^{3}H.[18] Therefore, on hydrogen-saturated surfaces, there is little probability for exchange with ^{3}H unless its specific activity is very high. Tritium enters plants through the roots, leaves, and stems, with high efficiency. It enters animals through ingestion, inhalation, and direct absorption through the skin. Unlike many radionuclides, however, tritium is seldom if ever concentrated in biological tissues to levels greater than those in the ambient air or water. There is, in fact, evidence that the steady-state concentration of tritium in biological tissues approaches but does not exceed the concentrations in ambient water or water vapor.[19,20] However, there is also some evidence that organically bound tritium may account for cases in which the T/H ratio in biota can slightly exceed the ratio in ambient water.[21] Tritium is lost from land plants through transpiration. It is lost from animals through exhalation, secretions, excretions, and surface evaporation.

Carbon-14, like tritium, is produced by cosmic rays, reactors, and nuclear explosions and is ubiquitously distributed in the biosphere. Carbon-14 decays with a half-life of about 5600 years through emission of 158 keV (maximum) beta particles. Natural formation and anthropogenic releases of ^{14}C are mainly atmospheric and the predominant chemical form in the atmosphere is $^{14}CO_2$. Dilution of atmospheric $^{14}CO_2$ with $^{12}CO_2$ occurs rapidly, and, under natural, steady-state conditions, there are about 15 dpm/g carbon. This represents about one atom of ^{14}C per 10^{12} atoms of ^{12}C. Carbon-14 enters food chains primarily through photosynthetic incorporation of CO_2. Complex transformations convert ^{14}C to many other chemical forms in plants and animals. When oxidized, compounds release ^{14}C back to the atmosphere as carbon dioxide. Carbon dioxide is also converted to carbonate and bicarbonate ions which may exist as inorganic carbon deposits.

In terms of bioaccumulation, ^{14}C is found in biota in roughly the same ratio to ^{12}C as it occurs in air or water. Since the total amount of carbon in organisms is roughly constant, except for growth phases, the amount of environmental ^{14}C that can accumulate is somewhat fixed. However, if the global ratio of $^{14}C/^{12}C$ were to increase through man's nuclear activities, the biological incorporation of ^{14}C would increase proportionately. On the other hand, burning of large quantities of fossil fuels, which are depleted of ^{14}C because of radioactive decay, tends to lower the $^{14}C/^{12}C$ ratio in the biosphere, thus reducing the incorporation of ^{14}C by plants and animals. There could be a slight fractionation between ^{14}C and ^{12}C as these nuclides pass through the chemical transformations in food chains because of the small isotopic mass difference. It is well known that ^{14}C forms a slightly stronger chemical bonds than ^{12}C and $^{12}CO_2$ diffuses slightly more rapidly than $^{14}CO_2$.[22] However, such isotopic effects do not appear sufficient to alter significantly the $^{14}C/^{12}C$ ratio through food chains.

Phosphorus is an essential component of biological systems, in that it is a building block of various kinds of tissues and a key element in many biochemical transformations, and in particular, those involving energy transduction. Phosphorus-32, one of six radioisotopes of phosphorus, is of considerable radioecological interest because it is frequently introduced into the environment, either as an unwanted contaminant or as a tracer used to study behavior of the element.[7] Produced mainly by neutron activation of stable phosphorus, ^{32}P has a half-life of about 14 days and a relatively energetic beta emission (1.71 MeV, maximum). Because of its short half-life and energetic beta particle, ^{32}P is attractive as a tracer as it is easy to measure and does not persist very long in the environment.

Phosphorus is a comparatively scarce element in the biosphere,[23] yet because of its chemistry, it is usually quite mobile in ecosystems and is readily assimilated by plants and animals in its more soluble forms, such as phosphates. Because of its vital role in biological tissues and its comparative scarcity in the environment, organisms can concentrate phosphorus to levels that greatly exceed the concentrations in ambient media. For instance, Polikarpov[13] gives a mean concentration factor of 24,000 for ^{32}P accumulation in freshwater plants and a mean concentration factor of 8000 for freshwater animals. In marine organisms, concentration factors for ^{32}P range from 10 to 1000.[13] Mammals assimilate around 75% of ingested ^{32}P and some 90% of the total body burden locates in bone.[6] Phosphorus in the body is excreted slowly, particularly that which is deposited in bone.

Iodine is of great interest to radioecologists because it too is an essential nutrient element and several of its radioisotopes are produced in copious quantities by nuclear fission. Environmental releases have enabled ready detection of ^{131}I in air, plants, and

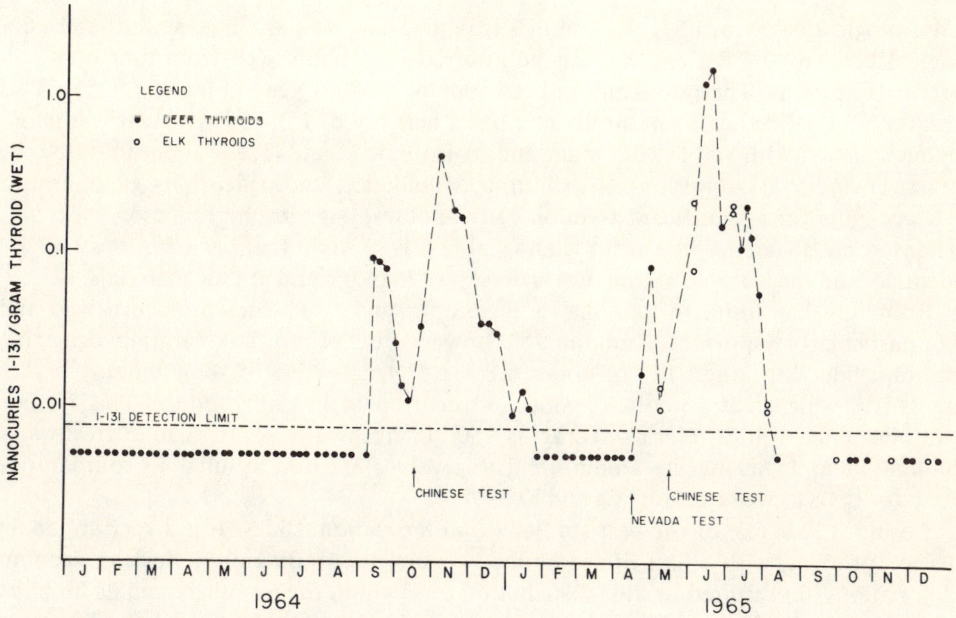

FIGURE 8. Iodine-131 concentrations of Colorado mule deer and elk thyroids in response to nuclear testing activities in 1964 and 1965. (From Whicker, F. W., Farris, G. C., and Dahl, A. H., *J. Wildl. Manage.*, 30(4), 781, 1966. With permission.)

animals on a global basis. Some 20 radioisotopes of iodine exist, but ^{131}I has received the greatest attention. Iodine-131 has a half-life of 8 days and has beta as well as gamma emissions. Iodine-129 has received attention in recent years because it is also produced by fission and has a very long half-life (1.7×10^7 years) which enables it to accumulate in the environment over long periods of time. The biological incorporation of radioisotopes of iodine is largely governed by the abundance and physiology of stable iodine.

Iodine readily enters biological systems. When assimilated by higher animals, it selectively concentrates in the thyroid gland. This gland has such a remarkable ability to concentrate iodine that it is an excellent bioindicator of ^{131}I fluctuations in the environment (Figure 8). Based upon ^{131}I uptake and retention data from mule deer, the thyroid ^{131}I concentration may exceed the forage ^{131}I concentration by three of four orders of magnitude under conditions of chronic intake.[25] Because of the short half-life of ^{131}I, the fact that it localizes in a small quantity of biological tissue, and that ingestion is usually a more important route of intake than inhalation, carnivore thyroids generally concentrate much less of the nuclide than herbivore thyroids. Iodine-131 exhibits concentration factors ranging from 10 to 1300 in freshwater organisms.[13] In marine organisms, iodine concentration factors range from 10 in fish tissues to over 1000 in sessile algae.[26] An excellent review of iodine in the environment was prepared by Hanson.[27]

The major commonality of the four radionuclides discussed under the nonmetal group is that they are all isotopes of essential biological nutrients. As such, their uptake and retention characteristics are largely controlled by the flux of these essential nutrients through biological processes. Radionuclides such as ^{40}K, ^{45}Ca, ^{54}Mn, ^{59}Fe, ^{60}Co, and ^{65}Zn, to be discussed later, are also isotopes of essential nutrients.

B. Light Metals

1. Group IA (K, Rb, Cs)

This group includes elements known as the "alkali metals". The ecological and physiological behavior of K, Rb, and Cs is remarkably similar. In general, the physiological behavior of Rb and Cs can be inferred reasonably well from that of K, an essential nutrient. The potassium radioisotope of greatest general interest is ^{40}K. This primordial nuclide occurs naturally and has a half-life of 1.3×10^9 years. It is homogenously mixed with stable potassium and its isotopic abundance is about 0.0119% by mass. Owing to its ubiquitous distribution, abundance, and penetrating gamma rays, ^{40}K accounts for a significant fraction of the natural background radiation exposure. The isotopic difference between ^{40}K and stable K is so slight that behavior is essentially identical and the isotopic abundance is preserved in biotic and abiotic materials.

Rubidium has some 16 possible radioisotopes, most of which are short-lived and not particularly important. Rubidium-87, however, is a primordial, naturally occurring radionuclide with a half-life of about 4.8×10^{10} years. Also of some interest is 18.6 day ^{86}Rb, which is a potential fission and neutron activation product. This nuclide exhibits concentration factors averaging 1500 in freshwater plants, 250 in freshwater animals, and 15 in marine animals.[13] These values are quite similar to comparative data for ^{137}Cs as well as stable Cs and K.

Cesium-137 is one of the best-studied of all the radionuclides. It is a very abundant fission product, with a half-life of about 30 years. Its production in nuclear weapons detonations has resulted in wide distribution throughout the biosphere and its mobility and physiological properties have led to detectable concentrations in essentially all organisms. The fact that ^{137}Cs emits energetic photons makes it simple to measure and, also, a potentially significant contributor to genetic dose in humans and other species. Because of its abundance, half-life and biological mobility, ^{137}Cs is a critical component of nuclear fission wastes.

As with most other radionuclides, cesium can enter plants by aerial deposition or surface adsorption and root uptake, and animals by inhalation, ingestion, and surface absorption or adsorption. As already stated, Cs is a chemical analog of K and behaves similarly. However, it does not behave identically. In fact, in specific food chains, Cs concentrations tend to increase with trophic level and its ratio to K can increase likewise.[14,28] In numerous studies the observed ratio (Cs/K) consumer ÷ (Cs/K) food varies from one to four and averages about three. The apparent explanation for this phenomenon is that while Cs and K are both assimilated to approximately the same extent, Cs is retained in the body longer. In fact, the biological half-time of Cs exceeds that of K by a factor of two to five in humans,[28] and this trend probably also holds for most other animals as well. We are not certain to what extent the trophic level effect is unique to Cs. This phenomenon has not been generally observed with other radionuclides. On the other hand, the trophic level effect has been observed for the pesticide DDT[29] and continuing research may well uncover other substances which concentrate with trophic level.

An important concept with regard to radionuclides such as ^{137}Cs, which are chemically analogous to major nutrient elements, is that the tendency for the radionuclide to accumulate in the biota is reduced if there is an abundance of the analogous element in the environment. Conversely, a scarcity of the analogous element usually leads to increased accumulation of the radionuclide in the biota. This pattern was illustrated for Sr and Ca (Figure 7) and a similar effect has also been shown for ^{137}Cs concentrations in fish from lakes in Finland which have a range of K concentrations in water.[15] This concept also applies to terrestrial ecosystems. For instance, increases in K intake in animals has been shown to reduce retention of Cs, which could effectively reduce the Cs body burden.[30,31] A comprehensive review of cesium and its relationships to potassium in ecosystems was prepared by Davis.[32]

As pointed out earlier in this chapter, the long-term availability of cesium and many other radionuclides depends heavily upon ecosystem characteristics, and in particular, soil properties. Soils and sediments of high clay content can effectively immobilize Cs by chemical binding. In such systems, the soil acts like a sink for Cs and in time very little of the nuclide is available for biological incorporation. Other systems have sandy soils with a low cation exchange capacity and larger quantities of Cs can be recycled through the biota of such systems for long periods of time.

2. Group IIA (Ca, Sr, Ba, Ra)

This group, termed the "alkaline earth metals", contains elements which have two valence electrons and an oxidation state of plus 2. These elements are chemically reactive and seldom occur in the free state. They commonly form salts such as carbonates, sulfates, and chlorides. Calcium is by far the most abundant of the alkaline earths in the biosphere and is an essential biological nutrient. The ecological and physiological behavior of Sr, Ba, and Ra can be inferred approximately from the behavior of calcium. Our principal interest in calcium, from the standpoint of radioecology, is the fact that its flow through ecological systems markedly affects important radionuclides such as ^{90}Sr, ^{140}Ba, and ^{226}Ra. The most important radioisotope of calcium, ^{45}Ca, is seldom produced by man or nature in sufficient quantities to be considered a biological hazard; rather, it is of greater interest as a tracer to study the behavior of Ca. A specific review of some of the radioecological aspects of calcium is available.[33]

The strontium radioisotopes ^{89}Sr and ^{90}Sr received great notoriety and study during the era of atmospheric nuclear testing in the 1950s and early 1960s. Strontium-90, like ^{137}Cs, is produced in high yields from nuclear fission and is persistent in the biosphere because of its 28-year half-life. Strontium has a tendency to form compounds that are comparatively soluble. Because of this and its chemical similarity to calcium, Sr isotopes are comparatively mobile in ecosystems and deposit in calcium-bearing tissues such as bone and shells, where they can remain for years. The beta particles of ^{90}Sr, and its radioactive daughter ^{90}Y, irradiate portions of the bone marrow, the site of blood cell formation, and thus are considered potential internal hazards.

Owing to its metabolic control by calcium, Sr uptake by organisms has often been expressed as an "observed ratio",[34] which is defined as the Sr/Ca ratio in an organism divided by the Sr/Ca ratio in the diet (or soil in the case of plant uptake). Observed ratios for soil to plants are usually close to unity, suggesting little discrimination between Sr and Ca by plants.[35] On the other hand, observed ratios describing food to animal tissue transfers are usually less than unity, indicating discrimination against Sr in favor of Ca.[35] Many studies have demonstrated that factors such as soil composition, diet, and age can substantially modify observed ratios. Therefore, the observed ratio is not a predictive panacea. Observed ratios describing Sr/Ca transport from water to fish muscle and bone also suggest discrimination against Sr.[36] Nevertheless, concentration factors for Sr in aquatic organisms can be very high.[13] In summary, strontium isotopes in the environment readily enter food chains, but tissue concentrations do not appear to increase with trophic level. High available calcium budgets in ecosystems tend to reduce the bioaccumulation of strontium.

The general behavior of isotopes of barium is not unlike that of strontium. Of several barium radioisotopes, ^{140}Ba is the most important.[37] A high-yield fission product, ^{140}Ba has a half-life of 12.8 days and emits both beta and gamma radiation. Because of the comparatively short life of ^{140}Ba, it will enter food chains primarily by the more direct pathways, such as foliar deposition and inhalation. Also because of its short half-life, ^{140}Ba generally does not build up in the calcium-bearing tissues to the same extent as, say, ^{90}Sr. If deposited in bone, ^{140}Ba will likely be removed mainly by radioactive decay rather than by excretion. In summary, ^{140}Ba is biologically mobile, but important for only a limited time following its entry into the environment.

Isotopes of radium constitute the most important naturally occurring radionuclides of the elements in Group IIA. Two are of overwhelming significance, namely, ^{226}Ra and ^{228}Ra. These radionuclides are generated in the decay chains of ^{238}U and ^{232}Th, respectively. Radium-226 is an alpha, gamma emitter with a 1620-year half-life while ^{228}Ra is a beta, gamma emitter with a 6.7 year half-life. Both are reasonably mobile in the environment, and, due to their chemical similarity to calcium, are considered bone seekers. The release of radium from bone is very slow, thus chronic intake can lead to comparatively high concentrations. Short-lived radioactive progeny of ^{226}Ra and ^{228}Ra can grow toward secular equilibrium with these nuclides during their residence in bony tissues, contributing substantially to radiation dose.[6]

Geological deposits serve as the major reservoirs of radium, but measurable quantities occur in soil, ground water, plants, and animals. The geographic distribution of Ra, like its primordial precursors U and Th, is ubiquitous yet heterogenous, with well-known "hot-spots" in many specific areas of the earth's surface. Although radium is obtained from soil by plants through root uptake, the degree of uptake appears somewhat less than for Sr or Ca.[38,39] The concentration of Ra in most plants is normally one or two orders of magnitude less than the soil upon which they grow,[39] but soil characteristics and species can modify uptake dramatically. An interesting example is the Brazil nut, which concentrates Ra to levels comparable to those found in the soil.[40] Ingested Ra is absorbed from the gastrointestinal tract to varying degrees; however, the International Commission on Radiological Protection (ICRP)[6] cites an assimilation fraction of 0.3. Polikarpov[13] indicates a general lack of information on the accumulation of Ra in aquatic organisms, but cites limited data which indicate concentration factors of 15 to 650 for marine animals and values of 100 to 2750 for plankton. A good review of the concentrations of ^{226}Ra in natural waters and foods is provided by Eisenbud.[41]

C. Noble Gases (A, Kr, Xe, Rn)

Elements of group VIIIA are called the "noble" or "inert gases". Under natural conditions, these elements exist in the gaseous state and because of their electronic configurations, do not enter into chemical combination with other elements. The inert gases therefore exist as monoatomic molecules. Each of the noble gases occurs naturally in the earth's atmosphere, with argon being the most abundant. Stable isotopes of He, Ne, A, Kr, and Xe occur in nature, whereas all isotopes of Rn are unstable and radioactive. Even though the longest-lived radon isotope, 3.8-day ^{222}Rn, is quite unstable, it is readily measurable in the atmosphere because it is continuously generated by the decay of ^{226}Ra in rock and soil.

From the standpoint of radioecological interest, radioisotopes of A, Kr, Xe, and Rn have dominated our attention. The argon radioisotopes of principal interest are ^{37}A and ^{41}A, both of which are produced by neutron activation of stable argon. Krypton-85 is the most important radioisotope of Kr because of its comparatively high fission yield and 10-year half-life. Xenon-133 has received some attention because it is also a common fission product and its 5.3-day half-life is sufficient for the nuclide to travel considerable distances in the atmosphere. As mentioned earlier in this volume, ^{222}Rn and ^{220}Rn are the most significant radon isotopes because they are produced in the decay chains of primordial ^{238}U and ^{232}Th, respectively.

In terms of biospheric transport, the noble gases exhibit similar behavior. They all behave as true gases and disperse through the atmosphere according to the laws of molecular and turbulent diffusion. These gases, unlike some of the diatomic or triatomic gases, are chemically inert and transport, therefore, is not altered by chemical transformations. The inert gases are not transported through food chains and they do not generally concentrate in biological tissues. An exception to the latter statement is

^{222}Rn, which can reside in fatty tissues if generated within an organism by the decay of ^{226}Ra. Also, atmospheric noble gases can permeate nonpolar substances, such as fatty compounds, but only to a limited extent. In almost all circumstances, the principal radiation exposure mode to plants and animals from the noble gases is submersion, or external irradiation from atoms decaying in the immediate atmosphere surrounding the organism. The radiation exposure from the noble gases is, therefore, controlled by atmospheric concentration, the types and energies of the radiations emitted, and the physical half-life of the radionuclide.

Radon-220 and 222 present a more complex radiation exposure problem than the A, Kr, and Xe isotopes discussed, in that $^{220,\,222}Rn$ generate a series of radioactive decay products. These decay products are not gaseous and tend to attach to small dust particles which can settle on or attach to biological surfaces. Of particular concern is the inhalation of radon progeny. Residence of these progeny in the lung leads to irradiation of lung tissues, which at sufficient levels has induced cancer in uranium miners.

D. Heavy Metals (Cr, Mn, Fe, Co, Zn, Zr, Tc, Ru, Pb, Po)

The heavy metals as a group exhibit complex and varied chemistry, therefore, few broad statements can be made for this group as a whole. Some of these elements play essential roles in biological processes, while others have no known biological function. Some of the important radioisotopes of these elements are naturally occurring, while others are strictly activation or fission products which result primarily from nuclear activities. In the following paragraphs, some of the more important radionuclides of the heavy metals are briefly discussed.

The only radioisotope of chromium of radioecological significance is 28-day ^{51}Cr, a neutron activation product of stable ^{50}Cr. In the natural environment, chromium occurs largely in the form of chromite deposits in rocks. It generally forms insoluble oxides and plays no known essential role in the physiology of organisms.[42] Very low concentrations of CrO_4^{--} ions occur in the hydrosphere. Chromium, especially in its more soluble chromate and dichromate forms, is considered a toxic metal. Chromium-51 has been used as a radiotracer in scientific investigations and is produced by reactors and nuclear detonations in quantities dependent on the amount of stable chromium accessible to thermal neutrons. In aquatic systems, ^{51}Cr exhibits concentration factors of 200 to 4000 in freshwater organisms[42] and values of 2 to 80 in marine biota.[13] The uptake of ^{51}Cr from soil by terrestrial plants is generally very low, and assimilation of the nuclide following ingestion by animals is also low, probably less than 0.5%.[6] Because of its short half-life, low biological mobility, and mostly weak radiations, ^{51}Cr is not generally considered an important biological hazard.

The most consequential radioisotope of manganese is 300-day ^{54}Mn, which is produced by nuclear weapons and reactors through the reactions $^{54}Cr(p,n)^{54}Mn$ and $^{54}Fe(n,p)^{54}Mn$.[41] Stable manganese, which occurs in igneous rock at a concentration of about 0.01%,[43] finds its way into the biosphere through geologic weathering and other geochemical processes. The chemistry of Mn is fairly complex, but large quantities in the environment tend to form insoluble MnO_2 in oxidizing conditions. Acidic or reducing conditions can convert oxide forms to Mn^{++} which can undergo exchange reactions with surfaces, especially clays, and which can be taken up by plants. Manganese can replace Ca in carbonates and probably in other compounds. Manganese is an essential element which is required for many biochemical processes in both plants and animals. Because of this, and certain of its chemical forms which are soluble, it enters and passes through food chains. The entry of ^{54}Mn into organisms is affected by availability of stable Mn, as well as other substances, and chemical form. It is assimilated by mammals to the extent of about 10%,[6] and the principal internal target organ is the liver.[6,43,44] Aquatic organisms also readily accumulate ^{54}Mn from the environment.[13]

Neutron activation of iron forms two radioisotopes commonly found in nuclear fallout debris and some reactor effluents, namely ^{55}Fe and ^{59}Fe. Iron-55 (2.6-year half-life) is somewhat longer-lived than ^{59}Fe and usually produced in greater quantities. Because of its mode of decay and weak radiations, ^{55}Fe is comparatively difficult to measure and not particularly hazardous. Its behavior in the environment is largely controlled by the geochemistry of stable iron, which is both abundant and biologically essential. The major biospheric reservoir of iron is the lithosphere, which contains the element at about 2 atom percent.[23] In general, the bioaccumulation of ^{55}Fe and ^{59}Fe is inversely related to the concentrations of biologically available iron in the environment. In the oceans, iron tends to form ferric hydroxide, an insoluble precipitate, resulting in very low concentrations of the element ($\sim$10 ppb) in sea water.[45] As a result, marine organisms accumulate substantial quantities of ^{55}Fe from worldwide fallout and certain marine food chains provide a significant vector of the nuclide to man.[46] Radioisotopes of iron are also accumulated readily by freshwater and terrestrial organisms. Persson,[47] who measured ^{55}Fe in the lichen-reindeer-human food chain, found readily measurable levels of the nuclide in each biotic component; however, the concentrations appeared to decline with trophic level. As one would expect, radioisotopes of iron tend to concentrate in red blood cells and the spleen is one of the critical organs in man.[6]

Several radioisotopes of cobalt are produced through neutron activation by reactors and nuclear detonations. Of these, the most commonly observed are 267-day ^{57}Co, 71-day ^{58}Co, and 5.2-year ^{60}Co. Cobalt-60 is probably of greatest overall interest because of its comparatively long half-life, penetrating gamma rays, and its uses in medicine, industry, and research. Cobalt-60 is produced mainly by neutron activation of stable ^{59}Co and ^{60}Ni. Cobalt is generally considered a trace element in the environment, but it occurs as a microconstituent of plants and animals and plays a role in certain biochemical reactions.[45] Cobalt radionuclides are readily accumulated from the environment by aquatic as well as terrestrial organisms. In marine ecosystems, cobalt concentration factors generally run on the order of 10 to 1000, while values of 100 to 100,000 have been observed in freshwater organisms.[13,48] Data on radiocobalt behavior are more limited for terrestrial than for aquatic ecosystems, but ^{60}Co concentration ratios of the order of 0.01 to 2.0 have been observed for plants/soil[49] and values of 0.1 to 100 were found for rat *(Rattus rattus)* tissues/soil at Eniwetak Atoll.[50] In mammals, cobalt is readily assimilated from the gut, and it is eliminated from the body comparatively rapidly. Major target organs for assimilated Co usually include the liver and kidney.

Most radionuclides of zinc are short-lived and seldom abundant in the environment. The major exception is 245-day ^{65}Zn, which is produced, sometimes in copious quantities, by neutron activation of stable Zn in nuclear reactors and detonations. Zinc-65 has also been used widely as an experimental tracer for stable zinc in biological and ecological studies. Stable zinc, which heavily influences the behavior of ^{65}Zn in the environment, is a comparatively rare constituent of the biosphere, yet it is considered essential for growth and function of organisms.[51] A substantial fraction of the Zn in both aquatic and terrestrial ecosystems ties up with soil and sediment where it is at least temporarily unavailable to all but those organisms which ingest or reside in soils and sediments. However, a measurable fraction of environmental Zn is accumulated by organisms and transported through aquatic and terrestrial food chains.[51] Zinc-65 is taken up through leaves and roots by plants, but uptake is strongly modified by chemical form of the nuclide and soil properties. When ingested by animals, ^{65}Zn

shows intermediate assimilation, but fairly long retention,[6] hence substantial tissue accumulation is possible under chronic intake. Liver and prostate represent critical organs in man. Weapons testing in the Pacific led to substantial ^{65}Zn contamination of marine biota, with concentration factors of up to 10^3 reported for algae and 10^5 for certain mollusks.[51] Polikarpov[13] cites ^{65}Zn concentration factors generally in the range of 10^3 to 10^4 for both marine and freshwater biota, but large variations in the values occur between specific organisms and circumstances. Because of its half-life and biological mobility, ^{65}Zn can be transported through food chains to man.[52]

Zirconium, though rare and lacking apparent biological function, is of interest because of 65-day ^{95}Zr, a high-yield fission product with energetic beta and gamma radiations. Most stable zirconium is associated with oxide compounds of the mineral zircon.[53] The chemistry of Zr in solutions is complicated by its tendency to form colloids and to undergo hydrolysis and polymerization. In general, the chemical forms of Zr in the environment are not very soluble and as a result, mobility of the nuclide in biological systems is low. Although ^{95}Zr has been measured in biological samples in areas which received fission product debris, the material is largely associated with external surfaces. Root uptake of ^{95}Zr from soil is very low and the assimilation fraction of the nuclide following ingestion is generally $<10^{-4}$, according to ICRP.[6] Because of its low assimilation, the critical organ for continuous ingestion of ^{95}Zr is considered to be the gastrointestinal tract. The small quantities of Zr which are assimilated, are generally retained for long periods of time. Despite low solubility and poor ability to enter organisms, reasonably high concentration factors, probably resulting from surface adsorption, have been observed for ^{95}Zr in aquatic biota. Concentration factors of the order of 10^2 to 10^4 have been documented for freshwater plants and animals, while values in the range of 1 to 10^3 in marine biota have been recorded.[13] Because of its high fission yield and penetrating gamma rays, ^{95}Zr and its 35-day daughter, ^{95}Nb, usually contribute importantly to external gamma fields following deposition of fresh fission debris.

Technicium is interesting because the element has no stable isotopes and ^{99}Tc is an abundant, long-lived fission product which appears to be very mobile in ecosystems. This element has received little attention until the last few years and few papers have appeared on the subject. Of substantial interest is the indication that the element is poorly retained by soil and that plant uptake is quite high.[54] Plant/soil concentration ratios as high as 10^3 have been reported for ^{99}Tc supplied to soil in the pertechnetate form.[55] Assimilation of Tc following ingestion appears high ($\sim$50%), but retention of the nuclide by animal tissues appears low, with biological half-times of 1 to 30 days, depending upon tissue type.[6] Since 2×10^5-year ^{99}Tc is a pure beta emitter, its importance as an external hazard is minimal. However, its abundance, long half-life, and ecological mobility may pose special problems, especially in regard to nuclear waste disposal.

Of the 10 or so radioisotopes of ruthenium, ^{103}Ru and ^{106}Ru are of greatest significance in that both are high-yield, intermediate-lived fission products which exhibit some mobility and biological concentration. Stable ruthenium is a rare element in the environment and it does not appear to have significant biological function. Therefore, radioisotopes of Ru are not considered as having essential nutrient analogs, even though they are absorbed to some extent by living tissues. Ruthenium can exhibit several oxidation states and form a wide variety of compounds and complexes, which complicates the matter of its uptake and retention by plants or animals.[56] In general, Ru appears to exhibit substantially greater mobility through soils and groundwater than other fission products; however, its mobility is strongly influenced by chemical form. Instances of substantial subsurface migration of ^{106}Ru have been documented at nuclear waste disposal sites. Ruthenium can be accumulated from the soil by plants, as shown by Auerbach and Olson who measured plant/soil concentration ratios up to 15 for bean leaves.[56] Ruthenium-106 has been found in substantial concentrations in muskrat (*Ondatra zibethica*) kidney tissue at Oak Ridge waste disposal areas and in measurable concentrations from fallout in deer liver.[44,56] Tabulated data indicate that ^{106}Ru has a moderate assimilation fraction in mammals and fairly rapid biological elimination. In terms of radiological assessment of ^{103}Ru and ^{106}Ru, account must be taken of the respective short-lived radioactive daughters, ^{103m}Rh and ^{106}Rh. These rhodium daughters rapidly reach secular equilibrium with ^{103,106}Ru and generally accompany the parent nuclides through ecological and biological systems.

Of the heavy metals discussed in this section, all but lead and polonium result primarily from the nuclear activities of man. The element lead consists of four stable isotopes of mass 204, 206, 207, and 208, and it has some 13 radioisotopes. The lead radioisotopes of mass 210, 211, 212, and 214 are all naturally occurring beta emitters generated by decay of radon isotopes. Of these four, ^{210}Pb has received the greatest attention because of its ubiquitous nature and 20-year half-life. One of the significant features of lead is that it is the heaviest element having stable isotopes and it marks the cutoff for alpha emitters, all of which are heavier than lead. Lead is present in all biospheric media, but concentrations vary widely according to sample type and location. It is not considered biologically essential, but it behaves in physiologically similar fashion to the alkaline earths and is incorporated in biological material.[40,41] The industrial activities of man have resulted in substantial global contamination by this potentially toxic element.[57] Radioactive ^{210}Pb is generated by decay of ^{222}Rn in the atmosphere and lithosphere. A major environmental pathway of this nuclide is aerial deposition, rates of 10^{11} atoms/m^2-year having been documented.[58] Measurable levels therefore occur on foliage,[59] particularly long-lived species such as lichens and mosses.[60] Herbivorous animals continually ingest ^{210}Pb and because it is moderately soluble and is tenaceously retained in the body, measurable body burdens have been documented in reindeer,[60] humans,[58] and other species. ^{210}Pb is also inhaled by animals and it has been documented that people who smoke have considerably higher body burdens than nonsmokers.[61] Assimilated lead primarily lodges in bone, but the kidney and liver are also important target organs. Though comparatively little data on the subject are available, the senior author has documented concentration factors of 10^2 to 10^4 for stable lead in freshwater biota.

Several radioisotopes of polonium are generated by the natural decay series of ^{232}Th, ^{238}U, and ^{235}U, but most are very short-lived and therefore exhibit behavior that is controlled largely by that of the parent nuclides. The important exception is 138-day ^{210}Po, which is generated as a radon daughter in the ^{238}U decay series. Because of its half-life, abundance, physiological behavior, and alpha emission, ^{210}Po is considered "very hazardous" in comparison to most other radionuclides.[62,63] Since there are no stable isotopes of polonium we do not have this usual yardstick for understanding its geochemistry or biology. As with ^{210}Pb, ^{210}Po enters the biosphere largely through aerial deposition on foliage and ingestion or inhalation by terrestrial animals. In long-lived biological tissues, such as lichens, ^{210}Po activity usually approaches secular equilibrium with ^{210}Pb. While absorption of ^{210}Po from the gut is comparable to lead, the distribution is more heavily oriented to the soft tissues, notably kidney, spleen, and liver.[6] Because of this, ^{210}Po has greater opportunity than lead to enter higher trophic levels of food chains. Since ^{210}Po accumulates on foilage, it is present in tobacco and smokers contain up to 3 times the lung burden of this nuclide as nonsmokers.[58] Polonium-210 is also accumulated by aquatic organisms and concentration factors of the order of 10^3 have been reported for marine biota.[13]

E. Rare Earths (Y, La, Ce, Pr, Pm)

The rare earth elements include the lanthanide series (atomic numbers 57 through 71) and because of their similar chemical properties, scandium and yttrium. The principal radioisotopes of interest in this group include ^{90,91}Y, ^{140}La, ^{141,144}Ce, ^{144}Pr, and ^{147}Pm. These radionuclides are produced as principal fission products and as daughters of other fission products. They share numerous important chemical and biological attributes. The stable isotopes of the rare earths occur in trace quantities in nature, and apparently no essential biological role for them has been demonstrated. In most aqueous solutions, the rare earths usually form +3 and +2 ions.[64] In normal aquatic environments, however, the rare earths mainly form hydroxides and occur as insoluble particulate forms. One of the chemical characteristics of the lanthanide series is that electrons can enter the 4f inner shell in addition to the valence shell. A comprehensive review of the chemistry of the rare earths is available.[65]

The rare earth radionuclides in the environment have largely been formed by nuclear weapons detonations. The initial chemical form of such material was principally the oxides, which are comparatively insoluble. Because of the insolubility and lack of biological function, the rare earth nuclides generally exhibit low biotic uptake and minimal food chain transport.[64] These nuclides, however, attach readily to biological surfaces and may be ingested in significant quantities by filter-feeding aquatic organisms, but concentrations usually diminish markedly at higher trophic levels. Rare earths are generally retained strongly by soils and uptake by terrestrial plants is very low, usually lower than uptake of strontium, cesium, and ruthenium. Measurable levels may, nevertheless, occur on terrestrial vegetation following aerial deposition of fallout debris. As a group, the rare earths are expected to be assimilated from the gut in animals to an extent of less than 0.01%.[6] However, the small fraction assimilated seeks bone, liver, and kidney as principal target organs, and excretion is very slow, with biological half-times of the order of 10^2 to 10^4 days. Because of low assimilation, the critical organ for continuous ingestion is usually the gastrointestinal tract. For inhalation of insoluble particles, the lung is sometimes listed as the critical organ.

Yttrium-90 and 91 are of interest because ^{91}Y is a fairly abundant radionuclide at a few months post-fission and ^{90}Y is a daughter of ^{90}Sr. Both have energetic beta emissions. Yttrium-90 contributes little to the total radioactivity until several years post-fission because its production is largely from 28-year ^{90}Sr, with which it grows into secular equilibrium. Because of this, the biological importance of ^{90}Y is largely dependent upon the ecological persistence and mobility of ^{90}Sr. In bone, the critical organ for ^{90}Sr, most of the radiation dose is actually delivered by the higher energy beta particles of ^{90}Y. By itself, ^{90}Y is neither persistent nor mobile in biological systems. Being longer-lived, 58-day ^{91}Y can accumulate on biological surfaces. Concentration factors ranging from 10 to over 10^4 have been measured for this nuclide in aquatic organisms.[13]

Lanthanum-140 is the only radionuclide of La that deserves discussion, and it is of importance primarily because it is generated by the decay of 13-day ^{140}Ba. The 40-hr half-life of ^{140}La precludes significant environmental persistence as a principal fission product. Since ^{140}Ba is also comparatively short-lived, ^{140}La is of relative significance only for a month or two post-fission. Similar to the situation for ^{90}Sr-^{90}Y, ^{140}La behavior and importance is largely dependent on its longer-lived parent; however, most of the radiation dose from ^{140}Ba actually arises from the energetic radiations of ^{140}La.

Cerium has two abundant fission-produced radioisotopes, 33-day ^{141}Ce and 285-day ^{144}Ce. As one would expect, ^{141}Ce is of relative importance from a few weeks to a few months post-fission, while ^{144}Ce peaks in relative importance between 10^2 and 10^3 days post-fission. During the latter period, ^{144}Ce and its 17-minute daughter, ^{144}Pr, contribute more of the total fission product beta activity than any other nuclide.[56] The ^{144}Ce-

^{144}Pr pair emits a mixture of beta particles and gamma rays, but the 3 MeV beta of ^{144}Pr is the dominant contributor to dose from internally deposited ^{144}Ce. Despite the generally low solubility and mobility of ^{144}Ce, the nuclide has been found in fish tissues following U.S. nuclear testing in the Pacific,[64] and in mule deer livers from worldwide fallout.[44] In the latter case, 71% of the estimated dose to deer liver resulting from several fallout radionuclides, was contributed by ^{144}Ce-^{144}Pr. In view of the very low assimilation expected for ^{144}Ce, the only reasonable explanation for significant liver burdens would be chronic ingestion for a long period, coupled with tenaceous retention by the organ. In aquatic ecosystems, ^{144}Ce concentration factors range from 1 to 10^4, with somewhat higher values for plants than animals and generally higher values for freshwater than marine biota.[13]

Promethium has 10 radioisotopes, but no stable isotopes. The isotope ^{147}Pm is a comparatively high-yield fission product which contributes a significant fraction of the total beta activity 2 to 10 years post-fission. Its peak relative abundance occurs 5 years after fission. Since ^{147}Pm is essentially a pure beta emitter, it does not contribute significantly to external exposure from fallout fields and is mainly of interest as an internal emitter. Its detection and measurement normally requires chemical separation, therefore it has not received as much study as some of the other rare earths. Like most of the other rare earths, the fraction of ingested ^{147}Pm assimilated is $<10^{-4}$, and 80% of the body burden is likely to be found in bone, with about 6% in the liver and 2% in the kidney.[6] Mean concentration factors of roughly 4000 and 700 have been reported for freshwater plants and animals, respectively.[13]

F. Actinides (Th, U, Pu)

The actinide elements include number 89 (actinium) and higher numbered elements. Included are the naturally occurring elements Ac, Th, Pa, and U, as well as the transuranium elements, which are largely produced by artificial neutron bombardment of natural uranium. All actinide elements can exist in the (III) oxidation state, but this state is most stable only for Ac, Am, and higher-numbered elements.[66] The (IV) oxidation state is most stable for Th, Np, and Pu, while the (VI) state is most stable for U. Elements 91 through 95 can exist in several oxidation states, complicating their chemistry. The actinides are chemically similar to the rare earths in a number of ways which can make their separation difficult. Like the rare earths, electrons can be added to a subshell of actinides, in this case the 5f subshell.

A very important feature of the actinides is that they are all radioactive, yet half-lives of the actinide nuclides vary widely. It is noteworthy that isotopes of all actinides can arise from the decay or neutron capture of primordial uranium. Certain isotopes of Ac, Th, Pa, and U arise in the decay chains of long-lived, primordial ^{235}U, ^{238}U, and ^{232}Th. Although some of the less massive transuranium nuclides, such as ^{239}Np and ^{239}Pu, are formed naturally in small quantities through neutron capture of ^{238}U, the overwhelming production of transuranics is the result of man's activities. Elements heavier than uranium arise principally through neutron capture and through beta decay, which effectively adds a proton to the nucleus. The high neutron fluxes associated with nuclear reactors and nuclear detonations can lead to successive neutron captures by heavy nuclei, leading to yet heavier, neutron-rich nuclides. Greater stability of such neutron-rich nuclides is frequently achieved through beta decay and the formation of a new element of higher atomic number. Prior to 1940, the transuranium elements were unknown. Early investigations were conducted with microgram quantities of these elements, and nowadays very large quantities are produced by the nuclear industry.

Many of the actinide nuclides emit alpha particles, which increases their potential hazard as internal emitters. In addition to presenting potential radiation hazards, several of the longer-lived actinides such as ^{232}Th, ^{238}U, and ^{237}Np can occur in sufficient mass quantities to be chemically toxic. Beta particle and photon emissions are also common among the actinides and a few nuclides, such as ^{242}Cm and ^{244}Cm, undergo spontaneous fission, which leads to emission of neutrons.

In general, the actinide nuclides form comparatively insoluble compounds in the environment and are therefore not considered biologically mobile. Furthermore, their interaction with biological systems is largely passive, and the heavier actinides, at least, have no known essential biological function. The actinides are transported in ecosystems mainly by physical and sometimes chemical processes. They tend to attach, sometimes strongly, to surfaces; and they tend to accumulate in soils and sediments which ultimately serve as storage reservoirs. Subsequent movement is largely associated with geological processes such as erosion and sometimes leaching. Organisms can effect actinide movement to a minor extent, usually through their own movements which cause rearrangement of abiotic materials.

Organisms are exposed to the actinides largely through surface contact, inhalation, and ingestion. Ordinarily, very little of these substances passes biological membranes to lodge internally, unless inhalation is the route of exposure. Uptake of the actinides by terrestrial plants from soil is generally considered to be low, especially if one considers plutonium, the element for which most data exist. Plant/soil concentration ratios for true uptake of Pu generally run of the order of 10^{-4} or less, especially for the oxide and hydroxide forms which usually occur in the environment.[67] There is scattered evidence, however, that a somewhat higher plant/soil concentration ratios exist for U and Am.[67,68] There is considerable variation in plant uptake of actinides, according to the specific element, soil characteristics, and plant species. Uranium is apparently taken up to a greater extent than thorium and some plant species accumulate uranium to the extent that they may serve as bioindicators of uranium ore deposits.[40] In field situations, actinides associated with vegetation are frequently attached to the surfaces to a much greater extent than they are biologically incorporated.

In the case of terrestrial animals, the actinides are taken in mainly through ingestion and inhalation, the latter pathway generally being regarded as the more important of the two. This is generally true for recent environmental deposits which can be readily resuspended in the airstream. As deposits age and weather into the soil, however, the ingestion pathway may become relatively more significant. For submicron particles, the fraction of inhaled actinides which enters the blood exceeds the fraction of ingested material which enters the blood. Furthermore, submicron, insoluble radioactive particles may reside in the deep lung for considerable periods of time. Larger particles are usually swallowed following inhalation and thus enter the gastrointestinal tract where they encounter the same physiological environment as ingested material. Assimilation of all actinide elements from the gastrointestinal tract is assumed to be less than 0.01%.[6] In addition, the actinides deposit within the body in similar fashion and based on animal experiments, some 80% or more of the total body burden is expected to be found in bone, with 1 to 10% in both the kidney and liver.[6] Retention of all the actinides except uranium is normally long, with whole body biological half-times of 10^4 days or greater assumed for predictive purposes. Uranium is excreted somewhat more rapidly from the body, as indicated by a biological half-time of the order of 10^2 days. Under conditions of continuous ingestion of most actinides, bone or the gastrointestinal tract are usually considered critical organs for dosimetry purposes. For inhalation of small, insoluble particles, the lung is usually the critical organ.

In aquatic systems, the actinides are readily adsorbed on the surfaces of plants and small animals such as zooplankton, and thus they enter the food chain. However, concentration factors generally decline substantially with trophic level and in the internal as compared to external tissues by virtue of membrane discrimination. Concentration factors for Th, U, Pu, Am, and Cm have been reported, but most data are for ^{239}Pu. Noshkin[69] summarized published ^{239}Pu concentration factors for marine environments, and most values for invertebrates and algae were in the range of 10^2 to 10^4. Concentration factors for vertebrate tissues ranged from 1 to 10^3. Concentration factors for other actinides can generally be expected to fall into these ranges.

Because of their high natural abundance in certain localities and ubiquitous presence in the biosphere, Th and U deserve some additional discussion. So does plutonium, because it has been studied to a far greater extent than the other transuranics. In terms of total quantities of transuranic radioactivity generated by the nuclear industry, several radionuclides, including ^{239}Np, ^{241}Pu, ^{241}Am, and ^{242}Cm are considerably more abundant than the well-studied fissile nuclide ^{239}Pu.[70] However, the environmental behavior of these nuclides is not well studied, but some general inferences to them can be made, with caution, from data on ^{239}Pu.

Thorium occurs ubiquitously throughout the lithosphere and is found at an average concentration of about 10 ppm in crustal material.[66] Its geographic distribution, however, shows many ''hot spots'', the more notable of which occur in Brazil and India.[41] Such hot spots are usually characterized by monazite-bearing sands, or certain types of volcanic intrusives. Thorium has a potential role in the nuclear fuel cycle because neutron capture by primordial ^{232}Th produces ^{233}Th, which decays through ^{233}Pa to form fissile ^{233}U. In addition, thorium has certain industrial uses, such as in the manufacture of incandescent gas mantles and welding rods. Of the 13 isotopes of Th, three are very long-lived with half-lives $>>$1000 years. The isotopes of predominant importance are ^{232}Th and ^{228}Th of the natural thorium series, and ^{234}Th and ^{230}Th, which are decay products in the natural uranium series (see Table 2 Chapter 4).

Mobility of Th in biological systems is extremely low. Plant uptake of the element is essentially negligible,[40,68] as is its absorption by animals, and there is apparently no evidence for measurable quantities of ^{232}Th in foods or in marine biota.[71,72] However, there is evidence for small quantities of ^{228}Th in plant and animal tissues.[72,73] This nuclide probably is generated within biological tissues by the decay of ^{228}Ra, which is taken up by plants and animals to a far greater extent than thorium. This may be of some significance because Th is retained tenaceously by bone and the carcinogenic effects of ^{228}Th at higher levels appear to be substantially greater than ^{226}Ra, and perhaps even greater than ^{239}Pu.[74]

Uranium is undoubtedly the most significant element of the nuclear era. It is the raw material from which is generated, either through fission or neutron activation, the majority of radionuclides discussed in this book. Natural uranium is present in the earth's crust at an average concentration of roughly 4 ppm and high grade ores in the Belgian Congo and Canada contain up to 1 to 4% U_3O_8.[66] Medium grade ores, ranging from 0.05 to 1% U_3O_8 occur more frequently in the U.S. and elsewhere. Natural uranium is 99.27% ^{238}U, which is fertile and 0.72% ^{235}U, which is fissile and therefore a primary fuel for nuclear reactors. Because of the long (4.5×10^9 year) half-life of ^{238}U, it is of low specific activity. Because of this, and its generally low biological mobility in ecological systems, it has received less attention from a radioecological point of view than it probably deserves.

Although of generally ''low'' biological mobility, uranium is taken up by plants to a considerably higher degree than thorium. In particular, certain perennial plants concentrate U in excess of 100 ppm, with plant/soil concentration ratios ranging from 10^{-4} to values in excess of 10^{-1}, and considerable effort has been spent investigating the use of certain species in prospecting for uranium.[75-77] The availability of U in soil apparently varies over a wide range according to soil chemistry and this undoubtedly contributes to high variability in plant uptake. There is some evidence that uranium stimulates plant growth at low concentrations, that it becomes toxic at slightly higher levels, and that certain species appear to be associated with soils containing uranium, selenium, and sulfur.[75] Through a comprehensive series of studies in both aquatic and terrestrial ecosystems, Kovalsky and co-workers found that uranium occurs in detectable concentrations in a variety of biota, but concentrations generally diminish with trophic level.[78] Uranium is present in the human diet to the extent that some 1 to 2 μg/day may be ingested per person.[39] In some localities, where spring waters contain elevated levels of U, 10 to 300 μg/day can be ingested by people.[39] Concentration factors ranging from 10 to 400 have been reported for U in marine biota.[13] Natural uranium is chemically toxic because of its low specific activity, and it is frequently difficult to distinguish radiation and chemical effects from the element.

Some 14 isotopes of plutonium have been documented, but most interest has centered around fissile ^{239}Pu, with secondary attention focused upon ^{238}Pu and ^{241}Pu. Because ^{239}Pu is a long-lived, bone-seeking alpha emitter of comparatively high radiotoxicity, and because of its potential importance as a nuclear fuel for reactors as well as a component of nuclear weapons, it has generated much respect, and from the public, much fear. It has been labeled as "one of the most toxic substances known to man". This has led to widespread adoption of the phrase, "the most toxic substance known to man", and frequent hysteria in connection with the politics of nuclear power and national defense. In actuality, while ^{239}Pu is comparatively hazardous under certain circumstances, it is no more toxic than naturally occurring ^{228}Th,[74] and probably far less toxic than arsenic, or certain biological toxins.[79] Despite much general misunderstanding of plutonium, "perhaps no single element has ever been so intensively studied,"[80] and consequently, a great deal is known about its physics, chemistry,[81] and biology.[82] The ecology of plutonium has also been investigated rather intensively, particularly within the last decade,[67] and a number of recent reviews on the subject are available.[69,83-87]

A fundamental reason for the attractiveness of ^{239}Pu as a nuclear fuel is that it can be chemically separated from its precursor, ^{238}U, in rather pure form. In contrast, the other fissile isotopes, ^{233}U and ^{235}U, require much more effort to separate from ^{238}U and to obtain in sufficent purity to serve as a reactor fuel or weapon component. Although plutonium can be chemically isolated from other elements, its chemistry is extremely complex. This complexity is due in large measure to the ability of Pu to exist in four oxidation states (III, IV, V, and VI), frequently simultaneously, in appreciable concentrations in aqueous systems.[81] It also occurs in six allotropic forms as a metal. In general, plutonium in the environment can be expected to undergo hydrolysis and oxidation, with PuO_2 a more common form. However, Pu can form complexes with a variety of organic and inorganic compounds.[88,89] When complexed with chelating agents, such as with DTPA for example, plutonium solubility in soil, uptake by plants, and general biological mobility is significantly increased.[88,90] Under most environmental conditions,however, plutonium occurs in forms which are comparatively insoluble and which are poorly transferred across biological membranes.

As a result of nuclear weapons testing, some 0.3 to 0.5 million curies of plutonium have been introduced into the biosphere.[82] This has resulted in very low levels of Pu, ubiquitously distributed in the environment. Additionally, accidents, inadvertent releases, and experiments have caused higher level Pu contamination in local areas. Examples include U.S. Air Force bomber accidents in Spain and Greenland, atmospheric reentry of a ^{238}Pu-powered satellite, releases at Rocky Flats, Oak Ridge, Hanford, and Mound Laboratory, and experiments at the Nevada Test Site.[91] Cleanup operations were instituted at several sites having the higher levels of Pu contamination. Residual contamination in undisturbed sites has provided opportunity to study the ecological behavior of plutonium in several kinds of environments.

In terrestrial ecosystems, well over 90%, and usually over 99% of the plutonium is found in the soil.[92] The remainder is distributed among the litter and biotic components of the ecosystem. An exception to this trend may be noted immediately following a contamination event, when a large fraction may be associated with vegetation. Distribution of Pu in aquatic systems is normally comparable, with sediments rapidly becoming the dominant reservoir of the material. Movement of Pu from soil and sediments to plants and animals is greatly inhibited by its insolubility and strong discrimination at biological membranes. As a rule of thumb, 10^{-4} is a reasonable discrimination factor for Pu to be applied at each step in the soil-plant-animal mineral chain. Concentrations of Pu in plants are normally of the order of 10^{-6} to 10^{-3} the concentrations in the soil in which they grow, when aerial deposition does not contribute to the plant activity.[90] Resuspension of soil particles and subsequent deposition on foliage surfaces can result in much higher plant/soil concentration ratios, up to 10^{-1} in some cases. Normally, less than 10^{-4} of the plutonium ingested by consumers is absorbed from the intestinal tract.[82] If inhaled, however, a somewhat larger fraction of the Pu can enter the body. For instance, some 5% of a pulmonary deposit can be absorbed into the blood stream and 15% can enter the lymphatic system.[93]

Plutonium deposits in the body are retained tenaceously.[93] For instance, pulmonary and lymph node deposits are cleared with half-times of the order of 500 and 1000 days, respectively. Skeletal deposit retention half-times increase with body weight, and range from roughly 1 year in small mammals to over 100 years in humans. The distribution of plutonium in biological tissues tends to be nonuniform, and "hot spots" are commonly observed. This complicates the dosimetry and has led to considerable controversy regarding the biological consequences of internal deposits.[94]

It is important to be aware of the fact that a tremendous amount of research has been conducted on the biological effects of internally deposited plutonium, and more is known about the biomedical aspects of this element than most other hazardous materials. An excellent review on this is available.[82] From an ecological viewpoint, the isotopes of plutonium do not appear to be as hazardous as some of the more biologically mobile radionuclides such as ^{137}Cs or ^{90}Sr. Based upon substantial research efforts, no specific physical injury to plants, animals, or man has been shown to be caused by plutonium, even though such exposures have occurred.

G. Summary: General Properties of Selected Radionuclides

A general idea of the ecological behavior of specific radionuclides can be obtained through tabulation of various attributes. Such attributes include general chemistry, half-life, distribution in ecosystems and organisms, modes of exposure to organisms, and absorption and retention by organisms. Table 2 represents an attempt at such a tabulation. Because of the wide variations in the quantification of such attributes, qualitative or semiquantitative descriptions are used, with the intent of giving generalized expectations for selected radionuclides. The literature should be consulted for more specific data which are to be applied to specific cases.

Much lumping and oversimplification are necessary for any generalized discussion of radionuclide behavior. For example, ecological transport is usually difficult to describe in terms that apply to all steps in a food chain, which apply to aquatic as well as terrestrial ecosystems, or which cover all chemical forms of the radionuclide in question. With such caveats in mind, we can proceed. In Table 2, important exposure modes or processes are described for animals and plants, both terrestrial and aquatic, with the kinds of organisms being implied by the name of the process. The categories describing degree of food chain transport and successive trophic level concentration are particularly subject to variations between specific trophic levels and between

Table 2
GENERAL ECOLOGICAL PROPERTIES OF SELECTED RADIONUCLIDES

Radionuclide ($T_{1/2}$)	Sources	Nutrient analogs	Principal biospheric reservoirs	Important exposure modes	Degree of food chain transport	Successive trophic level concentration	Critical organs (vertebrates)	Assimilation	Retention(T_{biol})
Nonmetals									
^{3}H (12 year)	Cosmic Fission Activation	H	Hydrosphere (HTO)	Ingestion, uptake, absorption, inhalation	High	Approaches unity	Total body	Complete	Low (days)
^{14}C (5600 year)	Cosmic Activation	C	Atmosphere (CO_2), lithosphere (C, CO_3)	Ingestion, absorption, uptake	High	Approaches unity	Fat, total body	Complete	Low(days)
^{32}P (14 day)	Activation	P	Biota, soil	Ingestion, uptake	High	<1.0	Bone	High	Low-high (days-years)
^{131}I (8 day)	Fission	I	Biota, soil	Ingestion, absorption, inhalation	High	$\sim 10^{3}$(thyroid/ plants)	Thyroid	High	Moderate (weeks-months)
Light metals (Group IA)									
^{40}K(1.3×10^{9} year)	Primordial	K	Lithosphere	Ingestion, absorption, uptake, external γ	High	Approaches unity	Total body	High	Moderate (weeks)
^{86}Rb (19 day)	Fission, activation	K	Biota, soil	Ingestion, absorption, uptake	High	Approaches unity	Total body	High	Moderate (weeks)
^{137}Cs(30 year)	Fission	K	Soil, sediments	Ingestion, absorption, external γ	High	Approaches 3.0	Total body	High	Moderate (weeks-months)

Table 2 (continued)
GENERAL ECOLOGICAL PROPERTIES OF SELECTED RADIONUCLIDES

Radionuclide ($T_{1/2}$)	Sources	Nutrient analogs	Principal biospheric reservoirs	Important exposure modes	Degree of food chain transport	Successive trophic level concentration	Critical organs (vertebrates)	Assimilation	Retention(T_{biol})
Light metals (Group IIA)									
^{45}Ca (160 day)	Activation	Ca	Biota, soil	Ingestion, absorption, uptake	High	<1.0	Bone	High	Moderate-high (weeks-years)
^{90}Sr (28 year)	Fission	Ca	Soil, biota	Ingestion, absorption, uptake	High	<1.0	Bone	Moderate	High (years)
^{140}Ba (13 day)	Fission	Ca	Biota, soil	Ingestion, absorption, uptake	High	<1.0	GI, bone	Moderate	Moderate (weeks)
^{226}Ra(1600 year)	^{238}U decay series	Ca	Lithosphere	Ingestion, absorption, uptake, external γ	Moderate	<1.0	Bone	Moderate	High (years)
Noble gases									
^{37}A (35 day)	Activation	None	Atmosphere	Submersion	Negligible	Negligible	Skin	Negligible	Negligible
^{85}Kr (11 years)	Fission	None	Atmosphere	Submersion	Negligible	Negligible	Skin	Negligible	Negligible
^{133}Xe (5.3day)	Fission	None	Atmosphere	Submersion	Negligible	Negligible	Whole body	Negligible	Negligible
^{222}Rn (3.8 day)	^{238}U decay series	None	Lithosphere, atmosphere	Inhalation of daughters	Negligible	Negligible	Lung (from daughters)	Negligible	Negligible
Heavy metals									
^{51}Cr (28 day)	Activation	None	Soil, sediment	Adsorption, inhalation	Very low	$<10^{-2}$	Lung, GI	Very low	Moderate-high (years)
^{54}Mn (300 day)	Activation	Mn	Soil, sediment	Ingestion, inhalation,	Moderate-high	<1.0	GI, liver, lung	Moderate	Moderate (days-weeks)

				adsorption, external γ					
^{55}Fe (2.6 years)	Activation	Fe	Soil, sediment	Ingestion, inhalation, adsorption	Moderate-high	<1.0	Spleen, lung	Moderate	Moderate-high (years)
^{60}Co (5.2 years)	Activation	Co	Soil sediment	Ingestion, adsorption, inhalation, external γ	Moderate-high	<1.0—10^2	GI, total body, lung	Moderate	Low (days)
^{65}Zn (245 day)	Activation	Zn	Soil, sediment	Ingestion inhalation adsorption external γ	Moderate-high	<1.0	Total body, liver, lung	Moderate	Moderate-high (months-years)
^{95}Zr (65 day)	Fission	None	Soil, sediment	Adsorption inhalation external γ	Very low	<10^{-2}	GI, lung	Very low-negligible	Moderate-high (months-years)
^{99}Tc (2×10^5 years)	Fission	None	Biota, soil	Ingestion inhalation uptake	High	<1.0	GI, lung	High	Low (days)
^{106}Ru (1 years)	Fission	None	Soil, sediment	Ingestion inhalation adsorption external γ	High	<1.0	GI, lung	Low-moderate	Low (days)
^{210}Pb (20 years)	^{238}U decay series	None	Soil, sediment	Ingestion inhalation adsorption	High	<1.0—10	Kidney, lung	Moderate	High (years)
^{210}Po (138 day)	^{238}U decay series	None	Soil, sediment	Ingestion inhalation adsorption	High	<1.0—10	Spleen, kidney, lung	Moderate	Moderate (weeks)
				Rare Earths					
^{91}Y (58 day)	Fission	None	Soil, sediment	Ingestion inhalation adsorption	Low	<0.1	GI, bone, lung	Very low-negligible	High (years)
^{140}La (40 hr)	Fission	None	Soil, sediments, biota	Ingestion inhalation adsorption external γ	Low	<0.1	GI	Very low-negligible	Moderate (1—3 years)

Table 2 (continued)
GENERAL ECOLOGICAL PROPERTIES OF SELECTED RADIONUCLIDES

Radionuclide ($T_{1/2}$)	Sources	Nutrient analogs	Principal biospheric reservoirs	Important exposure modes	Degree of food chain transport	Successive trophic level concentration	Critical organs (vertebrates)	Assimilation	Retention(T_{biol})
Rare Earths									
^{144}Ce (285 day)	Fission	None	Soil, sediment	Ingestion inhalation adsorption external γ	Low-moderate	<0.1	GI, bone, lung, liver	Very low-negligible	Moderate (1—5 years)
^{147}Pm(2.6 year)	Fission	None	Soil, sediment	Ingestion inhalation adsorption	Low	<0.1	GI, bone, lung	Very low-negligible	Moderate (2—5 years)
Actinides									
^{232}Th (1.4×10^{10} year)	Natural	None	Lithosphere	Ingestion inhalation external γ	Very low	$<10^{-2}$	Bone, lung	Very low-negligible	High (years)
^{238}U (4.5×10^{9} year)	Natural	S, Se?	Lithosphere	Ingestion inhalation uptake external γ	Low-moderate	<1.0	GI, kidney, lung	Very low	Moderate (months)
^{239}Pu (2.4×10^{4} year)	Activation	None	Soil, sediment	Ingestion inhalation adsorption	Very low	$<10^{-2}$	Bone, lung	Very low-negligible	High (years)

aquatic and terrestrial systems. The categories describing critical organs, assimilation, and retention are based mostly on laboratory experiments with mammals. Extensions to other animal groups can only be made in a very general way.

A number of general patterns can be gleaned from examination of Table 2. For instance, the radionuclides listed in the "nonmetals" category all have essential nutrient analogs and their assimilation and propensity to be transported through food chains is high. The principal biospheric reservoirs vary for the nonmetals and reflect the chemical and physical properties of the preferred chemical compounds. Physiologic distribution of this group is also varied, and governed by the distributions of the respective nutrient analogs. In general, the retention in mammals and many other animal groups is comparatively low, with biological half-times measured in terms of days to months.

The light metals listed have either potassium or calcium listed as nutrient analogs. Those which behave like potassium show fairly uniform distribution in the soft tissues of animals and therefore the total body is usually considered the critical organ. Those nuclides which simulate the physiologic behavior of calcium tend to wind up in hard tissues such as bone or exoskeleton. The light metals exhibit moderate to high mobility in food chains and in the special case of Cs, concentrations may in some instances increase with trophic level. The bone-seeking group (IIA) light metals are generally absorbed to a smaller extent but retained more tenaceously than the group (1A) light metals. The light metals include several long-lived gamma-emitting radionuclides which can expose organisms to external irradiation.

As a group, the noble gases stand apart from the other groups in terms of ecological behavior. They have no nutrient analogs and are not absorbed by organisms or transported through food chains to any significant degree. They reside in the atmosphere until decay and expose terrestrial biota largely through submersion and external irradiation. Radon-222 can present an inhalation problem by virtue of its radioactive, nongaseous progeny.

The somewhat larger number of radionuclides listed under the heavy metals exhibit quite variable behavior, especially in terms of nutrient analogs, food chain transport, critical organs, assimilation, and retention. Nutrient analogs of the heavy metals are generally considered micronutrients which perform specific biochemical functions. It is interesting to note that while radionuclides having nutrient analogs show moderate to high food chain mobility, as might be expected, some which have no nutrient analogs can also show relatively high mobility. Isotopes of chromium and zirconium show very low food chain mobility because of their general insolubility and very low biological assimilation. A general commonality of the heavy metals is that they all tend to wind up largely in soils or sediments after their introduction into the environment. The critical organs listed for the heavy metals are highly dependent upon mode of exposure.[6]

The rare earths listed are all fission products and none have nutrient analogs. They also exhibit considerable uniformity in characteristically poor assimilation and low food chain transport. In ecosystems, these nuclides are largely associated with soils and sediments, and in the vertebrate body, they tend to seek bony tissue. Retention in bone is generally high, with biological half-times in the order of a year to many years. The short physical half-lives of ^{91}Y and ^{140}La, however, preclude a long effective residence in biological tissues.

The actinides discussed are poorly absorbed and generally exhibit low biological mobility. However, it is important to note that uranium is considerably more mobile in biological system than thorium or plutonium. Thorium and uranium are naturally occurring primordial elements and their principal reservoir is the lithosphere. The principal reservoir of anthropogenic plutonium is soil or sediment. Ingestion and inhalation may constitute important exposure pathways for the actinides, but their relative importance varies with circumstances. The critical organ is dependent upon the principal mode of exposure. External gamma exposures arise from natural uranium and thorium in rock and soil and their radioactive progeny.

If one were to attempt even greater lumping of radionuclides to form more or less distinct groups, four general categories might result. These might include the nonmetals, the light metals, the noble gases, and a fourth group. The fourth group could include the heavy metals, the rare earths, and the actinides, because many similarities among these classes exist. However, the similarities between the rare earths and the actinides are perhaps more consistent than those within the heavy metals. The practice of trying to form groupings within which reasonable similarity exists is mainly useful for general understanding. Experience has shown that radionuclide behavior in ecosystems is usually a highly site-specific and nuclide-specific thing and quantitative predictions must have a greater basis than mere generalities.

IV. BEHAVIOR OF RADIONUCLIDES IN SPECIFIC ENVIRONMENTS

Because the behavior of radionuclides is strongly affected by ecosystem processes, one would expect considerable variation among different ecological systems in the cycling of specific radionuclides. This expectation is largely in accord with actual data which have been obtained from different environments. Ideally, a thorough discussion of the behavior of representative radionuclides in each of the biomes would be desirable. However, neither existing knowledge nor space is available to cover the scope of possibilities. Because of this, we have chosen to discuss a few well-studied environments which are particularly interesting and which illustrate the importance of some basic environmental factors. Some understanding of these factors and their effects in selected environments can be used to draw at least general inferences to other ecosystems.

Most ecosystem-scale studies in natural environments have focused on abundant, readily measurable radionuclides. In particular, the fission product ^{137}Cs, distributed throughout the biosphere in measurable quantities by the testing of nuclear weapons, has received the majority of attention over the years. For this reason, the discussions which follow will center around ^{137}Cs, with secondary attention to ^{90}Sr and a few other radionuclides. Since ^{137}Cs is to be our primary example radionuclide, the discussions are pertinent mostly to biologically mobile materials which seek soft tissues and are readily transported through food chains. Certain principles will also apply to biologically immobile radionuclides, but many will not. The authors will attempt to point out such differences as they arise.

A. Terrestrial Ecosystems

1. Temperate Mountains

A number of extensive radionuclide studies have been conducted in the temperate mountain ranges in North America. In the west, investigations have primarily centered around the Cascades in Washington, the Sierras in California, and the Rockies in Colorado and Utah. These mountain ranges afford the opportunity to study the effects of elevation, topography, physiognomy, and environmental processes unique to montane and alpine ecosystems on the distribution and movement of fallout radionuclides.

The Cascade, Sierra, and Rocky Mountains have received considerable quantities of worldwide fallout from nuclear weapons testing, particularly in higher elevation areas which receive copious precipitation, facilitating accurate radionuclide measurements in various kinds of environmental samples. The facts that these mountain ranges are within the latitudinal belt which receives maximal worldwide fallout, and that these mountains form orographic barriers which cause increased air turbulence and precipitation, explain the occurrence of relatively high fallout deposition. Proximity of the Rockies to the Nevada Test Site has, on occasion, explained high fallout deposition in the Wasatch and Uinta Mountains in Utah.

Evidence that fallout deposition generally increases with elevation in mountain regions is plentiful. For instance, there is substantial indication of increasing concentrations of ^{137}Cs with elevation in soils, vegetation, and mule deer collected in the Front Range of the Rocky Mountains.[8,95] Fallout deposition in the Colorado Front Range appears to increase two- to threefold between 5,000 and 10,000 ft elevation. Precipitation, which also increases with elevation in the Front Range to about the same degree, is primarily responsible for increased fallout. Orographic uplifting of air by mountains is a prime mechanism which frequently explains increased precipitation with elevation.[96] Increased ^{137}Cs in forest litter with elevation in the Washington Cascades has been reported, with a likewise strong correspondence to annual precipitation.[97] In that study, litter on the western slope of the Cascades was found to be higher in ^{137}Cs than litter at corresponding elevations on the eastern slope which receives considerably less precipitation. Similar findings of increasing concentrations of ^{137}Cs in soil with elevation in the Great Smoky Mountains have been reported, again in apparent response to increased rainfall at higher elevations.[98]

In addition to causing increased fallout deposition through orographically induced precipitation, mountains also induce numerous types of air motion perturbations[99] which may also enhance fallout deposition. Such perturbations consist of vertical air flows and flows counter to the prevailing winds which are caused by differential heating of the landscape and by terrain barriers. Air fountains, sinks, downdrafts, cascades, and eddies are a few of the terms used to describe air turbulence which is caused by mountainous terrain. It would seem that the greater the air turbulence, the greater the quantity of suspended particles which can be presented to the air/ground surface interface. Upon contact with soil, rock or vegetation, suspended particles may lodge on the surface through impaction. Complicating the situation is the fact that turbulent air motions may also cause resuspension and redistribution of previously deposited fallout particles.[100]

Another phenomenon of potential importance for fallout deposition in mountainous regions has been described.[101] Large orographic barriers, such as high, steep-sided mountain ranges, may induce large amplitude lee waves which apparently can disrupt the tropopause sufficiently to allow stratospheric air to reach the surface. Evidence for this is the measurement of ozone, presumably of stratospheric origin, at the surface along the Front Range of the Rocky Mountains in Colorado during a lee-waving phenomenon. There is good reason to believe that stratospheric fallout could reach the surface through the same mechanism.

Another important mechanism for fallout deposition at higher altitudes is the contribution of snowfall to annual precipitation. Snowfall is apparently more efficient in scrubbing suspended debris from the atmosphere than a similar quantity of rainfall. Snowflakes present a much greater surface area per unit water volume; they can be exposed to the atmosphere longer than raindrops, since their rate of descent is slower, and their horizontal pathway is longer if the storm is accompanied by winds. Snowfall is the dominant form of precipitation at high altitudes and in the Colorado Rockies, for example, snowfall is frequently recorded even during the summer months. In a watershed in the Colorado Front Range, which has a mean elevation of about 11,200 ft, Osburn reported that about two thirds of the annual precipitation was snowfall.[10] Stratospheric debris enters the troposphere to the greatest extent in the spring, while snowfall is still the dominant form of precipitation in the western mountains. Snow is extremely important in terms of fallout distribution because of its susceptability to wind drifting which causes amplified fallout deposition in depressions, lees of barriers and other snow accumulation areas, and decreased fallout deposition in sites exposed to the full forces of the wind.

Physiography, microterrain, and microclimate are probably more variable and have more influence on fallout distribution in mountainous areas than in any other major type of landscape. Whicker and Loveless measured fallout radioactivity within a 2 ha mountain study area in Colorado and found significant differences in fallout within the area.[102] A timbered portion of the study area had lower air concentrations of dust and radioactivity and lower soil concentrations of radioactivity than open, adjacent shrub communities. Fallout impaction on vertical gummed surfaces was much higher at exposed, wind-blown sites than in a sheltered draw or under a dense Douglas fir *(Pseudotsuga menziesii)* canopy. Hubbard studied ^{137}Cs distribution in an alpine watershed in Colorado and found the highest concentrations in alpine bog soils, microchannels and microdepressions, and, in general, in all areas in which snow and wind- or waterborne debris tend to accumulate.[95] Variation in topography and microclimate are manifest in biotic communities which vary conspicuously in mountain ecosystems. The same factors which are operative in the distribution of biotic communities also effect the distribution of radioactive debris and other airborne contamination.[103]

The fate of radioactive materials following deposition on a mountain ecosystem depends upon the nature of the system, its driving variables, and the properties of the radionuclides in question. Vegetation is the first biotic component of the ecosystem likely to contact significant quantities of fallout. Most higher plants in mountain areas do not seem to have particularly high fallout interception efficiencies, nor do they retain radionuclides for long periods of time. Osburn measured fallout interception efficiencies of 8 to 29% in *Kobresia,* 21 to 77% in *Deschampsia caespitosa,* and yet smaller efficiencies in *Carex scopulorum.*[10] We are not aware of any direct measurements of radionuclide retention in mountain vegetation, however, measurements of fallout ^{137}Cs in vegetation and mule deer since 1963 by the senior author in Colorado indicate relatively short ($\ll$1 year) residence times of ^{137}Cs since concentrations of the radionuclide have generally decreased at the same rate as the annual fallout increment.

Mosses and lichens, in contrast, appear to retain radionuclides for much longer periods of time. Furthermore, in sites having a high percentage ground cover with moss or lichens, the fallout interception efficiency may approach 100%. Measurements of fallout collection efficiency of *Sphagnum* bogs in the Colorado Front Range in 1962 to 1963 indicated that essentially all the ^{137}Cs and ^{90}Sr deposited was accounted for as an increment in the *Sphagnum* strata.[104] This suggests that *Sphagnum* cores obtained from undisturbed sites may give reliable estimates for cumulative deposition of long-lived radionuclides.

Within a few days or weeks after a single fallout deposition event, the largest proportion of the debris is likely to be found in standing vegetation, particularly in ecosystems having a dense vegetative canopy. However, in time, most of the fallout debris reaches the litter and upper soil horizons through leaf-fall, leaching, and other weathering processes. Since most radionuclides, and ^{137}Cs in particular, bind rather tenaciously to litter, humus, and small inorganic particles such as clay, they do not generally tend to move very rapidly into the deeper soil layers. Thus, the major zone of radionuclide accumulation is the litter-humus horizon or the moss-lichen carpet, if existent.

Radionuclides in snowmelt and other forms of surface runoff are sorbed rapidly and efficiently by moss carpets, litter, humus, and small particles of mineral soil. Extremely high concentrations of fallout debris can be found in litter and soil situated at the bases of large snowfields and through which passes large quantities of snowmelt. Nanocurie (10^{-9} Ci) to millicurie (10^{-3} Ci) quantities of gross beta radioactivity per square decimeter were found in certain "hot spots" in 1963.[10] Cesium-137 concentrations up to 1000 pCi/g have been measured in Engleman spruce *(Picea engelmannii)* litter which was found in a depression on bare rock in the Colorado Front Range.[93] Unusually high concentrations of fallout radioactivity in moss which filtered the water of a small stream which drained a barren, alpine watershed in the Washington Cascades have been reported.[105] A sample of this moss contained 400, 170, 510, and 440 pCi/g dry weight of ^{144}Ce, ^{137}Cs, ^{106}Ru, and ^{95}Zr, respectively. The senior author measured fallout ^{137}Cs and ^{239}Pu in soil at the base of a large rock slab in the Park Range in northern Colorado and found levels approximately 100 times the average in the vicinity.

Mountain vegetation in the more xeric sites is generally lower in radionuclide concentrations than vegetation in wet sites. Nevertheless, concentrations of ^{137}Cs and other fallout radionuclides were readily measurable in all types of mountain vegetation in the early 1960s. Eberhardt and colleagues reported ^{137}Cs concentrations in the range from about 2 to over 50 pCi/g (dry) in Douglas fir and various grass species collected in mountainous areas of the Pacific Northwest.[106] These studies indicated considerable seasonal and geographic variations, with much of the variation related to different precipitation regimes. Measured ^{137}Cs concentrations in 12 plant species collected within a single drainage system in the northern Colorado Front Range in 1963 and 1964 generally varied between 1 and 12 pCi/g (dry) with species collected above 8500 ft averaging about twice the values of species collected below that elevation.[8,107] Considerable variation between species growing in the same sampling sites was noted. For example, blueberry *(Vaccinium scoparium)* contained about four to seven times the ^{137}Cs concentration of aspen *(Populus tremuloides)* from sampling sites less than 10 m apart. These differences can be largely ascribed to variations in morphology. The prostrate form and fine branching of blueberry allows a much more intimate contact with the melting winter snow pack and direct transfer of radionuclides. The large variations in radionuclide concentrations among species of vegetation points to the importance of food habits in the accumulation of radionuclides by herbivores.

Deer are frequently one of the most abundant large herbivores in mountainous areas and constitute an important vector of radionuclide transfer to carnivores and a potential pathway to man. For these reasons, fallout accumulation in deer populations has been studied rather extensively.[8,44,108-113] Measurements of ^{137}Cs in mule deer collected weekly in north-central Colorado from February, 1962 through January 1965 gave values ranging from 200 to over 3000 pCi/kg fresh muscle.[8] An interesting seasonal pattern was evident with maximum values occurring during the summer months in deer collected at summer range elevations above 8500 ft and minimum values occurring during the winter in deer collected at lower elevations. The same seasonal pattern was evident in the important forage species being consumed by deer and the increased ^{137}Cs concentrations in summer range forage was attributed to greater precipitation and fallout at higher elevations. Somewhat the opposite seasonal pattern was found in mule deer from the Blue Mountains of Washington, in that wintertime ^{137}Cs values were higher than summertime values.[106] The higher ^{137}Cs concentrations in winter were attributed to increased intake of older plant parts which had been exposed to fallout for a longer period than the new, succulent vegetation consumed during summer. More recent observations on Colorado deer indicate that the seasonal pattern during the late 1960s was reversed from the 1962 to 1964 pattern, with higher ^{137}Cs values occurring in winter than in summer. This suggests that the high fallout rate in 1962 to 1963 caused higher deposition on summer range forage, but that ^{137}Cs retention is likely greater on the more woody winter range forage. Thus, in the late 1960s, when fallout deposition was greatly reduced because of the limited nuclear test ban treaty of 1962, the summer range forage contained less ^{137}Cs than winter range plants because of a differential in retention.

Comparisons indicate that deer flesh in Colorado during 1962 to 1964 contained five to ten times the concentrations of ^{137}Cs as beef or pork.[114] Thus, chronic consumption of deer meat could lead to considerably increased ^{137}Cs burdens in humans. Strontium-90 and ^{131}I would not significantly enter human bodies from consumption of deer meat since these radionuclides locate primarily in bone and thyroid tissues, respectively. However, consumption of deer liver, which is considered by many as a delicacy, could lead to the intake of fallout radionuclides such as ^{144}Ce, ^{137}Cs, ^{54}Mn, and ^{106}Ru.[44]

Lung tissue of deer may be convenient for monitoring insoluble aerosols such as ^{239}Pu and ^{95}Zr,[115] and the deer thyroid is an excellent collector of environmental ^{131}I.[24,110] The deer thyroid can concentrate fallout ^{131}I to several thousand times the concentration in vegetation[25] and the radionuclide may be detectable in thyroid tissues during periods when air, vegetation, or milk samples contain quantities below detection limits. Bone tissues of deer, including antlers, may be useful as a gauge of environmental contamination with ^{90}Sr, particularly if the age effect can be sorted out.[113] In terms of long-term radiation doses to deer from fallout, ^{90}Sr is an important radionuclide to consider. In 1963, for example, yearling mule deer in Colorado received an estimated annual dose of 330 mrad to bone from ^{90}Sr in comparison to about 12 mrad from ^{137}Cs and about 300 mrad from external sources, including natural background.[114] The annual ^{90}Sr dose of 330 mrad received by mule deer in 1963 exceeded the reported ^{90}Sr doses to young people in the U.S. by a factor of about 35, which points to the potential of radiation effects in wildlife populations before similar effects might arise in the human population.

The transfer of ^{137}Cs from mule deer to a principal predator, the mountain lion *(Felis concolor)*, has received some study.[28,116] Data suggest a threefold increase in ^{137}Cs concentration from deer to mountain lion. Increasing ^{137}Cs concentrations with trophic level have also been observed between caribou (*Rangifer arcticus*) and wolves (*Canis lupus*)[9] and between prey and predatory trout.[14] Such trophic level increases depend upon the relationships between cesium and potassium, where assimilation of the two elements from the diet is similar, and where the excretion rate of potassium is more rapid than for cesium. Slower excretion of cesium in comparison to potassium leads to a higher steady-state concentration relative to intake, if other possible variables are discounted. In order for a predator to reach a higher ^{137}Cs concentration than deer, for example, either deer must constitute the major fraction of the total diet, or other dietary items must approach the ^{137}Cs content of deer flesh.

Investigations in the variable ecosystems of mountainous regions have elucidated several principles concerning radionuclide behavior in the environment. For instance, these studies have demonstrated that:

1. Fallout radionuclide distribution becomes highly variable in time and space, particularly when under the influence of topographic, climatic, and biotic interactions.
2. Physical and biological forces can interact to cause extremely high concentrations of fallout contaminants in local ecosystem components, both biotic and abiotic, relative to the general, regional level of contamination.
3. There is a basic need to understand fundamental ecosystem structure and processes before radionuclide behavior can be interpreted or predicted.
4. The naturally occurring biotic resources found in mountain regions may be more susceptible to the potential direct effects of large-scale atmospheric contamination with radionuclides than the human population.

2. Arctic Regions

The arctic and subarctic regions of the northern hemisphere are extremely interesting from a radioecological standpoint because relatively high concentrations of radionuclides, notably the longer-lived ones such as ^{137}Cs, ^{90}Sr, ^{55}Fe, ^{210}Pb, and ^{210}Po, accumulate in some of the major plant and animal species, including man. This occurs in spite of comparatively low fallout deposition in these regions. Increasingly crowded conditions for humans in temperate regions and a growing scarcity of natural resources are illuminating the potential importance of resources in arctic and subarctic regions. These regions, which constitute about one fifth of the land area of the northern hemisphere, support approximately 5 million reindeer and caribou, which in turn provide an important food source for a portion of the some one million people living in these regions.[117] Major radioecological studies on terrestrial ecosystems of these regions have been carried out in Alaska,[118] Sweden,[119] Finland,[117] the U.S.S.R.,[120] and Norway.[121]

One of the earliest clues that food chains in arctic regions concentrate unusually high quantities of radionuclides and provide an important vector to man were the findings of Liden.[122] Professor Liden, in the course of measuring body burdens of radionuclides in Scandinavian people in 1960 to 1961, found that persons who had eaten significant quantities of reindeer meat had much higher than average body burdens of ^{137}Cs. Two reindeer-raising Laplanders from northern Sweden contained 30 and 45 times more ^{137}Cs than the average from a control group in Lund, near the southern tip of Sweden. Liden also reported that a sample of reindeer meat from northern Sweden contained 28 nCi ^{137}Cs/kg while beef from the Lund area showed less than 0.1 nCi/kg. Hanson found high concentrations of ^{137}Cs in Alaskan lichens in 1959 and in caribou by 1960.[123] By 1962, Hanson had measured body burdens of ^{137}Cs in Alaskan Eskimos comparable to those of the reindeer-breeding Lapps from northern Scandinavia. Various ethnic groups of Alaskan natives contained body burdens of ^{137}Cs in direct proportion to the consumption of caribou and reindeer. Studies of ^{137}Cs in the lichen-reindeer-man food chain by Professor Miettinen's group in Finland were also getting underway by 1960 to 1961 and early results were similar to the findings of Liden in Sweden and Hanson in Alaska.[124] During the decade of the 1960s, work by these groups and also by the Russians[120] elucidated the major patterns of radionuclide movement and accumulation in arctic ecosystems and some of the mechanisms involved.

The major factor involved in the high concentrations of radionuclides in terrestrial arctic food chains is the remarkable ability of lichens to absorb and retain radioactive particles. This ability is related to the morphological and physiological characteristics of lichens which have evolved to enable these organisms to live in extreme environments and to obtain a large proportion of their nutrients directly from the atmosphere and from precipitation. Lichens in arctic and subarctic regions are generally prostrate or mat-like and cover a large amount of ground area relative to their biomass. Most lichens achieve a high surface area/weight ratio through branching (fruiticose forms), flattened thalli (foliose forms), or scabrous, flattened thalli (crustose forms). The large surface area, efficient ground coverage, and low profile all combine to enable lichens to intercept a large fraction of the fallout debris contained in air or precipitation. Once fallout debris is incorporated with the lichen thallus, it is lost at a very slow rate in comparison to other forms of vegetation, apparently because the lichen has evolved efficient mechanisms to conserve and retain nutrients that may be in scarce supply in areas where lichens are dominant. The effective retention half-time for ^{137}Cs in lichens, for example, has been estimated to be of the order of 10 years or longer.[119,125] Lichens are very long-lived organisms and Gorham was one of the first investigators to observe higher concentrations of radioactive materials in plants having persistent above-ground parts.[126] Many studies have compared rationuclide concentrations in lichens and seed plants growing in the same areas and, in general, lichens contain roughly two- to tenfold higher concentrations than the seed plants.[127,128]

The importance of lichens in the passage of radionuclides to reindeer and caribou is further promoted by their generally low mineral nutrient content. For example, Alaskan lichens appear to contain only 0.1 to 0.5 the potassium concentrations that higher plants, such as willow *(Salix* spp.), birch *(Betula* spp.), and sedges *(Carex* spp.), contain.[31] A reduced potassium intake of this magnitude, which probably occurs when reindeer and caribou change their feeding habits from higher plants to lichens, would increase ^{137}Cs retention appreciably, which in turn would enhance net ^{137}Cs accumulation in tissues.

Reindeer and caribou, which range over most of the arctic and subarctic regions of the northern hemisphere, are subject to significant seasonal migrations and corresponding changes in feeding habits. During the winter and early spring months these animals rely principally upon lichens for food, whereas, during the summer and early fall months, they prefer seed plants such as willow, birch, blueberry, and various monocots, especially while they are green and succulent. As a result of seasonal change in feeding habits, these animals undergo large fluctuations in ^{137}Cs body burdens through the year.[9] The seasonal differences in ^{137}Cs concentrations in caribou can range from four- to tenfold between spring maxima and summer minima. Because of their food link with lichens, caribou and reindeer contain up to tenfold higher concentrations of ^{90}Sr and ^{137}Cs than other large herbivores, such as moose (*Alces alces*), which occupy many of the same areas but exhibit different food preferences.

Major carnivores which rely upon caribou and reindeer for food include wolves and humans. Radionuclides which accumulate in the soft tissues of these herbivores, such as ^{137}Cs, ^{55}Fe, and ^{210}Po, are readily ingested by these carnivorous species. The high ^{137}Cs concentrations in caribou are reflected in even higher concentrations of the nuclide in wolf flesh. Wolves in apparent equilibrium with their food generally show a twofold increase in ^{137}Cs concentrations over caribou,[9] which strengthens the concept that ratios of ^{137}Cs to K tend to increase at progressively higher trophic levels.[28]

Since the ^{137}Cs intake by Lapps and Eskimos is essentially proportional to consumption of reindeer and caribou, and ^{137}Cs content of the meat, the large ^{137}Cs fluctuations and variations observed in these people are plausible. Seasonal food habits vary between native villages according to availability of various natural foods, and government programs and other external influences which have, in many cases, altered village food habits. The concentration of ^{137}Cs in caribou and reindeer meat is dependent upon the time of slaughter, the location and food habits of the animals prior to slaughter, and the disposition of the meat following slaughter.[129] Although Eskimos and Lapps have received substantially higher radiation doses from both natural and artificial radionuclides as a result of the lichen-reindeer/caribou-man food chain than peoples of temperate latitudes, the total radiation exposures received by these people fall within the range of variation of natural background radiation.[130]

In the preceeding discussion, the authors have briefly covered but one facet of the story of radionuclide behavior in arctic ecosystems. However, the studies mentioned have not only been of extreme interest in themselves, but they have demonstrated several fundamental concepts:

1. Ecosystem processes can operate to channel seemingly insignificant quantities of fallout deposition into measurable concentrations in major biological forms, including man.
2. The behavior of radionuclides as observed in arctic ecosystems reflects many important functional processes of ecosystems. The importance of factors such as

climate, vegetation, and food chain relationships of consumers is clearly demonstrated.

3. Relatively simple ecosystems, which have evolved under rigorous climatic conditions in nutrient-poor conditions, tend to conserve basic elements and fallout radionuclides and to channel these materials into a relatively small array of species and a relatively small biomass.

3. Lower Coastal Plain of the Southeastern U.S.

The worldwide and continental patterns of radioactive fallout deposition have been carefully documented.[131,132] Although threefold or slightly greater differences in fallout deposition have occurred within the contiguous U.S., much greater regional differences have been observed in the concentrations of fallout radionuclides in plants and animals. Furthermore, in some instances, fallout radionuclide burdens in biotic components of ecosystems show little relation to the magnitude of fallout deposition. An excellent example of this occurs in the lower coastal plain in portions of Florida, Georgia, South Carolina, and North Carolina. Although this region receives rather comparable fallout deposition in relation to the average for U.S., certain species of plants and animals, both domestic and native, accumulate ^{137}Cs and sometimes ^{90}Sr to levels which can be one or two orders of magnitude greater than counterparts in adjacent regions.

The lack of a relationship between fallout deposition and high bioaccumulation of ^{137}Cs in the lower coastal plain suggests that ecological characteristics unique to the region must account for the unusually high cesium values. Characteristics of soil and vegetation in the lower coastal plain appear to be key factors in this phenomenon. The region in which notably high bioaccumulation of ^{137}Cs occurs is characterized by soils which are sandy, low in clay and organic matter, low in exchangeable potassium, and low in cation-fixing capacity. Also, high water tables frequently exist in these areas. All of these soil characteristics have been shown by experimentation to enhance uptake of cesium by plants from the soil. Plants adapted to such nutrient-poor soils must efficiently obtain and retain nutrient elements and they do likewise with analogous radionuclides. Many of the important plant species involved in ^{137}Cs transport have long-lived parts which accumulate the radionuclide over significant periods of time.

According to data from the lower coastal plain near Tampa, Fla., five soils had exchangeable potassium values averaging 0.08 meq/100 g while seven agricultural soils from Alabama and Ohio had respective values averaging 0.32 meq/100 g.[133] Perhaps more importantly, the Florida soils displayed Cs^+ fixing capacities averaging only 0.04 meq/100 g while the Alabama and Ohio soils had respective values averaging 0.51 meq/100 g. These comparative soil properties would be expected to lead to greater ^{137}Cs uptake from the Florida soils by plants growing in them and this was indeed observed in experiments. The low Cs^+ fixing capacities of lower coastal plain soils would explain the observation that these soils contained only one sixth as much fallout ^{137}Cs as soils from the Georgia Piedmont.[134] These and other data in the literature support the general hypothesis that fallout ^{137}Cs deposited on the lower coastal plain of the southeastern United States is more readily available for entry into food chains than elsewhere because of the poor cation-fixing properties of soils in the region. Soils in most other regions usually act as a sink for ^{137}Cs, fixing the nuclide so that it is not readily available for uptake by vegetation.

A direct result of low Cs^+ fixing capacities and low available K^+ concentrations of lower coastal plain soils, is that vegetation is relatively high in ^{137}Cs and low in K^+. Data indicate that lower coastal plain vegetation was 2 to 10 times higher in ^{137}Cs and 4.5 times lower in K^+ than piedmont vegetation in Georgia.[134] Lichens, mushrooms, and ferns contain some of the highest ^{137}Cs values in the lower coastal plain, but the importance of these species in the diets of important herbivores is apparently not well quantified. Some of the Georgia lichens contain ^{137}Cs concentrations comparable to those growing in Alaska and elsewhere. The generally high ^{137}Cs, low K^+ content in lower coastal plain vegetation enhances ^{137}Cs accumulation in herbivores and higher members of the food chain, sometimes to a remarkable degree.

One of the most dramatic examples of high ^{137}Cs accumulation in lower coastal plain herbivores is in the whitetail deer *(Odocoileus virginianus)*. Jenkins and Fendley[135] analyzed muscle samples from 165 whitetail deer from various regions of the Southeast and found ^{137}Cs values ranging from less than 250 to over 150,000 pCi/kg. Lowest values were from samples collected in mountain, piedmont, and upper coastal plain regions. Mean values for the lower coastal plain ranged from 5,000 to over 120,000 pCi/kg. The relatively high ^{137}Cs values in lower coastal plain deer are attributed to high intake of the nuclide from vegetation, and a high assimilation factor and long retention time as a result of a low potassium intake. A seasonal change in food habits of lower coastal plain deer was indicated by fluctuating body burdens, with highest values occurring in late winter. Data from other sources indicate that mushrooms contain sufficient ^{137}Cs and are ingested by deer frequently enough to account for observed levels of the nuclide in Florida whitetails.[136]

Relatively high ^{137}Cs concentrations have also been found in many other species from the lower coastal plain. The list includes cottontail rabbits (*Sylvilagus* sp.), fox squirrels (*Sciurus niger*), cotton rats (*Sigmodon hispidus*), oppossums (*Didelphis virginiana*), raccoons (*Procyon lotor*), foxes (*Vulpes* sp.), and bobcats (*Lynx rufus*).[134,135] In addition, domestic food products such as beef and milk, particularly in central Florida, have noticeably elevated ^{137}Cs values.[137,138] As a result of high ^{137}Cs levels in meat, milk, and vegetables, Florida residents concentrate 2 to 3 times more ^{137}Cs than persons living elsewhere in the contiguous U.S.[139]

One of the more interesting features of the behavior of ^{137}Cs in lower coastal plain ecosystems is the apparent persistence of the nuclide in food webs. Roessler and coworkers reported an observed half-time of ^{137}Cs in Floridians of the order of 5 years during the period 1965-69, when fallout deposition was quite low and decreasing with time.[139] During the same period, ^{137}Cs concentrations in U.S. residents were decreasing with a half-time of about 1.5 years. Peak ^{137}Cs values in lower coastal plain deer occurred in 1968 to 1969, some 5 or 6 years following major nuclear testing and peak fallout rates in 1962 to 1963.[134] In other ecosystems in which ^{137}Cs soon finds its way into the soil where it is irreversibly retained, the nuclide concentrations in biota decreased rather markedly over the period 1964 to 1971. In these coastal plain ecosystems, ^{137}Cs appears to recycle to a significant degree, which explains its persistence in food webs. Another phenomenon which would help explain the persistence of cesium would be tenacious retention by long-lived plant parts.

In summary, data from studies conducted in the lower coastal plain of the Southeast suggest that ecosystems having sandy soils low in available nutrients and low in cation-fixing capacity can be expected to result in relatively high and persisting radiocesium contamination of plants and animals. This can be expected for other soluble, biologically mobile radionuclides also, but to varying degrees, depending upon the chemistry of the nuclide and the nature of the ecosystem.

B. Aquatic Ecosystems

A very large number of investigations have been conducted on the behavior of radionuclides in both freshwater and marine ecosystems. As is the case with terrestrial ecosystems, aquatic environments present seemingly unlimited variability in time and space. Such variability can elicit considerable complexity in the transport and accumulation of radionuclides. However, certain elements and features common to all aquatic systems provide a basis for general discussion. The obvious differences in chemistry, geometry, and other factors between freshwater and marine systems justify separate discussions for each.

1. Freshwater Ecosystems

All freshwater ecosystems, with exception of subterranean bodies of water, have received radioactive materials from worldwide fallout and the long-lived, biologically important nuclides, ^{137}Cs and ^{90}Sr, can be readily measured in components of such systems. Studies on the behavior of ^{137}Cs and ^{90}Sr are therefore plentiful and the authors shall give emphasis to these nuclides. It is recognized that many other radionuclides are of importance in certain lakes and streams exposed to local sources of contamination from reactors, nuclear fuel processing facilities, or other facilities and situations. However, many of the principles of radionuclide behavior in aquatic systems can be developed from observations on ^{137}Cs and ^{90}Sr, recognizing that quantitative differences can be expected for other nuclides.

One of the earlier papers giving both field and experimental data on the behavior of ^{137}Cs in freshwater ecosystems was published in 1958 by Pendleton and Hanson.[140] An experiment was conducted in which an artificial pond was acutely contaminated with ^{137}Cs and, subsequently, the components of the system were sampled and assayed for the nuclide. The results were similar to those obtained in many subsequent experiments by other investigators with the following general observations:

1. The nuclide concentration in water rapidly decreased with a simultaneous increase observed in biotic and other abiotic components.
2. The net rates of ^{137}Cs increase in biological materials varied significantly through time and between species.
3. After some time, all components of the system reached a reasonably stable concentration of ^{137}Cs, with all nonaqueous components containing a much higher concentration than that prevailing in water.

The concentration of a radionuclide in a component of an aquatic system relative to the concentration measured in water at the same point in time and space is commonly referred to as the concentration factor. The concentration factor, CF, is defined as

$$CF = \frac{C_s}{C_w} = \frac{q_s/M_s}{q_w/M_w} \tag{8}$$

where C_s and C_w are the concentrations of radionuclide in the substance of concern and water, respectively, q_s is the quantity of radionuclide in M_s grams of substance, and q_w is the quantity of radionuclide in M_w grams (mℓ) of water. The mass of substance, M_s, is normally expressed as fresh weight and q_w is normally measured in water which has been filtered to remove seston, which is composed of suspended debris, phytoplankton, and other microorganisms. Because of the inconsistency with which the CF has been calculated, it should be carefully defined whenever used. Immediately following introduction of a radionuclide into the water phase, the concentration factor will change rapidly with time as the radionuclide is differentially transferred between compartments. In time, the system's compartments will approach steady-state conditions with incomes balancing losses in each of the compartments. In this condition, the radionuclide concentration in a given compartment assumes a reasonably constant relationship to the concentrations in the other compartments and the CF assumes a fairly constant value. It is at this point that the CF can be useful for comparisons and predictions. Polikarpov[13] prepared a treatise on marine radioecology (with considerable reference to freshwaters), which is based largely upon reported concentration factors for a large array of radionuclides, aquatic forms, and environmental conditions.

In freshwaters, concentration factors for ^{137}Cs, and for nearly all other radionuclides as well, can exceed unity by orders of magnitude. Concentration factors for ^{137}Cs, for example, have ranged from less than 100 to over 25,000.[13] A notable feature of concentration factors is their tremendous variability. This variability may be associated with the radionuclide, the organism, the part of the organism assayed, and the chemical, physical, and biological attributes of the ecosystem.

Many examples can be found which illustrate the expected variability in the concentration factor between radionuclides. For instance, a concentration factor of about 0.5 was measured for tritium in the alga *Chlorella pyrenoidosa*,[141] while a value of 154 was found for ^{137}Cs in the same organism.[142] Concentration factors of up to 300,000 have been noted for ^{32}P in sessile algae.[143] While these experiments undoubtedly vary in factors other than the radionuclides studied, the range of variability noted is possible among different isotopes. Polikarpov[13] cites work by Timofeyeva-Resovskaya on the concentration factors for a matrix of 20 radionuclides in 34 freshwater plants. Over all species studied, mean concentration factors ranged from 290 for ^{45}Ca to 18,665 for ^{60}Co. A comparable range of variability between radionuclides occurs in freshwater animals.

An indication of variations among species in concentration factors can also be obtained from Polikarpov.[13] Concentration factors averaged over 20 radionuclides ranged from 22 for cattails *(Typha augustifolia)* to 29,260 for an alga *(Mougeotia* sp.). Williams and Swanson[142] studied ^{137}Cs concentration factors in nine species of freshwater algae under uniform experimental conditions and found values ranging from 52 in *Chlamydomonas* sp. to 1530 in *Rhizoclonium hieroglyphicum*. Differences in concentration factors between trophic groups of aquatic organisms are observed, however, such differences do not appear with sufficient pattern to warrant many generalizations. However, one generalization is frequently made regarding trophic level relationships for radiocesium and potassium: "The observed ratio Cs/K tends to increase with trophic level."[28] This rule, which appears true for aquatic as well as terrestrial ecosystems, seems most reliable in situations where ingestion is the primary mode of cesium transfer and where feeding habits and trophic relationships are well known. Exceptions to the rule can usually be explained on the basis of variations of the Cs/K ratio in the diet. Chemical and physical forms of cesium, which might reduce its solubility, could also result in a variance from the rule of trophic level increase.

One of the more frequently observed regularities is the tendency toward higher concentration factors in aquatic systems having low dissolved nutrient concentrations. Templeton and Brown[36] measured ^{90}Sr and Ca^{++} in brown trout *(Salmo trutta)* and water from various locations in the U.K. and found an inverse correlation between the ^{90}Sr concentration factors and Ca^{++} content of the water. At a Ca^{++} concentration of 1 mg/ℓ, the CF for ^{90}Sr in trout bone was about 4×10^4. At a Ca^{++} concentration of 100 mg/ℓ, the CF was only about 400. Similar results were obtained in Sweden by Agnedal[144] for ^{90}Sr and stable strontium in perch, pike, and roach, and for trout in the mountain lakes of Colorado.[14] Kolehmainen and co-workers[145] reported a decreasing ^{137}Cs concentration in pike *(Esox lucius)* flesh with increasing K^+ concentrations in Finnish lakes and Polikarpov[13] attributes the tenfold lower concentration factors of ^{137}Cs in marine organisms as compared to freshwater species to an approximate 100-fold higher K^+ concentration in seawater.

Cesium-137 accumulation in fish is of considerable interest because they provide a vector of transport from aquatic systems to man and other carnivores. It is of interest to compare concentrations of ^{137}Cs in fish to other human food items and to characterize the types of freshwater systems which yield fish that are relatively high in ^{137}Cs. Nelson and Whicker[146] measured ^{137}Cs in several species of game fish, mainly trout, in Colorado. They found that several species of trout from high mountain lakes contained ^{137}Cs concentrations up to several nCi/kg wet muscle, which was from 20 to 300 times the respective concentrations in domestic meat products. Consumption of a single trout meal in 1966 could have resulted in about 5 times the daily ^{137}Cs intake from the normal human diet. Similar ^{137}Cs values were reported by Gustafson for freshwater fish, notably northern pike from Minnesota,[147] and for fish inhabiting Finnish lakes.[148]

In general, it appears that fish inhabiting natural lakes accumulate higher concentrations of fallout ^{137}Cs than fish inhabiting streams. Limited data gathered for trout from mountain streams and lakes in Colorado suggest two- to tenfold lower values in stream fish. This pattern would seem likely for those situations in which stream water has been subjected to fallout removal processes for longer periods of time than lake waters. However, the reverse could be true if one were to compare an oligotrophic stream with a eutrophic lake. Artificial impoundments at lower elevations generally are comparatively high in mineral nutrients and receive mainly "older" water from feeder streams, hence they usually produce fish with relatively low ^{137}Cs concentrations.

According to the work of Kolehmainen and co-workers,[148] freshwater lakes which are oligotrophic in nature enhance radionuclide accumulation in fish and other biota, while more eutrophic lakes produce fish with lesser radionuclide concentrations. An exception can be found in some of the shallow, semidrainage lakes which, although having relatively high mineral nutrient and humic acid concentrations, and thus classified as dyseutrophic, produce fish high in ^{137}Cs. In one such lake in the Colorado mountains, the cause of high ^{137}Cs burdens in rainbow trout *(Salmo gairdnerii)* was attributed to the ingestion of bottom detritus and sediment by the fish as they pursued benthic organisms near the lake bottom.[11] The process of sediment ingestion by fish may be of considerable significance because sediments represent the accumulation or "sink" compartment for most radionuclides. Long-term buildup of long-lived radionuclides in such sediments may ultimately lead to high concentrations of radioactive material in fish and other organisms.

In mountainous regions, physiography of lake watersheds appears important with regard to accumulation of fallout radionuclides by fish. The senior author's work on high mountain lakes in Colorado suggests that lakes located in rocky, steep-sided glacial cirques can accumulate greater than average snow depths and thus receive relatively large inputs of fallout. Furthermore, a rocky, barren watershed will funnel snowmelt directly into the lake with minimal fallout removal from the meltwater. In contrast, a lake surrounded by well-developed soils and vegetation will receive minimal fallout from runnoff because radionuclides dissolved in snowmelt are efficiently adsorbed by soil, litter, and low-stature vegetation. Alpine lakes in barren, rocky cirques, are frequently very low in dissolved nutrients, which further enhances accumulation of ^{137}Cs and ^{90}Sr by fish and other aquatic organisms.

The general behavior of radionuclides in aquatic systems is very complex because of the large number of physical and biological processes which control radionuclide movement. Some of these processes are competitive, others mutually augmentative. While some processes tend to disperse radionuclides, others tend to concentrate them within certain ecosystem compartments. Such processes are likely to vary quantitatively through the seasons of the year, adding further complexity.

Dispersion of radionuclides from a point source in aquatic systems is accomplished through the processes of turbulent and molecular diffusion, dilution of the water medium with less contaminated water, and biologically-related dispersion. Competing with dispersion processes are those which cause "detention" of radionuclides within a given space. Detention processes involve incorporation of the radionuclide with ecosystem components which are at least temporarily fixed in a given position. Such components include sediment, rocks, periphyton (microscopic plants attached to the surfaces of rocks and other substrates), rooted vegetation, and animals which maintain their position through swimming, burrowing, or clinging. A significant portion of the radioactive material released into the aqueous phase may become incorporated with suspended, free-floating organisms such as phytoplankton and nonliving particulate matter, and in this form, the radionuclide is still subject to physical dispersion. Armstrong and Gloyna have discussed aquatic dispersion and detention processes in terms of a generalized dispersion equation.[149]

Possible modes of accumulation of radionuclides by aquatic organisms include physical surface adsorption, absorption through membranes exposed to the aqueous phase, and ingestion. While aquatic plants extract dissolved minerals directly from the water, aquatic mammals may obtain nutrients by ingestion as well as by direct adsorption from the aqueous phase.[150] Davis and Foster[151] reported that Columbia River fish were 100 times as radioactive as fish maintained in water having the same radioactivity but fed uncontaminated food. On the other hand, Gallegos[152] found that amphipods *(Gammarus lacustris)* maintained a rather constant CF for ^{134}Cs whether or not autotrophs or organic detritus were present as a food source. Thus, the ^{134}Cs concentration in the aqueous phase appeared to control the nuclide concentrations in amphipods, not the ^{134}Cs content of the amphipod's food. Literature suggests that adsorption and absorption, leading to direct equilibration with water, are the predominant mechanisms of accumulation in invertebrate animals and plants while ingestion becomes more important for vertebrates, particularly the larger carnivores.

The long-term behavior of persistent radionuclides is of considerable interest and importance. Observations on the ^{137}Cs content of freshwater fishes suggest that some systems contain irreversible sinks for the nuclide, causing a fairly rapid decrease in biotic radioactivity following a contaminating event while other systems, effectively lacking such sinks, recycle the nuclide. Significant recycling results in persistent levels of ^{137}Cs in the biota. Temporal patterns of ^{137}Cs in trout from mountain lakes in Colorado indicate that, in some lakes, ^{137}Cs values in fish correspond closely with the annual rates of fallout, suggesting a sink (most probably bottom sediments) with little recycling occurring.[14] The systems generally exhibiting this pattern included alpine and montane lakes and reservoirs with fine-grained clay sediments which contain little organic detritus. On the other hand, ^{137}Cs concentrations in trout from shallow, semidrainage dyseutrophic lakes appear to more closely parallel cumulative cesium deposition as measured in the sediments through time. In these lakes, mechanisms exist to effect transfer of ^{137}Cs from the sediments back into living biota. As mentioned previously, a likely mechanism for such a transfer was indicated to be the inadvertent ingestion of bottom detritus by fish while in pursuit of benthic animals.[11]

An intellectually intriguing approach to the behavior of radionuclides in aquatic ecosystems is systems analysis. In this approach an aquatic ecosystem is conceived as an array of compartments with intercompartmental couplings which allow the transfer of materials between compartments. Attempts are made through the techniques of systems analysis to quantify compartments and couplings in terms of mass content and flow rates, respectively. Whittaker[153] utilized this approach in a study of ^{32}P behavior in aquarium microcosms. Through a series of experiments in which ^{32}P was introduced into systems of known composition, and measured through time in the major compartments, Whittaker was able to quantitatively describe the kinetic behavior of radiophosphorus in the total system by determining intercompartmental rate constants. A similar effort for determination of the kinetics of cesium in a natural lake was undertaken by Hakonson.[154] If rate constants are known for a given system, then the concentrations of radioactivity can be calculated as a function of time after a hypothetical introduction, provided that the relative masses of compartments are known. Since specific rate constants and compartment masses vary with time and other factors, and indeed are difficult to measure in the first place, this approach must be used with caution for predictive purposes. Perhaps the greatest value of a system model is that it describes the whole system rather than just some of its parts and thus allows greater insights into the relative importance of specific compartments and couplings.

Since most aquatic ecosystems have reasonably well-defined boundaries, and a central compartment (water) which is in direct contact with essentially all other compartments, these systems are suitable for quantitative description of processes which control radionuclide behavior. Radionuclide studies in aquatic systems are important, not only because of man's needs for water, food, recreation, and other amenities derived from such systems, they are also important for the elucidation of radioecological principles common to other kinds of ecosystems. While we now know quite a good deal about the behavior of cesium and strontium in freshwater systems, we know little about the aquatic behavior of many other radionuclides. There appears, without question, to be much room for further contributions in freshwater radioecology.

2. Marine Ecosystems

A great deal of research has been carried out and published on the behavior of radionuclides in marine environments. In this section, the authors provide a general discussion of marine ecosystems and the ways that some of the important radionuclides behave. In addition, some of the more important differences between marine and freshwater systems in terms of occurrence and behavior of radionuclides are pointed out. One cannot overstress the importance of the oceans as they affect and help maintain the biosphere as a functioning whole, as they relate to human welfare, and as they constitute the ultimate, giant receptable of wastes generated on land as well as in the sea. Clearly, through time man is placing greater demands, as well as greater insults, upon marine and other ecosystems. Our ability to understand and deal with this problem can be measured in part by our knowledge of radionuclide transport phenomena.

As with terrestrial systems, the oceans in general have been contaminated with radioactive materials, primarily from the testing of nuclear weapons in the 1950s and early 1960s.[48] In addition, localized contamination of marine systems has occurred from reactor effluents, nuclear fuel processing facilities, disposal of radioactive wastes, accidents, and miscellaneous sources.[155]

Most of the nuclear weapons tests not conducted over continental land areas have been carried out in the Pacific. The U.S. detonated some 98 nuclear devices at Bikini and Eniwetak Atolls, Christmas, and Johnston Islands in the central Pacific between 1946 and 1962, while the U.K. was responsible for several other nuclear explosions in the Pacific between 1952 and 1958.[156] Many of the nuclear explosions in the Pacific created localized fallout in the near vicinity and some, conducted near or beneath the surface, caused direct contamination of the ocean with radionuclides. Nearly all of the larger nuclear devices detonated in the atmosphere, whether over the land or the sea, have produced worldwide fallout, the major fraction of which has settled somewhere into the sea. Several investigators have produced data which indicate that at comparable latitudes, the oceans have received greater fallout deposition than land areas,[157] and the oceans cover about 71% of the earth's surface.[158] Of course, in addition, some fallout deposited on land eventually reaches the ocean through atmospheric resuspension and land surface drainage systems.

One well-studied situation, involving oceanic contamination from a local source of radionuclides, is the mouth of the Columbia River which enters the Pacific Ocean at the Washington-Oregon border. It is reported that, at one time, about 1000 Ci/day entered the ocean from this river.[48] The origin of the radioactivity was the Hanford Atomic Products Operation, some 370 miles upstream from the mouth of the Columbia. The radioactivity consisted of about 60 different radionuclides, but only a few of these were predominant near the river mouth.[159]

Of the naturally occurring radionuclides present in seawater, ^{40}K is overwhelmingly predominant over the nuclides of uranium, thorium, radium, and rubidium.[48] The man-made radionuclides which have found their way into the oceans consist primarily of fission products and nuclides formed by neutron activation. A few of the more important fission products found in seawater and marine biota include ^{90}Sr, ^{90}Y, ^{137}Cs, ^{144}Ce, ^{95}Zr-Nb, $^{103,106}Ru$-Rh, and soon after fission, ^{131}I and ^{140}Ba. Some of the predominant activation products of importance in marine systems include $^{55,59}Fe$, ^{65}Zn, $^{57,60}Co$, ^{54}Mn, and ^{51}Cr. Carbon-14 and tritium are also present in low concentrations in seawater. Also present in the sea, but in very low concentrations, is ^{239}Pu and other transuranium radionuclides.

Physicochemical form of a radionuclide has a significant bearing upon its behavior in any natural environment. The large diversity of chemicals and substances present in seawater renders difficult the prediction of physical and chemical forms of most radionuclides. As a crude generalization, however, the literature suggests a tendency for Cs, Sr, and Zn to exist in ionic form, and a tendency for Ru, Ce, Zr, Y, Nb, and Fe to exist as suspended particulates and as colloids.[158,160] Radionuclides of these elements would presumably tend toward the same forms. There is evidence that the radionuclides $^{57,60}Co$, ^{54}Mn, and ^{59}Fe primarily exist as insoluble forms in seawater.[161] Depending upon the solubility product, nuclides in ion form may precipitate or co-precipitate with other ions. For example, soluble ^{90}Sr may co-precipitate with calcium and become irreversibly bound in $CaCO_3$ deposits.

A general understanding of the behavior of radionuclides in the ocean requires some review of the physics, chemistry, and biology of the sea. While the ocean is a relatively homogenous body in comparison to the innumerable independent bodies of freshwater, it consists of many distinctly different ecosystems. For instance, there are open oceans, abyssal depths, continental shelves, coral reefs, tidal pools, estuaries, and so on, each of which has many unique physical, chemical, and biological characteristics. The ocean as a whole is immense, covering some 60% of the northern and 80% of the southern hemispheres. Ocean depths average about 4000 m and contain a total water volume of about 1.4×10^9 km^3.[48] Unseen by the average voyager are gigantic, submerged mountain ranges, plains, and trenches up to 10,000 m in depth. Such vast dimensions must be considered in terms of potential dilution and dispersion of radionuclides. While the oceans and seas are ultimately connected to form the entire ocean, physical barriers more or less isolate individual bodies of seawater which have consequently developed unique ecological characteristics. The gently sloping continental shelves occupy only about 6% of the oceanic surface yet they are probably of greater importance in terms of seafood production than the open seas. In general, waters of the continental shelf are less salty but support higher biological productivity than the open seas. The inherent importance of the continental shelf waters, combined with their proximity to sources of man-generated pollution, underscores the importance of basic knowledge on the ecology of coastal waters.

Ocean waters are vertically stratified in response to a density gradient which is formed by variations with depth in temperature, pressure, and salt content. At temperate and tropical latitudes, abyssal waters are colder and saltier than surface waters. The surface water layer mixes fairly rapidly to the depth of the thermocline, pycnocline (referring to density), or halocline (referring to saltiness). This upper, mixed layer of seawater varies considerably, but averages about 75 m in thickness and can undergo rapid horizontal migrations. The surface currents of the ocean are driven primarily by winds. However, large-scale currents, upwelling, convergent sinking, land forms, spatial variations in temperature and salinity, and other factors may influence the direction and speed of these currents.

The deeper ocean layers below the pycnocline mix and move about, but much more slowly than the surface layers. Mixing of seawater and its contents between the upper and lower layers is retarded considerably by the pycnocline, yet some exchange does take place. The deeper layers are frequently richer in mineral nutrients than the surface layers, which tend to become depleted in nutrients from freshwater inputs and from the death and subsequent setting out of biotic substances. In many cases, the surface layers must rely upon some interchange with deeper water for replenishment of minerals needed for biological activities. One way this is frequently accomplished is through the upwelling of deeper waters along coastlines when the surface waters are driven offshore by winds.

The very processes which affect radionuclide behavior can be elucidated more clearly by measuring the dispersal of radionuclides which are placed within the system. For example, Revelle and Schaefer[158] cite studies in which the vertical mixing rate in the upper, "mixed" layers was measured by sampling water contaminated with debris from weapons tests. In one study it was found that the lower boundary of the radioactive water moved downward at the rate of about 0.1 cm/sec until it reached the thermocline, where it abruptly stopped. Other radioactive tracer studies are cited by these authors which have shown that horizontal dispersal is much more rapid than vertical. Seymour[162] estimated North Equatorial Current speeds of about 8 nmi/day between Eniwetak and Guam, based upon the arrival of fallout from the Hardtack weapons tests.

Another set of marine attributes which plays a major role in the behavior of radionuclides is the chemical composition of seawater. While considerable spatial variation can be expected in chemical composition, numerous attempts to summarize existing data have been made for purposes of general comparisons and predictions. An excellent summary of this kind was prepared by Krumholz et al.[160] Apart from hydrogen and oxygen, sodium and chlorine are the predominant elements in seawater, averaging some 10.5 and 19 g/ℓ, respectively. Potassium and calcium run about 380 and 400 ppm, respectively, in seawater, while in comparison these elements most frequently occur in pristine freshwaters at concentrations less than 10 ppm. Nelson, a colleague and limnologist with the Colorado Division of Wildlife, has measured potassium concentrations as low as 0.03 ppm in alpine lakes in Colorado and calcium values as low as a few tenths parts per million. This is one important reason why freshwater organisms, including fish, are usually much higher in ^{137}Cs and ^{90}Sr than their marine counterparts (Table 1). Another reason involves the greater volume for dilution of the radionuclides in the sea as compared to the generally shallower freshwaters.

Radionuclides, such as ^{137}Cs and ^{90}Sr, tend to be "carried" in the dissolved phase in seawater as a result of the high concentrations of chemically analogous carriers. Several kinds of studies have shown that these nuclides, existing primarily in the dissolved phase, disperse primarily according to the ocean currents, with total movement into biota and sediments being minimal and relatively slow.[163,164] In oceanic waters of average depths, sheer volume precludes rapid exchange of nuclides between the sediments and the major water fraction. Also, in the open, less productive areas of the ocean such as in the tropics, the standing crop of biomass may be so small that in spite of reasonably large concentration factors, the major fraction of radioactivity will likely reside in the water phase.[165,166] A large fraction of the radioactivity in the water phase will, in time, be found in the deeper layers where biological productivity is very low. Bowen and Sugihara[157] studied depth profiles of certain fission products in the Atlantic and showed, in 1958, that large fractions of these fission products occurred at depths greater than 100 m, and sometimes at depths exceeding 1000 m. Vertical profiles of particulate-associated nuclides such as ^{144}Ce and ^{147}Pr differed considerably from the profiles of ^{90}Sr, illustrating effects of physicochemical form.

Although seldom of overwhelming importance, biotic components nevertheless have some influence on the movement of radionuclides in the marine environment. The degree of biotic influence, as would be expected, is dependent upon the radionuclide and other considerations. Plankton and higher order consumers accumulate measurable quantities of radionuclides from seawater and the subsequent fate of these radionuclides, if not excreted, is determined by the fate of the organisms. Populations of small organisms, such as phytoplankton which undergo rapid turnover, carry considerable quantities of radionuclides to deeper layers and to the sediments as the dead organisms settle out. Fecal materials also carry radionuclides toward the bottom, as do dead cells which can still accumulate radionuclides from solution in spite of the fact that they are no longer alive.[167] Salmon, potentially, provide a good example of a pelagic species which can accumulate radionuclides from a large volume of seawater through the food chain and then release the material in small freshwater streams following the spawning ritual.

Concerning general accumulation patterns of radionuclides in marine biota, it appears that activation products and fission products which are found preferentially in the particulate or colloidal state are of generally greater concern than nuclides such as strontium and cesium. In freshwater and terrestrial systems, the reverse seems to hold true, with strontium and cesium being more predominantly accumulated. In spite of their insoluble form, particulate-form radionuclides such as Ru, Ce, and Zr are readily accumulated by marine filter-feeders including zooplankton and mollusks.[48] Radionuclides of zinc, iron, cobalt, and manganese also accumulate readily in marine plankton, but of these, only zinc and iron seem to be readily transferred to and accumulated by higher order consumers and predators.[168,169] Radionuclides of Zn, Fe, Al, C, and P can be expected to exhibit comparatively high concentration factors in marine food chains because the concentrations of these elements in seawater are generally very low. Coral is an unusual but not surprising example of a marine organism in which ^{90}Sr may be preferentially accumulated over other radionuclides.[5,169] Studies at Rongelap and Eniwetak Atolls in the Pacific further substantiate that marine systems handle the same initial spectrum of radionuclides much differently than terrestrial systems. Although exposed to the same fallout radionuclides, marine biota at Rongelap and Eniwetak selectively accumulated activation products and insoluble fission products while terrestrial forms preferentially accumulated ^{90}Sr and ^{137}Cs.[161,169]

Reasons for different radionuclide behavior between marine and other ecosystems should be reiterated. Firstly, as mentioned earlier, cesium and stronium ions are vastly more diluted with potassium and calcium ions in seawater than in freshwaters or terrestrial soils. While fallout on land is mixed vertically in only a few inches of soil and in freshwaters only a few tens of feet, that in the sea is mixed (rather rapidly) throughout the upper 75 m or so. Such differences in vertical dispersion, coupled with different nutrient ion concentrations, are manifest in much higher ^{90}Sr/Ca and ^{137}Cs/K ratios on land and in freshwaters than in the sea. Since potassium and calcium ions are very abundant in seawater, biological concentration factors are comparatively low for these elements and their analogs. In contrast, the element phosphorus for example, which is frequently found only in limited (and limiting) concentrations in seawater, may concentrate 2 million-fold over seawater in marine vertebrates.[160] Radionuclides of elements which are biologically essential, but in scarce supply, will accumulate impressively in organisms. In addition, it is possible for an element (such as nitrogen) to enhance the bioaccumulation of a radionuclide such as ^{32}P.[158] Food chains that include filter-feeding organisms, such as zooplankton and mollusks in marine systems, are exposed effectively to radionuclides which tend to occur in insoluble, particulate forms. In terrestrial ecosystems, such radionuclides tend to remain in the soil.

If interested in the more specific aspects of radionuclide behavior in the marine environment, the reader is encouraged to consult the treatise prepared by the National Academy of Sciences Panel on Radioactivity in the Marine Environment[155] and Polikarpov.[13] These works cover most of the important aspects of marine radioecology in depth.

REFERENCES

1. **Slade, D. H., Ed.,** Meterology and Atomic Energy 1968, U.S. AEC Rep. TID-24190, U.S. Atomic Energy Commission, Washington, D. C., 1968.
2. **Turner, D. B.,** Workbook on Atmospheric Dispersion Estimates (Revised 1970), Rep. AT-26, Office of Air Programs, U.S. Environmental Protection Agency, Research Triangle Park, N.C., 1971.
3. **Engelmann, R. J. and Sehmel, G. A.,** Coords., Atmosphere-Surface Exchange of Particulate and Gaseous Pollutants (1974), (ERDA Symp. Ser. 38), U.S. ERDA Rep. CONF-740921, U.S. Energy Research and Development Administration, Washington, D.C., 1976.
4. **Englemann, R. J. and Slinn, W. G.,** Coords., Precipitation Scavenging (1970) (AEC Symp. Ser. 22), U.S. AEC Rep. CONF-700601, U.S. Atomic Energy Commission, Washington, D.C., 1970.
5. **Robbins, W. W., Weier, T. E., and Stocking, C. R.,** *Botany — An Introduction to Plant Science,* 2nd ed., John Wiley & Sons, New York, 1957.
6. International Commission on Radiological Protection, Report of Committee II on permissible dose for internal radiation (1959), *Health Phys.,* 3, 1, 1960.
7. **Hanson, W. C. and Case, A. C.,** A method for measuring waterfowl dispersion utilizing phosphorus-32 and zinc-65, in *Radioecology,* Schultz, V. and Klement, A. W., Eds., Reinhold, New York, 1963, 451.
8. **Whicker, F. W., Farris, G. C., Remmenga, E. E., and Dahl, A. H.,** Factors influencing the accumulation of ^{137}Cs in Colorado mule deer, *Health Phys.,* 11(12), 1407, 1965.
9. **Hanson, W. C.,** Radioecological concentration processes characterizing arctic ecosystems, in *Radioecological Concentration Processes,* Åberg, B. and Hungate, F. P., Eds., Pergamon Press, New York, 1967, 183.
10. **Osburn, W. S.,** Ecological concentration of nuclear fallout in a Colorado mountain watershed, in *Radioecological Concentration Processes,* Åberg, B. and Hungate, F. P., Eds., Pergamon Press, New York, 1967, 675.
11. **Gallegos, A. F., Whicker, F. W., and Hakonson, T. E.,** Accumulation of radiocesium in rainbow trout via a non-food chain pathway, Proc. 5th Ann. Health Physics Society Midyear Topical Symp., *Health Physics Aspects of Nuclear Facility Siting,* Voilleque, P. G. and Baldwin, B. R., Comps., II, 477, 1971.
12. **Arthur, W. J., III,** Plutonium Intake by Mule Deer at Rocky Flats, Colorado, M.Sc. thesis, Colorado State University, Fort Collins, 1977.
13. **Polikarpov, G. G.,** *Radioecology of Aquatic Organisms* (English transl.), Schultz, V. and Klement, A. W., Jr., Eds., North-Holland, Amsterdam, 1966; *Radioecology of Marine Organisms,* Atomizdat, Moscow, 1964.
14. **Whicker, F. W., Nelson, W. C., and Gallegos, A. F.,** Fallout ^{137}Cs and ^{90}Sr in trout from mountain lakes in Colorado, *Health Phys.,* 23(4), 519, 1972.
15. **Kolehmainen, S., Hasanen, E., and Miettinen, J. K.,** ^{137}Cs in fish, plankton and plants in Finnish lakes during 1964-5, in *Radioecological Concentration Processes,* Åberg, B. and Hungate, F. P., Eds., Pergamon Press, New York, 1967, 913.
16. **Jacobs, D. G.,** Sources of Tritium and Its Behavior Upon Release to the Environment, U.S. AEC Rep. TID-24635, U.S. Atomic Energy Commission, Washington, D.C., 1968.
17. **Libby, W. F.,** Radioactive fallout, *Proc. Natl. Acad. Sci. U.S.A.,* 44(8), 800, 1958.
18. **Perkins, R . W. and Nielsen, J. M.,** Cosmic-ray produced radionuclides in the environment, *Health Phys.,* 11(12), 1297, 1965.

19. **Anspaugh, L. R., Koranda, J. J., and Robison, W. L.,** Environmental aspects of natural gas stimulation, in Radionuclides in Ecosystems, U.S. AEC Rep. CONF-710501-P1, Nelson, D. J., Ed., U.S. Atomic Energy Commission, Washington, D.C., 1973, 37.
20. **Bard, S. T.,** Environmental deposition of tritiated water vapor, Ph.D. dissertation, Colorado State University, Fort Collins, 1977; *Diss. Abstr.*, 38(7), 3073-B, 1977.
21. **Koranda, J. J. and Martin, J. R.,** Persistence of radionuclides at sites of nuclear detonations in Biological Implications of the Nuclear Age, (AEC Symp. Ser. 16), U.S. AEC Rep. CONF-690303, Goulden, A. M., Ed., U.S. Atomic Energy Commission, Washington, D.C., 1969, 159.
22. **Wolf, G.,** *Isotopes in Biology*, Academic Press, New York, 1964.
23. **Deevey, E. S., Jr.,** Mineral cycles, *Sci. Am.*, 223(3), 148, 1970.
24. **Whicker, F. W., Farris, G. C., and Dahl, A. H.,** Radioiodine in Colorado deer and elk thyroids during 1964-65, *J. Wildl. Manage.*, 30(4), 781, 1966.
25. **Gist, C. S. and Whicker, F. W.,** Radioiodine uptake and retention by the mule deer thyroid, *J. Wildl. Manage.*, 35(3), 461, 1971.
26. **Bowen, V. T., Olsen, J. S., Osterberg, C. L., and Ravera, J.,** Ecological interactions of marine radioactivity, in *Radioactivity in the Marine Environment*, Panel on Radioactivity in the Marine Environment, National Academy of Sciences, Washington, D.C., 1971, 200.
27. **Hanson, W. C.,** Iodine in the environment, in *Radioecology*, Schultz, V. and Klement, A. W., Jr., Eds., Reinhold, New York, 1963, 581.
28. **Pendleton, R. C., Mays, C. W., Lloyd, R. D., and Church, B. W.,** A trophic level effect on ^{137}Cs concentration, *Health Phys.*, 11(12), 1503, 1965.
29. **Woodwell, G. M.,** Toxic substances and ecological cycles, *Sci. Am.*, 216(3), 24, 1967.
30. **Wasserman, R. H. and Comar, C. L.,** The influence of dietary potassium on the retention of chronically ingested cesium-137 in the rat, *Radiat. Res.*, 15(1), 70, 1961.
31. **Holleman, D. F., Luick, J. R., and Whicker, F. W.,** Transfer of radiocesium from lichen to reindeer, *Health Phys.*, 21(5), 657, 1971.
32. **Davis, J. J.,** Cesium and its relationship to potassium in ecology, in *Radioecology*, Schultz, V. and Klement, A. W., Jr., Eds., Reinhold, New York, 1963, 539.
33. **Hsiao, S. C.,** The radioecology of calcium, in *Radioecology*, Schultz, V. and Klement, A. W., Jr., Eds., Reinhold, New York, 1963, 525.
34. **Comar, C. L. and Wasserman, R. H.,** Radioisotope absorption and methods of elimination: differential behavior of substances in metabolic pathways, in *Radioisotopes in the Biosphere*, Caldecott, R. S. and Snyder, L. A., Eds., University of Minnesota, Minneapolis, 1960, 526.
35. **Comar, C. L.,** Movement of fallout radionuclides through the biosphere and man, *Annu. Rev. Nucl. Sci.*, 15, 175, 1965.
36. **Templeton, W. L. and Brown, V. M.,** Accumulation of calcium and strontium by brown trout from waters in the United Kingdom, *Nature (London)*, 198(4876), 198, 1963.
37. **French, N. R.,** Review and discussion of barium, in *Radioecology*, Schultz, V. and Klement, A. W., Jr., Eds., Reinhold, New York, 1963, 557.
38. **Osburn, W. S.,** Primordial radionuclides: their distribution, movement, and possible effect within terrestrial ecosystems, *Health Phys.*, 11(12), 1275, 1965.
39. **Garner, R. J.,** *Transfer of Radioactive Materials from the Terrestrial Environment to Animals and Man*, CRC Press, Boca Raton, Fla., 1972.
40. **Russell, R. S. and Smith, K. A.,** Naturally occurring radioactive substances: the uranium and thorium series, in *Radioactivity and Human Diet*, Russell, R. S., Ed., Pergamon Press, New York, 1966, 367.
41. **Eisenbud, M.,** *Environmental Radioactivity*, 2nd ed., Academic Press, New York, 1973.
42. **Foster, R. F.,** Environmental behavior of chromium and neptunium, in *Radioecology*, Schultz, V. and Klement, A. W., Jr., Eds., Reinhold, New York, 1963, 569.
43. **Wangersky, P. J.,** Manganese in ecology, in *Radioecology*, Schultz, V. and Klement, A. W., Jr., Eds., Reinhold, New York, 1963, 499.
44. **Whicker, F. W., Walters, R. A., and Dahl, A. H.,** Fall-out radionuclides in Colorado deer liver, *Nature (London)*, 214(5087), 511, 1967.
45. **Lowman, F. G.,** Iron and cobalt in ecology, in *Radioecology*, Schultz, V. and Klement, A. W., Jr., Eds., Reinhold, New York, 1963, 561.
46. **Palmer, H. E. and Beasley, T. M.,** ^{55}Fe in the marine environment and in people who consume ocean fish, in *Radioecological Concentration Processes*, Åberg, B. and Hungate, F. P., Eds., Pergamon Press, New York, 1967, 259.
47. **Persson, B.,** ^{55}Fe from fallout in lichen, reindeer and Lapps, in *Radioecological Concentration Processes*, Åberg, B. and Hungate, F. P., Eds., Pergamon Press, New York, 1967, 253.
48. **Chipman, W. A.,** Food chains in the sea, in *Radioactivity and Human Diet*, Russell, R. S., Ed., Pergamon Press, New York, 1966, 421.

49. **Grummitt, W. E.,** Transfer of cobalt-60 to plants from soils treated with sewage sludge, in *Radioecology and Energy Resources,* Cushing, C. E., Jr. et al., Eds., Dowden, Hutchinson & Ross, Stroudsburg, Pa., 1976, 331.
50. **Bastian, R. K. and Jackson, W. B.,** ^{137}Cs and ^{60}Co in a terrestrial community at Enewetak Atoll, in *Radioecology and Energy Resources,* Cushing, C. E., Jr. et al., Eds., Dowden, Hutchinson & Ross, Stroudsburg, Pa., 1976, 314.
51. **Rice, T. R.,** Review of zinc in ecology, in *Radioecology,* Schultz, V. and Klement, A. W., Jr., Eds., Reinhold, New York, 1963, 619.
52. **Perkins, R. W., Nielsen, J. M., Roesch, W. C., and McCall, R. C.,** Zinc-65 and chromium-51 in foods and people, *Science,* 132(3443), 1895, 1960.
53. **Held, E. E.,** Some aspects of the biology of zirconium-95, in *Radioecology,* Schultz, V. and Klement, A. W., Jr., Eds., Reinhold, New York, 1963, 577.
54. **Rogers, L. E. and Rickard, W. H., Eds.,** Ecology of the 200 Area Plateau Waste Management Environs: A Status Report, U.S. DOE Rep. PNL-2253, Battelle Pacific Northwest Laboratories, Richland, Wash., 1977.
55. **Wildung, R. E., Garland, T. R., and Cataldo, D. A.,** Preliminary studies on the uptake of technetium by soybeans, in Pacific Northwest Laboratory Annu. Rep. for 1974 to the USAEC Division of Biomedical and Environmental Research. Part 2. Ecological Sciences, U.S. AEC Rep. BNWL-1950 Pt. 2, Vaughn, B. E. and Staff, Eds., Pacific Northwest Laboratory, Richland, Wash., 1974, 41.
56. **Auerbach, S. I. and Olson, J. S.,** Biological and environmental behavior of ruthenium and rhodium, in *Radioecology,* Schultz, V. and Klement, A. W., Jr., Eds., Reinhold, New York, 1963, 509.
57. **Goldberg, E. D.,** The chemical invasion of the oceans by man, in *Global Effects of Environmental Pollution,* Singer, S. F., Ed., Springer-Verlag, New York, 1970, 178.
58. **Blanchard, R. L.,** Relationship between ^{210}Po and ^{210}Pb in man and his environment, in *Radioecological Concentration Processes,* Åberg, B. and Hungate, F. P., Eds., Pergamon Press, New York, 1967, 281.
59. **Hill, C. R.,** Lead-210 and plonium-210 in grass, *Nature (London),* 187(4733), 211, 1960.
60. **Kauranen, P. and Miettinen, J. K.,** ^{210}Po and ^{210}Pb in environmental samples in Finland, in *Radioecological Concentration Processes,* Åberg, B. and Hungate, F. P., Eds., Pergamon Press, New York, 1967, 275.
61. **Holtzman, R. B. and Ilcewicz, F. H.,** Lead-210 and polonium-210 in tissues of cigarette smokers, *Science,* 153(3741), 1259, 1966.
62. **Morgan, K. Z., Snyder, W. S., and Ford, M. R.,** Relative health hazard of the various radioactive materials, *Health Phys.,* 10(3), 151, 1964.
63. **Stannard, J. N. and Casarett, G. W., Eds.,** Metabolism and biological effects of an alpha particle emitter, polonium-210, *Radiat. Res.,* Suppl. 5, 1964.
64. **Palumbo, R. F.,** Factors controlling the distribution of the rare earths in the environment and in living organisms, in *Radioecology,* Schultz, V. and Klement, A. W., Jr., Eds., Reinhold, New York, 1963, 533.
65. **Stevenson, P. C. and Nervik, W. E.,** The Radiochemistry of the Rare Earths, Scandium, Yttrium, and Actinium, Nucl. Sci. Ser. NAS-NS-3020, National Academy of Sciences, National Research Council, Washington, D.C., 1961.
66. **Glasstone, S.,** *Sourcebook on Atomic Energy,* 3rd ed., D Van Nostrand, New York, 1967.
67. **U.S. Energy Research and Development Administration,** Workshop on Environmental Research for Transuranic Elements, Rep. ERDA-76/134, U.S. Energy Research and Development Administration, Washington, D.C., 1976.
68. **Verkhovskaya, I. N., Vavilov, P. P., and Maslov, V . I.,** The migration of natural radioactive elements under natural conditions and their distribution according to biotic and abiotic environmental components, in *Radioecological Concentration Processes,* Åberg, B. and Hungate, F. P., Eds., Pergamon Press, New York, 1967, 313.
69. **Noshkin, V. E.,** Ecological aspects of plutonium dissemination in aquatic environments, *Health Phys.,* 22(6), 537, 1972.
70. **Thomas, C. W. and Perkins, R. W.,** Transuranium Elements in the Atmosphere, U.S. AEC Rep. BNWL-1881, Battelle Pacific Northwest Laboratories, Richland, Wash., 1974.
71. Food Protection Committee, *Radionuclides in Foods,* National Research Council, National Academy of Sciences, Washington, D.C., 1973.
72. **I.A.E.A.,** Effects of Ionizing Radiation on Aquatic Organisms and Ecosystems, Tech. Rep. Ser. 172, International Atomic Energy Agency, Vienna, 1976.
73. **Gopal-Ayengar, A. R. and Mistry, K. B.,** On the radioactivity of plants from the high radiation areas of the Keral Coast and adjoining regions. I. Studies on the uptake of gamma-emitting radionuclides, in Radioisotopes in Soil-Plant Nutrition Studies, Publ. STI/PUB/55, International Atomic Energy Agency, Vienna, 1962, 3.

74. **Mays, C. W. et al.**, Radiation induced bone cancer in beagles, in *Delayed Effects of Bone-seeking Radionuclides,* Mays, C. W., Lee, S. S., Lloyd, R. D., Stover, B. J., Doughtery, J. H., and Taylor, G. N., Eds., University of Utah Press, Salt Lake City, 1969, 387.
75. **Cannon, H. L.**, The effect of uranium-vanadium deposits on the vegetation of the Colorado Plateau, *Am. J. Sci.,* 250(10), 735, 1952.
76. **Cannon, H. L.**, The development of botanical methods of prospecting for uranium on the Colorado Plateau, *Geol. Surv. U.S. Bull.,* 1085-A, 1960.
77. **Froelich, A. J. and Kleinhampl, F. J.**, Botanical prospecting for uranium in the Deer Flat area, White Canyon District, San Juan County, Utah, *Geol. Surv. U.S. Bull.,* 1085-B, 1960.
78. **Kovalsky, V. V., Vorotnitskaya, I. E., and Lekarev, V. S.**, Biogeochemical food chains of uranium in aquatic and terraneous organisms, in *Radioecological Concentration Processes,* Åberg, B. and Hungate, F. P., Eds., Pergamon Press, New York, 1967, 329.
79. **Beckmann, P.**, *The Health Hazards of Not Going Nuclear,* Golem Press, Boulder, Colo., 1976.
80. **Olafson, J. H. and Larson, K. H.**, Plutonium, its biology and environmental persistence, in *Radioecology,* Schultz, V. and Klement, A. W., Jr., Eds., Reinhold, New York, 1963, 633.
81. **Seaborg, G. T.**, *The Transuranium Elements,* Yale University Press, New Haven, 1958.
82. **Bair, W. J. and Thompson, R. C.**, Plutonium: biomedical research, *Science,* 183(4126), 715, 1974.
83. **Francis, C. W.**, Plutonium mobility in soil and uptake in plants: a review, *J. Environ. Qual.,* 2(1), 67, 1973.
84. **Price, K. R.**, A review of transuranic elements in soils, plants, and animals, *J. Environ. Qual.,* 2(1), 62, 1973.
85. **Romney, E. M. and Davis, J. J.**, Ecological aspects of plutonium dissemination in terrestrial environments, *Health Phys.,* 22(6), 551, 1972.
86. **Hanson, W. C.**, Ecological considerations of the behavior of plutonium in the environment, *Health Phys.,* 28(5), 529, 1975.
87. **Hakonson, T. E.**, Environmental pathways of plutonium into terrestrial plants and animals, *Health Phys.,* 29(4), 583, 1975.
88. **Wildung, R. E., Drucker, E., and Au, F. H. F.**, The relationship of microbial processes to the fate of transuranic elements in soil, in Transuranics in Natural Environments, U.S. ERDA Rep. NVO-178, White, M. G. and Dunaway, P. B., Eds., U.S. Energy Research and Development Administration, Washington, D.C., 1977, 127.
89. **Bondietti, E. H. and Sweeton, F. H.**, Transuranic speciation in the environment, in Transuranics in Natural Environments, U.S. ERDA Rep. NVO-178, White, M. G. and Dunaway, P. B., Eds., U.S. Energy Research and Development Administration, Washington, D.C., 1977, 449.
90. **Schulz, R. K.**, Root uptake of transuranic elements, in Transuranics in Natural Environments, U.S. ERDA Rep. NVO-178, White, M. G. and Dunaway, P. B., Eds., U.S. Energy Research and Development Administration, Washington, D.C., 1977, 321.
91. **Yoder, R. E., Jr.**, Potential source terms and control measures, in Plutonium and Other Transuranium Elements: Sources, Environmental Distribution and Biomedical Effects, U.S. AEC Rep. WASH-1359, Wachholz, B. W., Comp., U.S. Atomic Energy Commission, Washington, D.C., 1974, 13.
92. **Whicker, F. W., Little, C. A., and Winsor, T. F.**, Plutonium behavior in the terrestrial environs of the Rocky Flats installation, in Environmental Surveillance Around Nuclear Installations, Vol. 2, Publ. STI/PUB/353, International Atomic Energy Agency, Vienna, 1974, 89.
93. **International Commission on Radiological Protection**, *The Metabolism of Compounds of Plutonium and Other Actinides,* ICRP Publ. 19, Pergamon Press, New York, 1972.
94. **Bair, W. J., Richmond, C. R., and Wachholz, B. W.**, Radiobiological Assessment on the Spatial Distribution of Radiation Dose from Inhaled Plutonium, U.S. AEC Rep. WASH-1320, U.S. Atomic Energy Commission, Washington, D.C., 1974.
95. **Hubbard, J. E.**, Cesium-137 in an Alpine Watershed, Ph.D. dissertation, Colorado State University, Fort Collins, 1968; *Nucl. Sci. Abstr.,* 23(20), 40862, 1968.
96. **Williams, P., Jr. and Peck, E. L.**, Terrain influences on precipitation in the intermountain west as related to synoptic situations, *J. Appl. Meterol.,* 1(3), 343, 1962.
97. **Rickard, W. H.**, Accumulation of ^{137}Cs in litter and understory plants of forest stands from various climatic zones of Washington, in *Radioecological Concentration Processes,* Åberg, B. and Hungate, F. P., Eds., Pergamon Press, New York, 1967, 527.
98. **Ritchie, J. C., Clebsch, E. E. C., and Rudolph, W. K.**, Distribution of fallout and natural gamma radionuclides in litter, humus and surface mineral soil layers under natural vegetation in the Great Smoky Mountains, North Carolina - Tennessee, *Health Phys.,* 18(5), 479, 1970.
99. **Humphreys, W. J.**, *Physics of the Air,* 2nd ed., McGraw-Hill, New York, 1929.
100. **Healy, J. W. and Fuquay, J. J.**, Wind pickup of radioactive particles from the ground, *Proc. 2nd Int. Conf. Peaceful Uses of Atomic Energy,* 18, 291, 1958.

101. **Lovill, J. E.**, Transport Processes in Orographically Inducted Gravity Waves as Indicated by Atmospheric Ozone, Paper No. 135, Department of Atmospheric Science, Colorado State University, Fort Collins, 1969.
102. **Whicker, F. W. and Loveless, C. M.**, Relationships of physiography and microclimate to fallout deposition, *Ecology*, 49(2), 363, 1968.
103. **Osburn, W. S., Jr.**, The dynamics of fallout distribution in a Colorado alpine tundra snow accumulation ecosystem, in *Radioecology*, Schultz, V. and Klement, A. W., Jr., Eds., Reinhold, New York, 1963, 51.
104. **Osburn, W. S., Jr.**, Accountability of nuclear fallout deposited in a Colorado high mountain bog, in Symposium on Radioecology, U.S. AEC Rep. CONF-670503, Nelson, D. J. and Evans, F. C., Eds., U.S. Atomic Energy Commission, Washington, D.C., 1969, 578.
105. **Davis, J. J., Hanson, W. C., and Watson, D. G.**, Some effects of environmental factors upon accumulation of worldwide fallout in natural populations, in *Radioecology*, Schultz, V. and Klement, A. W., Jr., Eds., Reinhold, New York, 1963, 35.
106. **Eberhardt, L. L., Rickard, W. H., Cushing, C. E., Watson, D. G., and Hanson, W. C.**, A study of fallout cesium-137 in the Pacific Northwest, *J. Wildl. Manage.*, 33(1), 103, 1969.
107. **Remmenga, E. E. and Whicker, F. W.**, Sampling variability in radionuclide concentrations in plants native to the Colorado Front Range, *Health Phys.*, 13(9), 977, 1967.
108. **Hanson, W. C., Dahl, A. H., Whicker, F. W., Longhurst, W. M., Flyger, V., Davey, S. P., and Greer, K. R.**, Thyroidal radioiodine concentrations in North American deer following 1961-1963 nuclear weapons tests, *Health Phys.*, 9(12), 1235, 1963.
109. **Schultz, V. and Flyger, V.**, Relationship of sex and age to strontium-90 accumulation in white-tailed deer mandibles, *J. Wildl. Manage.*, 29(1), 39, 1965.
110. **Whicker, F. W., Farris, G. C., and Dahl, A. H.**, Concentration patterns of ^{90}Sr, ^{137}Cs and ^{131}I in a wild deer population and environment, in *Radioecological Concentration Processes*, Åberg, B. and Hungate, F. P., Eds., Pergamon Press, New York, 1967, 621.
111. **Longhurst, W. M., Goldman, M., and Della Rosa, R. J.**, Comparison of the environmental and biological factors affecting the accumulation of ^{90}Sr and ^{137}Cs in deer and sheep, in *Radioecological Concentration Processes*, Åberg, B. and Hungate, F. P., Eds., Pergamon Press, New York, 1967, 635.
112. **French, N. R. and Bissell, H. D.**, Stronium-90 in California mule deer, *Health Phys.*, 14(5), 489, 1968.
113. **Farris, G. C., Whicker, F. W., and Dahl, A. H.**, Strontium-90 levels in mule deer and forage plants, in Symposium on Radioecology, U.S. AEC Rep. CONF-670503, Nelson, D. J. and Evans, F. C., Eds., U.S. Atomic Energy Commission, Washington, D.C., 1969, 602.
114. **Whicker, F. W., Farris, G. C., and Dahl, A. H.**, Wild deer as a source of radionuclide intake by humans and as indicators of fallout hazards, in *Radiation Protection. Part 2*, Snyder, W. S. et al., Eds., Pergamon Press, New York, 1968, 1105.
115. **Pendleton, R. C., Mays, C. W., Lloyd, R. D., and Hancock, N. V.**, Fall-out plutonium-239 and zirconium-95 in the lungs of deer, *Nature (London)*, 202(4933), 715, 1964.
116. **Pendleton, R. C., Lloyd, R. D., Mays, C. W., and Church, B. W.**, Trophic level effect on the accumulation of caesium-137 in cougars feeding on mule deer, *Nature (London)*, 204(4959), 708, 1964.
117. **Miettinen, J. K.**, Enrichment of radioactivity by arctic ecosystems in Finnish Lapland, in Symposium on Radioecology, U.S. AEC Rep. CONF-670503, Nelson, D. J. and Evans, F. C., Eds., U.S. Atomic Energy Commission, Washington, D.C., 1969, 23.
118. **Hanson, W. C., Watson, D. G., and Perkins, R. W.**, Concentration and retention of fallout radionuclides in Alaskan arctic ecosystems, in *Radioecological Concentration Processes*, Åberg, B. and Hungate, F. P., Eds., Pergamon Press, New York, 1967, 233.
119. **Liden, K. and Gustafsson, M.**, Relationships and seasonal variation of ^{137}Cs in lichen, reindeer and man in northern Sweden 1961-1965, in *Radioecological Concentration Processes*, Åberg, B. and Hungate, F. P., Eds., Pergamon Press, New York, 1967, 193.
120. **Nevstrueva, M. A., Ramzaev, P. V., Moiseer, A. A., Ibatullin, M. S., and Teplykh, L. A.**, The nature of ^{137}Cs and ^{90}Sr transport over the lichen-reindeer-man food chain, in *Radioecological Concentration Processes*, Åberg, B. and Hungate, F. P., Eds., Pergamon Press, New York, 1967, 209.
121. **Hvinden, T. and Lillegraven, A.**, Caesium-137 and strontium-90 in precipitation, soil and animals in Norway, *Nature (London)*, 192(4808), 1144, 1961.
122. **Liden, K.**, Cesium-137 burdens in Swedish Laplanders and reindeer, *Acta Radiol.*, 56(3), 237, 1961.
123. **Hanson, W. C.**, Fallout radionuclides in Alaskan food chains, *Am. J. Vet. Res.*, 27(116), 359, 1966.
124. **Miettinen, J. K. and Hasanen, E.**, ^{137}Cs in Finnish Lapps and other Finns in 1962-6, in *Radioecological Concentration Processes*, Åberg, B., Hungate, F. P., Eds., Pergamon Press, New York, 1967, 221.

125. **Hanson, W. C. and Eberhardt, L. L.**, Effective half-times of radionuclides in Alaskan lichens and eskimos, in Symposium on Radioecology, U.S. AEC Rep. CONF-670503, Nelson, D. J. and Evans, F. C., Eds., U.S. Atomic Energy Commission, Washington, D.C., 1969, 627.

126. **Gorham, E.**, A comparison of lower and higher plants as accumulators of radioactive fallout, *Can. J. Bot.*, 37(2), 327, 1959.

127. **Rickard, W. H., Davis, J. J., Hanson, W. C., and Watson, D. G.**, Gamma-emitting radionuclides in Alaskan tundra vegetation 1959, 1960, 1961, *Ecology*, 46(3), 352, 1965.

128. **Potter, L. D. and Barr, M.**, Cesium-137 concentrations in Alaskan arctic tundra vegetation, 1967, *Arct. Alp. Res.*, 1(3), 147, 1969.

129. **Hanson, W. C.**, ^{137}Cs: seasonal patterns in native residents of three contrasting Alaskan villages, *Health Phys.*, 20(6), 585, 1971.

130. **Gustafsson, M.**, External and internal irradiation of Swedish reindeer breeders 1950-1970, *Health Phys.*, 17(1), 19, 1969.

131. **Hardy, E. P., Jr., List, R. L., Machata, L., Alexander, L. T., Allen, S. S., and Meyer, M. W.**, Strontium-90 on the Earth's Surface. II. Summary and Interpretation of a World-Wide Soil Sampling Program: 1960—1961 Results, U.S. AEC Rep. TID-17090, U.S. Atomic Energy Commission, Washington, D.C., 1962.

132. **List, R. J., Machta, L., Alexander, L. T., Allen, J. S., Meyer, M. W., Valassis, V. T., and Hardy, E. P., Jr.**, Strontium-90 on the earth's surface. III. in Radioactiive Fallout from Nuclear Weapons Tests, (AEC Symp. Ser. 5), U.S. AEC Rep. CONF-765, Klement, A. W., Jr., Ed., U.S. Atomic Energy Commission, Washington, D.C., 1965, 359.

133. **Cummings, S. L., Bankert, L., Garrett, A. R., Jr., and Regnier, J. E.**, ^{137}Cesium uptake by oat plants as related to the soil fixing capacity, *Health Phys.*, 17(1), 145, 1969.

134. **Jenkins, J. H.**, Evaluation of the factors involved in bioaccumulation of gamma-emitting radionuclides in white-tailed deer *(Odocoileus virginianus)*, 3rd Tech. Progr. Rep. February 1, 1971 - January 31, 1972, U.S. AEC Contract AT(38-1)-642, U.S. AEC Rep. SRO-642-1, University of Georgia, Athens, 1972; *Nucl. Sci. Abstr.*, 26(3), 4720.

135. **Jenkins, J. H. and Fendley, T. T.**, The extent of contamination, detection, and health significance of high accumulations of radioactivity in southeastern game populations, *Proc. Annu. Conf. Southeast. Assoc. Game Fish Comm.*, 22, 89, 1968.

136. **Johnson, W. and Nayfield, C. L.**, Elevated levels of Cs-137 in common mushrooms (Agaricaceae) with possible relationship to high levels of Cs-137 in whitetail deer, 1968-69, *Radiol. Health Data Rep.*, 11(10), 527, 1970.

137. **Roessler, C. E., Dunavant, B. G., and Bevis, H. A.**, Investigations of unusual cesium ecology in Florida — cesium-137 levels in feed-lot beef, *Health Phys.*, 16(6), 691, 1969.

138. **Porter, C. R., Phillips, C. R., Carter, M. W., and Kahn, B.**, The cause of relatively high ^{137}Cs concentrations in Tampa, Florida, milk, in *Radioecological Concentration Processes*, Åberg, B. and Hungate, F. P., Eds., Pergamon Press, New York, 1967, 95.

139. **Roessler, G. S., Dunavant, B. G., and Roessler, C. E.**, Cesium-137 body burdens in Florida residents, *Health Phys.*, 16(6), 673, 1969.

140. **Pendleton, R. C. and Hanson, W. C.**, Absorption of cesium-137 by components of an aquatic community, *Proc. 2nd Int. Conf. Peaceful Uses of Atomic Energy*, 18, 419, 1958.

141. **Weinberger, D. and Porter, J. W.**, Metabolism of hydrogen isotopes by rapidly growing *Chlorella pyrenoidosa* cells, *Arch. Biochem. Biophys.*, 50(1), 160, 1954.

142. **Williams, L. G. and Swanson, H. D.**, Concentration of cesium-137 by algae, *Science*, 127(3291), 187, 1958.

143. **Kornberg, H. A. and Davis, J. J.**, Food chains in fresh water, in *Radioactivity and Human Diet*, Russell, R. S., Ed., Pergamon Press, New York, 1966, 385.

144. **Agnedal, P. O.**, Calcium and strontium in Swedish waters and fish and accumulation of ^{90}Sr, in *Radioecological Concentration Processes*, Åberg, B. and Hungate, F. P., Eds., Pergamon Press, New York, 1967, 879.

145. **Kolehmainen, S., Hasanen, E., and Miettinen, J. K.**, ^{137}Cs in the plants, plankton and fish of the Finnish lakes and factors affecting its accumulation, in *Radiation Protection, Part 1*, Snyder, W. S. et al., Eds., Pergamon Press, New York, 1968, 407.

146. **Nelson, W. C. and Whicker, F. W.**, Cesium-137 in some Colorado game fish, 1965-66, in Symposium on Radioecology, U.S. AEC Rep. CONF-670503, Nelson, D. J. and Evans, F. C., Eds., U.S. Atomic Energy Commission, Washington, D.C., 1969, 258.

147. **Gustafson, P. F.**, Cesium-137 in freshwater fish during 1954-1965, in Symposium on Radioecology, U.S. AEC Rep. CONF-670503, Nelson, D. J. and Evans, F. C., Eds., U.S. Atomic Energy Commission, Washington, D.C., 1969, 249.

148. **Kolehmainen, S. E., Hasanen, E., and Miettinen, J. K.**, ^{137}Cs levels in fish of different limnological types of lakes in Finland during 1963, *Health Phys.*, 12(7), 917, 1966.

149. **Armstrong, N. E. and Gloyna, E. F.,** Mathematical models for the dispersion of radionuclides in aquatic systems, in Symposium on Radioecology, U.S. AEC Rep. CONF-670503, Nelson, D. J. and Evans, F. C., Eds., U.S. Atomic Energy Commission, Washington, D.C., 1969, 329.
150. **Polikarpov, G. G.,** Regularities of uptake and accumulation of radionuclides in aquatic organisms, in *Radioecological Concentration Processes,* Åberg, B. and Hungate, F. P., Eds., Pergamon Press, New York, 1967, 819.
151. **Davis, J. J. and Foster, R. F.,** Bioaccumulation of radioisotopes through aquatic food chains, *Ecology,* 39(3), 530, 1958.
152. **Gallegos, A. F.,** Radiocesium Kinetics in the Components of a Montane Lake Ecosystem, Ph.D. dissertation, Colorado State University, Fort Collins, 1969, *Nucl. Sci. Abstr.,* 24(14), 27561.
153. **Whittaker, R. H.,** Experiments with radiophosphorus tracer in aquarium microcosms, *Ecol. Monogr.,* 31(2), 157, 1961.
154. **Hakonson, T. E.,** Cesium Kinetics in a Montane Lake Ecosystem, Ph.D. dissertation, Colorado State University, Fort Collins, 1972; *Nucl. Sci. Abstr.,* 27(9), 19836.
155. **Panel on Radioactivity in the Marine Environment,** *Radioactivity in the Marine Environment,* National Academy of Sciences, Washington, D.C., 1971.
156. **U.S. Weather Bureau,** Announced Nuclear Detonations 1945-1962. United States, United Kingdom, Republic of France, Union of Soviet Socialist Republics, in Fallout Program Quarterly Summary Rep. (September 1, 1963 Through December 1, 1963), U.S. AEC Rep. HASL-142, Hardy, E. P., Jr., Rivera, J., and Collins, W. R., Jr., Eds./Comps., Health Safety Laboratory, U.S. Atomic Energy Commission, Washington, D.C., 1964, 218.
157. **Bowen, V. T. and Sugihara, T. T.,** Cycling and levels of strontium-90, cerium-144, and promethium-147 in the Atlantic Ocean, in *Radioecology,* Schultz, V. and Klement, A. W., Jr., Eds., Reinhold, New York, 1963, 135.
158. **Revelle, R. and Schaefer, M. B.,** General considerations concerning the ocean as a receptacle for artificially radioative materials, in The Effects of Atomic Radiation on Oceanography and Fisheries, Publ. 551, Committee on Effects of Atomic Radiation on Oceanography and Fisheries, National Academy of Sciences-National Research Council, Washington, D.C., 1957, 1.
159. **Watson, D. G., Davis, J. J., and Hanson, W. C.,** Interspecies differences in accumulation of gamma emitters by marine organisms near the Columbia River mouth, *Limnol. Oceanogr.,* 8(2), 305, 1963.
160. **Krumholz, L. A., Goldberg, E. D., and Boroughs, H.,** Ecological factors involved in the uptake, accumulation, and loss of radionuclides by aquatic organisms, in The Effects of Atomic Radiation on Oceanography and Fisheries, Publ. 551, Committee on Effects of Atomic Radiation on Oceanography and Fisheries, National Academy of Sciences-National Research Council, Washington, D.C., 1957, 69.
161. **Palumbo, R. F.,** The difference in uptake of radioisotopes by marine and terrestrial organisms, in *Recent Advances in Botany,* Vol. 2, University of Toronto Press, Canada, 1961, 1367.
162. **Seymour, A. H.,** Radioactivity of marine organisms from Guam, Palau and the Gulf of Siam, 1958-1959, in *Radioecology,* Schultz, V. and Klement, A. W., Jr., Eds., Reinhold, New York, 1963, 151.
163. **Voipio, A. and Salo, A.,** On the balances of ^{90}Sr and ^{137}Cs in the Baltic Sea, *Nord. Hydrol.,* 2(1), 57, 1971.
164. **Burton, J. D.,** Radioactive nuclides in sea water, marine sediments and marine organisms, in *Chemical Oceanography,* Vol. 2, Riley, J. P. and Skirrow, G., Eds., Academic Press, New York, 1965, 425.
165. **Lowman, F. G.,** Radionuclides in plankton and tuna from the central Pacific, in *Radioecology,* Schultz, V. and Klement, A. W., Jr., Eds., Reinhold, New York, 1963, 145.
166. **Ketchum, B. H.,** The effects of the ecological system on the transport of elements in the sea, in The Effects of Atomic Radiation on Oceanography and Fisheries, Publ. 551, Committee on Effects of Atomic Radiation on Oceanography and Fisheries, National Academy of Sciences-National Research Council, Washington, D.C., 1957, 52.
167. **Corcoran, E. F. and Kimball, J. F., Jr.,** The uptake, accumulation and exchange of strontium-90 by open sea phytoplankton, in *Radioecology,* Schultz, V. and Klement, A. W., Jr., Eds., Reinhold, New York, 1963, 187.
168. **Lowman, F. G., Rice, T. R., and Richards, F. A.,** Accumulation and redistribution of radionuclides by marine organisms, in *Radioactivity in the Marine Environment,* Panel on Radioactivity in the Marine Environment, National Academy of Sciences, Washington, D.C., 1971, 161.
169. **Held, E. E.,** Qualitative distribution of radionuclides at Rongelap Atoll, in *Radioecology,* Schultz, V. and Klement, A. W., Jr., Eds., Reinhold, New York, 1963, 167.

ADDITIONAL READINGS

Aleksakhin, R. M. and Tikhomirov, F. A., Radionuclide migration in forest biogeocenoses, in *Contemporary Problems of Radiobiology,* Vol. 2, Klechkovskiy, V. M., Polikarpov, G. G., and Aleksakhin, R. M., Eds., Atomizdat, Moscow, 1971, 145; *Radioecology,* Greenberg, D ., Ed., John Wiley & Sons, New York, 1973, 126; (English transl.).

Auerbach, S. I., Kaye, S. V., Nelson, D. J., Reichle, D. E., Dunaway, P. B., and Booth, R. S., Understanding the dynamic behavior of radionuclides released to the environment, *Proc. 4th Int. Conf. Peaceful Uses of Atomic Energy,* 11, 575, 1972.

Baranova, D. D. and Polikarpov, G. G., Sorption and desorption of radionuclides by shallow-water sediments of the Black Sea, *Okeanologiya,* 8(3), 427, 1968, *Oceanology,* 8(3), 328; (English transl.).

Berg, G. G. and Ginsberg, E., Partition of ruthenium-106 between the freshwater environment and crayfish, *Health Phys.,* 30(4), 329, 1976.

Book, S. A., Connolly, G. E., and Longhurst, W. M., Fallout ^{137}Cs accumulation in two adjacent populations of northern California deer, *Health Phys.,* 22(4), 379, 1972.

Bovard, P., Grauby, A., Saas, A., and Schaeffer, R., Effect of the pollutant burden of water on the behaviour of radionuclides (in French; English summary), in Combined Effects of Radioactive, Chemical and Thermal Releases to the Environment, Publ. STI/PUB/404, International Atomic Energy Agency, Vienna, 1975, 179.

Brisbin, I. L., Jr., Geiger, R. A., and Smith, M. H., Accumulation and redistribution of radiocaesium by migratory waterfowl inhabiting a reactor cooling reservoir, in Environmental Behaviour of Radionuclides Released in the Nuclear Industry, Publ. STI/PUB/345, International Atomic Energy Agency, Vienna, 1973, 373.

Brisbin, I. L., Jr., Staton, M. A., Pinder, J. E., III, and Geiger, R. A., Radiocesium concentrations of snakes from contaminated and non-contaminated habitats of the AEC Savannah River Plant, *Copeia,* 1974(2), 501, 1974.

Brisbin, I. L., Jr., Beyers, R. J., Dapson, R. W., Geiger, R. A., Gentry, J. B., Gibbons, J. W., Smith, M. H., and Woods, S. K., Patterns of radiocesium in the sediments of a stream channel contaminated by production reactor effluents, *Health Phys.,* 27(1), 19, 1974.

Bryan, G. W., The accumulation of ^{137}Cs by brackish water invertebrates and its relation to the regulation of potassium and sodium, *J. Mar. Biol. Assoc. U.K.,* 43(2), 541, 1963.

Bryan, G. W., Preston, A., and Templeton, W. L., Accumulation of radionuclides by aquatic organisms of economic importance in the United Kingdom, in Disposal of Radioactive Wastes into Seas, Oceans and Surface Water, Publ. STI/PUB/126, International Atomic Energy Agency, Vienna, 1966, 623.

Crossley, D. A., Jr. and Reichle, D. E., Analysis of transient behavior of radioisotopes in insect food chains, *BioScience,* 19(4), 341, 1969.

Cushing, C. E., Periphyton productivity and radionuclide accumulation in the Columbia River, Washington, U. S. A., *Hydrobiologia,* 29(1/2), 125, 1967.

Cushing, C. E. and Watson, D. G., Cycling of zinc-65 in a simple food web, in Radionuclides in Ecosystems, U.S. AEC Rep. CONF-710501-Pl, Nelson, D. J., Ed., U.S. Atomic Energy Commission, Washington, D.C., 1973, 318.

Dahlman, R. C., Prediction of radionuclide contamination of grass from fallout-particle retention and behavior, in Survival of Food Crops and Livestock in the Event of Nuclear War, (AEC Symp. Ser. 24), U.S. AEC Rep. CONF-700909, Benson, D. W. and Sparrow, A. H., Eds., U.S. Atomic Energy Commission, Washington, D.C., 1971, 492.

Dahlman, R. C. and Van Voris, P., Cycling of ^{137}Cs in soil and vegetation of a flood plain 30 years after initial contamination, in *Radioecology and Energy Resources,* Cushing, C. E., Jr. et al., Eds., Dowden, Hutchinson & Ross, Stroudsburg, Pa., 1976, 291.

Davis, J. J., The effects of environmental factors upon the accumulation of radioisotopes by ecological systems, *Tex. Eng. Exp. Stn., Misc. Publ.,* E-72-60, 37, 1960; *Nucl. Sci. Abstr.,* 14(24), 25209.

Domby, A. H., Paine, D., and McFarlane, R. W., Radiocesium dynamics in herons inhabiting a contaminated reservoir system, *Health Phys.,* 33(6), 523, 1977.

Evans, D. W. and Cutshall, N. H., Effects of ocean water on the soluble-suspended distribution of Columbia River radionuclides, in Radioactive Contamination of the Marine Environment, Publ. STI/PUB/313, International Atomic Energy Agency, Vienna, 1973, 125.

Fendley, T. T., Manlove, M. N., and Brisbin, I. L., Jr., The accumulation and elimination of radiocesium by naturally contaminated wood ducks, *Health Phys.,* 32(5), 415, 1977.

Fleishman, D. G., Accumulation of artificial radionuclides in freshwater fish, in *Contemporary Problems of Radiobiology,* Vol. 2, Klechkovskiy, V. M., Polikarpov, G. G., and Aleksakhin, R. M., Eds., Atomizdat, Moscow, 1971, 395; *Radioecology,* Greenberg, D., Ed., John Wiley & Sons, New York, 1973, 347; (English transl.).

Forster, W. O., Radioactive and stable nuclides in the Columbia River and adjacent northeast Pacific Ocean, in *The Columbia River Estuary and Adjacent Ocean Waters: Bioenvironmental Studies,* Pruter, A. T. and Alverson, D. L., Eds., University of Washington Press, Seattle, 1972, 663.

Forster, W. O., Radionuclide distribution in Columbia River adjacent Pacific shelf sediments, in *The Columbia River Estuary and Adjacent Ocean Waters: Bioenvironmental Studies,* Pruter, A. T. and Alverson, D. L., Eds., University of Washington Press, Seattle, 1972, 701.

Francis, C. W. and Tamura, T., Cesium-137 soil inventory of a tagged *Liriodendron* forest 1962 and 1969, in Radionuclides in Ecosystems, U.S. AEC Rep. CONF-710501-Pl, Nelson, D. J., Ed., U.S. Atomic Energy Commission, Washington, D.C., 1973, 140.

French, N. R., Comparison of radioisotope assimilation by granivorous and herbivorous mammals, in *Radioecological Concentration Processes,* Åberg, B. and Hungate, F. P., Eds., Pergamon Press, New York, 1967, 665.

Garten, C. T., Jr., Radiocesium uptake by a population of cotton rats *(Sigmodon hispidus)* inhabiting the banks of a radioactive liquid waste pond, *Health Phys.,* 36(1), 39, 1979.

Gilat, E., Laichter, Y., and Shafrir, N. H., Behaviour of caesium-137 in the marine environment, in Impacts of Nuclear Releases into the Aquatic Environment, Publ. STI/PUB/406, International Atomic Energy Agency, Vienna, 1975, 63.

Glaser, R., The radioactivity of several submerged water plants, *Limnologica,* 1(1), 51, 1962; AEC-tr-5753, U.S. Atomic Energy Commission, Washington, D.C., 1963; (English transl.).

Gustafson, P. F., Comments on radionuclides in aquatic ecosystems, in *Radioecological Concentration Processes,* Åberg, B. and Hungate, F. P., Eds., Pergamon Press, New York, 1967, 853.

Hakonson, T. E. and Johnson, L. J., Distribution of environmental plutonium in the Trinity Site ecosystem after 27 years, in Proc. 3rd Int. Conf. Int. Radiation Protection Association, U.S. AEC Rep. CONF-730907-Pl, Snyder, W. S., Ed., U.S. Atomic Energy Commission, Washington, D.C., 1974, 242.

Hoffman, G. R., The accumulation of cesium-137 by cryptogams in a *Liriodendron tulipifera* forest, *Bot. Gaz. (Chicago),* 133(2), 107, 1972.

Holleman, D. F. and Luick, J. R., Radiocesium kinetics in arctic carnivores, *Health Phys.,* 30(2), 241, 1976.

Ijuin, M., Picat, P., Sass, A., and Grauby, A., Determination of the diffusion coefficient of radioelements in the Rhone sediments, *Health Phys.,* 24(6), 665, 1973.

Il'yenko, A. I., Radioecology of freshwater fishes, *Vopr. Ikhtiol.,* 55, 1969; *Probl. Ichthyol.,* 9(2), 249; (English transl.).

Il'yenko, A. I., Radioecology of wild animals, in *Contemporary Problems of Radiobiology,* Vol. 2, Klechkovskiy, V. M., Polikarpov, G. G., and Aleksakhin, R. M., Eds., Atomizdat, Moscow, 1971, 279; *Radioecology,* Greenberg, D., Ed., John Wiley & Sons, New York, 1973, 241; (English transl.).

Jordan, C. F. and Kline, J. R., Strontium-90 in a tropical rain forest: 12th-yr validation of a 32-yr prediction, *Health Phys.,* 30(2), 199, 1976.

Kalninya, Z. and Polikarpov, G., Strontium-90 concentration factors of lake plankton, macrophytes, and substrates, *Science,* 164(3887), 1517, 1969.

Kaye, S. V. and Nelson, D. J., Analysis of specific-activity content concept as related to environmental concentration of radionuclides, *Nucl. Saf.,* 9(1), 53, 1968.

Kimura, Y. and Honda, Y., Uptake and elimination of some radionuclides by eggs and fry of rainbow trout. I, *J. Radiat. Res.,* 18(3), 170, 1977.

Kimura, Y. and Honda, Y., Uptake and elimination of some radionuclides by eggs and fry of rainbow trout. II, *J. Radiat. Res.,* 18(3), 182, 1977.

Kirkham, M. B., Adriano, D. C., and Corey, J. C., Comparison of plutonium concentrations in deer from the southwestern United States and in deer from an integrated nuclear fuel cycle facility, *Health Phys.,* 36(4), 516, 1979.

Kitchings, T., DiGregorio, D., and Van Voris, P., Review of the ecological parameters of radionuclide turnover in vertebrate food chains, in *Radioecology and Energy Resources,* Cushing, C. E., Jr. et al., Eds., Dowden, Hutchinson & Ross, Stroudsburg, Pa., 1975, 304.

Klepper, B., Rogers, L. E., Hedlund, J. D., Schreckhise, R. G., and Price, K. R., Radiocesium movement in a gray rabbit brush community, in Environmental Chemistry and Cycling Processes, (DOE Symp. Ser. 45), U.S. DOE Rep. CONF-760429, Adriano, D. C. and Brisbin, I. L., Jr., Eds., U.S. Department of Energy, Washington, D.C., 1978, 725.

Kline, J. R., Retention of fallout radionuclides by tropical forest vegetation, in A Tropical Rain Forest: A study of Irradiation and Ecology at El Verde, Puerto Rico, U.S. AEC Rep. TID-24270, Odum, H. T. and Pigeon, R. F., Eds., U.S. Atomic Energy Commission, Washington, D.C., 1970, H191.

Kolehmainen, S. E., The balances of ^{137}Cs, stable cesium and potassium of bluegill *(Lepomis macrochirus* Raf.) and other fish in White Oak Lake, *Health Phys.,* 23(3), 301, 1972.

Koyanagi, T., Nakahara, M. and Iimura, M., Absorption of sediment-bound radionuclides through the digestive tract of marine demersal fishes, *J. Radiat. Res.,* 19(4), 295, 1978.

Kulikov, N. V., Radioecology of freshwater plants and animals, in *Contemporary Problems of Radiobiology,* Vol. 2, Klechkovskiy, V. M., Polikarpov, G. G., and Aleksakhin, R. M., Eds., Atomizdat, Moscow, 1971, 367; *Radioecology,* Greenberg, D., Ed., John Wiley & Sons, New York, 1973, 323; (English transl.).

Livingston, H. D. and Bowen, V. T., Americium in the marine environment — relationships to plutonium, in *Environmental Toxicity of Aquatic Radionuclides: Models and Mechanisms,* Miller, M. W. and Stannard, J. N., Eds., Ann Arbor Science, Mich., 1976, 107.

Lowman, F. G., Marine biological investigations at the Eniwetok Test Site, in Disposal of Radioactive Wastes, Vol. 2, Publ. STI/PUB/18, International Atomic Energy Agency, Vienna, 1960, 106.

Maslov, V. I., Gruzdev, B. I., Maslova, K. I., Nikiforov, V. S., and Verkhovskaya, I. N., The role of muriform rodents in the biogenic migration of uranium, radium and thorium under different radioecological conditions, in Proc. Int. Symp. on Radioecology, Vol. 2, Centre d'Etudes Nucleaires de Cadarache, Commissariat a l'Energie Atomique, France, 1970, 923; *Nucl. Sci. Abstr.,* 24(2), 41787.

Mattsson, L. J. S. and Persson, R. B. R., Radioecology of ^{22}Na, ^{210}Po and ^{210}Pb in the unique food-chain: lichen-reindeer-man, in Proc. Int. Symp. Radioecology Applied to the Protection of Man and His Environment, Vol. 2, Rep. EUR 4800 d-f-i-e, Commission of the European Communities, Luxembourg, 1971, 1135; *Nucl. Sci. Abstr.,* 27(3), 5285.

Miyake, Y., Artificial radioactivity in the sea, in *The Sea,* Vol. 2, Hill, M. N., Ed., Interscience, New York, 1963, 78.

Nakahara, M., Koyanagi, T., and Saiki, M., Temperature effect on the concentration of radionuclides by marine organisms, *J. Radiat. Res.,* 18(2), 122, 1977.

Nelson, D. J., Kaye, S. V., and Booth, R. S., Radionuclides in river systems, in *River Ecology and Man,* Oglesby, R. T., Carlson, C. A., and McCann, J. A., Eds., Academic Press, New York, 1972, 367.

Nishiwaki, Y., Honda, Y., Kimura, Y., Morishima, H., Koga, T., Miyaguchi, Y., and Kawai, H., Behaviour and distribution of radioactive substances in coastal and estuarine waters, in Radioactive Contamination of the Marine Environment, Publ. STI/PUB/313, International Atomic Energy Agency, Vienna, 1973, 177.

O'Farrell, T. P. and Gilbert, R. O., Transport of radioactive materials by jackrabbits on the Hanford Reservation, *Health Phys.,* 29(1), 9, 1975.

Osterberg, C., Pearcy, W. G., and Curl, H., Jr., Radioactivity and its relationship to oceanic food chains, *J. Mar. Res.,* 22(1), 2, 1964.

Ovchenkov, V. Ya., Radium migration from sites of its concentration under natural conditions (in Russian), in *Radioecological Investigations in Natural Biogeocenoses,* Verkhovskaya, I. N., Ed., Nauka, Moscow, 1972, 147, *Nucl. Sci. Abstr.,* 26(23), 56112.

Patel, B., Mulay, C., and Ganguly, A. K., Radioecology of Bombay harbour - a tidal estuary, *Estuarine Coastal Mar. Sci.,* 3(1), 13, 1975.

Pentreath, R. J., The roles of food and water in the accumulation of radionuclides by marine teleost and elasmobranch fish, in Radioactive Contamination of the Marine Environment, Publ. STI/PUB/313, International Atomic Energy Agency, Vienna, 1973, 421.

Polikarpov, G. G., Trends in accumulation of radioactive substances by aquatic organisms in various seas in World Ocean, in *Marine Radioecology,* Polikarpov, G. G., Ed., Naukova Dumka, Kiev, 1970, 248; AEC-tr-7299, U.S. Atomic Energy Commission, Washington, D.C., 1972, 307; (English transl.).

Polikarpov, G. G., Generalizations in marine radioecology, in *Marine Radioecology,* Polikarpov, G. G., Ed., Naukova Dumka, Kiev, 1970, 255; AEC-tr-7299, U.S. Atomic Energy Commission, Washington, D. C., 1972, 318; (English transl.).

Pritchard, D. W., Reid, R. O., Okubo, A., and Carter, H. H., Physical processes of water movement and mixing, in *Radioactivity in the Marine Environment,* Panel on Radioactivity in the Marine Environment, National Academy of Sciences, Washington, D.C., 1971, 90.

Ragsdale, H. I. and Shure, D. J., Flood plain transfer and accumulation of ^{137}Cs from a reactor effluent stream, in Environmental Behaviour of Radionuclides Released in the Nuclear Industry, Publ. STI/PUB/345, International Atomic Energy Agency, Vienna, 1973, 243.

Reichle, D. E., Radioisotope turnover and energy flow in terrestrial isopod populations, *Ecology,* 48(3), 351, 1967.

Reichle, D. E. and Crossley, D. A., Jr., Radiocesium dispersion in a cryptozoan food web, *Health Phys.,* 11(12), 1375, 1965.

Reichle, D. E., Dunaway, P. B., and Nelson, D. J., Turnover and concentration of radionuclides in food chains, *Nucl. Saf.,* 11(1), 43, 1970.

Renfro, W. C., Forster, W. O., and Osterberg, C. L., Seasonal and areal distributions of radionuclides in the biota of the Columbia River Estuary, in *The Columbia River Estuary and Adjacent Ocean Waters: Bioenvironmental Studies,* Pruter, A. T. and Alverson, D. L., Eds., University of Washington Press, Seattle, 1972, 777.

Rice, T. R. and Wolfe, D. A., Radioactivity - chemical and biological aspects, in *Impingement of Man on the Oceans,* Hood, D. W., Ed., John Wiley & Sons, New York, 1971, 325.

Schelske, C. L., Wolfe, D. A., and Hoss, D. E., Ecological implications of fallout radioactivity accumulated by estuarine fishes and mollusks, in Radionuclides in Ecosystems, U.S. AEC Rep. CONF-710501-P2, Nelson, D. J., Ed., U.S. Atomic Energy Commission, Washington, D.C., 1973, 791.

Seymour, A. H. and Nelson, V. A., Decline of ^{65}Zn in marine mussels following the shutdown of Hanford reactors, in Radioactive Contamination of the Marine Environment, Publ. STI/PUB/313, International Atomic Energy Agency, Vienna, Austria, 1973, 277.

Seymour, A. H. et al., Aquatic environment, in *Long-term Worldwide Effects of Nuclear—Weapons Detonations,* National Research Council, National Academy of Sciences, Washington, D.C., 1975, 103.

Shure, D. J. and Gottschalk, M. R., Cesium-137 dynamics within a reactor effluent stream in South Carolina, in *Radioecology and Energy Resources,* Cushing, C. E., Jr. et al., Eds., Dowden, Hutchinson & Ross, Stroudsburg, Pa., 1975, 234.

Springer, J. T., ^{90}Sr and ^{137}Cs in coyote scats from the Hanford Reservation, *Health Phys.,* 36(1), 31, 1979.

Stara, J. F., Nelson, N. S., Della Rosa, R. J., and Bustad, L. K., Comparative metabolism of radionuclides in mammals: a review, *Health Phys.,* 20(2), 113, 1971.

Straney, D. O., Beaman, B., Brisbin, I. L., Jr., and Smith, M. H., Radiocesium in birds of the Savannah River Plant, *Health Phys.,* 28(4), 341, 1975.

Ueda, T., Nakamura, R., and Suzuki, Y., Comparison of influence of sediments and sea water on accumulation of radionuclides by worms, *J. Radiat. Res.,* 18(1), 84, 1977.

Ueda, T., Nakamura, R., and Suzuki, Y., Comparison of influences of sediments and sea water on accumulation of radionuclides by marine organisms, *J. Radiat. Res.,* 19(1), 93, 1978.

Verkhovskaya, I. N., Gruzdev, B. I., Maslov, V. I., Maslova, K. I., Nikoforov, V. S., Ovchenkov, V. J., Popova, O. N., Rubtsov, D. M., and Vavilov, P. P., The content and translocation of natural radioactive elements in the system "soil-plants-animals" under natural and experimental conditions, in Proc. Int. Symp. on Radioecology, Vol. 2, Centre d'Etudes Nucleaires de Cadarache, Commissariat a l'Energie Atomique, France, 1969, 781; *Nucl. Sci. Abstr.,* 24(20), 41785.

Wahlgren, M. A. and Marshall, J. S., The behaviour of plutonium and other long-lived radionuclides in Lake Michigan. I. Biological transport, seasonal cycling, and residence times in the water column, in Impacts of Nuclear Releases into the Aquatic Environment, Publ. STI/PUB/406, International Atomic Energy Agency, Vienna, 1975, 227.

Watson, D. G., Cushing, C. E., Coutant, C. C., and Templeton, W. L., Cycling of radionuclides in Columbia River biota, in *Trace Substances in Environmental Health —IV,* Hemphill, D. D., Ed., University of Missouri, Columbia, 1971, 144; *Nucl. Sci. Abstr.,* 26(7), 15205.

Willard, W. K., Avian uptake of fission products from an area contaminated by low-level atomic wastes, *Science,* 132(3420), 148, 1960.

Witherspoon, J. P., Jr., Cycling of cesium-134 in white oak trees, *Ecol. Monogr.,* 34(4), 403, 1964.

Witkamp, M. and Merchant, V. A., Effects of light, temperature, and soil fertility on distribution of manganese-54 and cesium-137 in producer-consumer microcosms, in Radionuclides in Ecosystems, U.S. AEC Rep. CONF-710501-P1, Nelson, D. J., Ed., U.S. Atomic Energy Commission, Washington, D.C., 1973, 204.

Wolfe, D. A. and Rice, T. R., Cycling of elements in estuaries, *Fishery Bull. (U.S.),* 70(3), 959, 1972.

Zaytsev, Yu. P. and Polikarpov, G. G., The problems of radioecology of hyponeuston, *Okeanologiya,* 4(3), 423, 1964; JPRS 25, 966, 1964, 28 (English transl.); *Nucl. Sci. Abstr.,* 19(1), 22.

INDEX

A

A, I: 152
Aberrations in chromosomes, II: 125, 143
Abiotic compartments, I: 132
Abiotic features of environment, I: 23
Abiotic formation of amino acids, II: 183
Absorbed dose, II: 120
Absorption, I: 31, 180; II: 30
  alpha particle, I: 46
  beta particle, I: 47
  energy, I: 63
  photoelectric, I: 50
  physical, II: 30
  self-, I: 59
Ac, I: 158
Acceptable risk, II: 186
Accidental releases of radioactivity, II: 187
Accidents
  nuclear, II: 142
  public fear of, II: 187
  Urals, II: 188
  uranium mill tailings dam failure, II: 188
Accumulation
  compartmentalization of, II: 66
  factors affecting, I: 133
Actinides, I: 158
Activation of neutrons, I: 135
Activation products, I: 69, 93
Active growth phases, II: 123
Activity median aerodynamic diameter (AMAD), II: 46
Actual birth rate, I: 17
Actual death rate, I: 17
Acute dose, II: 50
Acute gamma radiation, II: 137
Acute irradiation, II: 119
  investigations of, II: 131
Acute lethality, II: 125
Adaptation to stress, II: 123
Adsorption, I: 30; II: 10, 30
  chemical, I: 130
  surface, I: 180
Advection, II: 11
Advection-diffusion equations, II: 9
AEC, see Atomic Energy Commission
Aerodynamic drag force, II: 14
Aerosols, I: 142
  inhalation of, I: 132
Aging, II: 122, 125, 181
Agreement States, II: 186
Airborne fallout, I: 111
Airborne transport, II: 27
Air turbulence, I: 169
ALARA, see As low as reasonably achievable
Algae, I: 178
Alkaline metals, I: 149, 151
Allocation of resources, I: 21
Alpha emission, II: 130
  theory of, I: 37
Alpha particles, I: 37
  absorption of, I: 46
Alpine lakes, I: 179
Altitude, I: 76
Am, I: 159
241-Am, I: 160
AMAD, see Activity median aerodynamic diameter
Amchitka Island, I: 105
Amensalism, I: 33
Americium, see Am
Amino acids, abiotic formation of, II: 183
Ammonium diuranate, I: 91
Amphibians, II: 135
Amphipods, I: 180
Animal/plant concentration ratios,II: 63
Animal populations, II: 162
Annihilation radiations, I: 48
Anspaugh's model, II: 29
Anthropogenic releases of radioactivity, II: 185
Antinuclear sentiment, II: 181
Ants, II: 163, 164
Aphids, II: 124, 163
Apparatus for measurement of radiation, I: 52
Applications of nuclear explosives, I: 117
Applied Fisheries Laboratory, I: 4
Approximation, numerical, II: 109
Aquatic ecosystems, I: 24, 142, 176
Aqueous dispersion, II: 9
Arctic regions, I: 173
Argon, see A
Argonne National Laboratory, I: 6
Arms race, II: 189
Arthropods, II: 149
Artificial ecosystems, I: 13
Asexual reproduction, II: 141
As low as reasonably achievable (ALARA), II: 186
Assimilation, I: 133
  fraction of, II: 37, 39, 108
  of minerals, I: 14
Atmospheric dispersion, II: 1
Atmospheric fallout transport, I: 114
Atmospheric stability, II: 4
Atmospheric tracers, I: 77
Atom bomb, I: 106
Atomic Energy Commission (AEC), I: 4—6; II: 130
Atomic number, I: 64
Attenuation
  coefficient of, I: 49
  photon, I: 51
  total, I: 48
Attraction, molecular, II: 30
Autecology, I: 19
Autoradiograph, I: 57
Autotrophic, I: 14, 19

B

Ba, I: 151; II: 35
Background radiation, I: 73
  high natural, II: 143, 149
  natural, II: 143, 149, 182
Barium, see Ba
Barn swallows, II: 104
Basal metabolic rate, II: 42
Batemann equations, II: 95
Battelle Pacific Northwest Laboratories, I: 5
Bedrock leaching, I: 81
Beef, I: 173
Behavior, I: 140
  of radionuclides, I: 144
Beliefs about harmfulness of radiation, I: 73
Benthic animals, I: 180
Beta particles, I: 38, 39, 67
  absorption of, I: 47
Big Thompson Canyon flood of 1976, I: 30
Bioaccumulation, I: 175; II: 61
Biochemical lesions, II: 125
Biological effects, I: 64, 129
  of ionizing radiation, II: 120
Biological half-lives, I: 135—136
Biological half-times, II: 57—60
Biological membranes, II: 34
Biological risk, II: 121
Biological surfaces, I: 137
Biological uptake, II: 10
Biology
  cellular, I: 15
  molecular, I: 15
Biomass, I: 26, 183; II: 159
  community, I: 146
Biomes, I: 2, 16, 24
Biosphere, I: 2, 16, 24
  nuclear-powered, II: 182
Biota, I: 184
Biotic communities, I: 19
Birch, I: 174; II: 161
Birds, II: 136, 164
Birth rates, I: 17
Black oak, II: 160
Blood changes, II: 139, 152
Blueberry, I: 171
Blue crab, II: 134
Bluc gills, II: 135
Bobcats, I: 176
Bombs, I: 106
Bone, I: 156
Bone seekers, I: 139
Borel's Theorem, II: 82
Boundaries of ecosystem, I: 24
Breathing rate, II: 45
Breeder reactors, I: 86
Breeding, I: 86
Bremsstrahlung X-rays, I: 42
Brewer-Dobson model, I: 114
Brine shrimp, II: 133, 134
Brookhaven National Laboratory, II: 124, 155, 156, 158, 162, 192
Bryophytes, II: 150
Buggy, I: 119
Buildup curves, II: 102
Burdens, radionuclide, I: 141
Burrowing, I: 141

C

C, I: 29, 147
14-C, I: 29, 76, 148
Ca, I: 136, 151
45-Ca, I: 149
Cabriolet, I: 119
Calcium, see Ca
Caloric value of food, II:43
Cancer, II: 125, 181
Capillary forces, II: 31
Carbon, see C
Carbon cycle, I: 29; II: 182
Caribou, I: 172, 173
Carnivores, I: 14, 26
Carrying capacity, I: 17, 18
Cation exchange capacity, I: 145; II: 33
Cation-fixing capacity, I: 175
Cattails, I: 178
CC, see Coefficient of community
Ce, I: 157
Cells
  differentiation of, II: 122
  division rate of, II: 122
  lesions in, II: 125
  replication in, II: 182
Cellular biology, I: 15
Central nervous system (CNS) damage, II: 125
Centrifugation, I: 91
Centromeres, II: 141
CEQ, see Council on Environmental Quality
Cerium, see Ce
Cesium, see Cs
Chain reactions, I: 69
  spontaneously triggered, I: 79
Characteristic X-rays, I: 42
Charged primary particles, I: 76
Chariot, I: 121
Chelating agents, I: 145, 161
Chelyabinsk, II: 188
Chemical adsorption, I: 130
Chemical analog, I: 150
Chemical conversion of uranium, I: 85
Chemical effluents, I: 90
Chemical energy, I: 13, 14, 25
Chemical factors, I: 142, 144
Chemical sorption, I: 130; II: 31
Chemical toxicity, I: 129
Chemical toxins, II: 125
Chemisorption, see Chemical sorption
Chinook salmon, I: 4
Chipmunck, II: 139
Chromium, see Cr
Chromosomal polymorphism, II: 151

Chromosomes
aberrations of, II: 125, 143
damage to, II: 122
volume of, II: 135
Chronic ingestion, II: 41
Chronic radiation stress and exploitation, II: 148
Chronic radiation studies, II: 142, 165
Cirques, I: 142
Climatate, I: 24
Climate, I: 19, 141
changes in, II: 193
Climax community, I: 22
Cloud depletion, II: 7, 22
242-Cm, I: 159, 160
244-Cm, I: 159
CNS, see Central nervous system
Co, I: 153
60-Co, I: 149
Coal mines, I: 88
Coastal plains, I: 175
Coastal waters, I: 182
Cobalt, see Co
Code of Federal Regulations, II: 185
Coefficient
of attenuation, I: 49
of community (CC), I: 20
of dispersion, II: 4
of distribution, II: 32
of rainout, II: 26
Coefficient matrix, II: 93
Collection of samples for radiation counting, I: 58
Colligative properties, I: 136
Commensal relationship, I: 33
Communities, I: 2, 16
biotic, I: 19
climax, I: 22
coefficient of, I: 20
composition of, I: 19; II: 156, 158
function of, II: 153
investigation of effects on, II: 126
ionizing radiation effects on, II: 119
island, II: 149
plant, II: 153
productivity of, II: 159
stable, I: 22
structure of, II: 153
Community biomass, I: 146
Comparative radiosensitivities, II: 162
Compartmental inventories, I: 133
Compartments, I: 180
accumulation, II: 66
defined, II: 66
in first-order systems, II: 90
in three-compartment systems, II: 90
in two-compartment systems, II: 54, 83
single, II: 70, 74, 78
Compartment systems, kinetics of, II: 1, 64
Competition, I: 33; II: 133, 142
for resources, I: 17
interspecific, II: 123
intraspecific, II: 123
Composition of community, I: 19; II: 156, 158
Compton effect, I: 50
Compton scattering, I: 68
Computers, II: 106
Concentration factor, I: 143, 177; II: 61
Concentration ratios, II: 65, 108
animal/plant, II: 63
empirical, II: 61
plant/soil, II: 62
Concentrations, equilibrium, I: 131
Conceptual models, II: 67
Consumer organisms, I: 14
Consumption, I: 138; II: 124
rate of, II: 42
Containment, I: 98
Contamination
cross-, I: 58
levels of, I: 3
Continental shelves, I: 182
Conversion of uranium ore, I: 90
Convolution integral applied to single compartments, II: 78
Coral reefs, I: 182
Core meltdown, II: 187
Cosmic flux, I: 76
Cosmic rays, I: 1, 84; II: 183
energy from, I: 76
hard component of, I: 73
intensity of, I: 75, 76
interactions with matter and, I: 76
primary, II: 74, 75
secondary, I: 74
Cosmogenic radionuclides, I: 76, 77
Cotton rats, I: 176; II: 131, 137, 164
Coulombic repulsion, I: 47
Council on Environmental Quality (CEQ), I: 11
Counting error, I: 61
Counting statistics, I: 61
Counting yield, I: 59
Count rate, I: 59
Cr, I: 153
51-Cr, II: 146
Crabs, II: 134
Cramer's Rule, II: 92, 93
Crater-forming explosions, I: 109
Cratering, I: 117; II: 127
Creep, II: 27
Critical mass, I: 106
Crop, standing, I: 26
Cross-contamination, I: 58
Cs, I: 137—150; II: 35
137-Cs, II: 109, 152
Curie, I: 46
Curium, see Cm
Curve-fitting, II: 54, 99
Cycles
carbon, I: 29; II: 182
mineral, I: 15, 22, 26, 30, 31
nuclear fuel, I: 84, 85, 101, 121; II: 185
processes of, I: 28
radionuclide, I: 22
water, I: 28

D

Damage
biologic, I: 64
chromosomal, II: 122
recovery from, II: 161
Daphinia pulex, II: 147
Dating, I: 77
Daughter radionuclides, I: 78
DE, see Dose equivalent
Death rates, I: 17
Decay, I: 31, 43, 156
Decomposers, I: 14, 15
Deer, I: 171, 176
lung tissue of, I: 172
Deer mice, II: 131, 137, 148, 151
Deforestation, II: 193
Degradation of photon energy, I: 68
Delayed fallout, I: 112
Department of Energy (DOE), I: 5—7, 96
Depletion of soil, I: 81
Deposition, I: 30, 130
dry, II: 13, 17
internal, I: 80
of fallout, I: 169
processes of, II: 13
rate of, II: 17
velocity of, II: 7, 17, 19, 108
wet, II: 24
Depressions, I: 142
Depth dose, I: 67
Descent times, II: 17
Desorption, II: 51
Destruction of habitat, I: 18—19
Detection
of gamma radiation, I: 52
of X-radiation, I: 52
Detectors
Geiger-Mueller, I: 40, 53, 54
NaI, I: 56
proportional, I: 53
scintillation, I: 55
semiconductor, I: 55
thermoluminescent, I: 57
Detention processes, I: 180
Determinant, II: 92
Deterministic models, II: 68
Detonation, II: 127, 190
Detritus, I: 133
organic, I: 180
Detritus food chain, I: 26
Detritus reservoir, I: 133
Diffusion, I: 179
equation for, II: 2, 9
gaseous, I: 90, 91
molecular, II: 2, 25
theory of, II: 2
turbulent, II: 2
vertical, II: 6
Digital computers, II: 106
Dilution, I: 97, 179; II: 12
Diodes, I: 59
Discharges from reactors, I: 93
Discrimination, I: 133
Dispersion, I: 130, 142, 179
aqueous, II: 9
atmospheric, II: 1
coefficients of, II: 4
hydrospheric, II: 9
Disposal
geologic, I: 101
spent fuel, I: 85
waste, I: 95; II: 127, 150
Distribution
absorbed dose, II: 121
coefficient of, II: 32
Gaussian, II: 2
Poisson, I: 61
spatial, of dose, II: 120
tissue, I: 137
Diversity, II: 149
ecosystem, I: 23, 146
functional, I: 20, 26
index of, II: 152
old-field, II: 158
DNA strand breakage, II: 124, 125
DOE, see Department of Energy
Dominance, II: 133
Dormancy, II: 123
Dose, see also Exposure; Irradiation, I: 3; II: 120, 121
acute, II: 50
calculation of, I: 66
concept of, I: 63
depth, I: 67
estimation of, II: 108, 131
evaluation of, I: 67
external, I: 67
from natural radiation, I: 82
lethal, II: 123, 142
limits on, II: 185, 186
low, II: 125, 142
median lethal, II: 131
spatial distribution of, II: 121
sublethal, II: 142
to the public, I: 101
Dose-dependent syndromes, II: 125
Dose equivalent (DE), I: 64
Dose-response curve, II: 120
Dosimeters
glass, II: 152
implantation of, II: 131
lithium fluoride, I: 65
pocket, I: 52
thermoluminescent, I: 65; II: 130, 149
Dosimetry, I: 63; II: 129
field, I: 68
Double standard for humans vs. other populations, II: 125
Drag force, II: 14
Drosophila, see Fruit flies
Dry deposition, II: 13, 17
Dry fallout, I: 115
Ducks, II: 136

Duhamel's integral, II: 79
Dumping in the ocean, I: 98
Duration of exposure, I: 73
Dust, extraterrestrial, I: 77
Dynamics of populations, I: 16
Dyseutrophic, defined, I: 179

E

Earth's heat content, I: 73
Ecological principles applied to radioecology, I: 11—36
Ecological properties of selected radionuclides, I: 163—166
Ecological systems, radionuclide movement within, I: 2
Ecologists, I: 12
Ecology, I: 11
  basic premise of, I: 12
  definition of, I: 12
  recognition of as a bona fide science, I: 12
Ecosystems, I: 2
  aquatic, I: 24, 142
  artificial, I: 13
  attributes of, I: 23, 141, 145
  biological attributes of, I: 145
  boundaries of, I: 12, 24
  categorized, I: 24
  concept of, I: 12
  diversity of, I: 23, 146
  energy flow in, I: 25
  freshwater, I: 24
  function of, I: 25, 26, 31
  functional relationships within, I: 25
  investigation of effects on, II: 126
  ionizing radiation effects on, II: 119
  key elements of, I: 13
  marine, I: 13, 24
  mineral cycles of, I: 26
  net production of, I: 26
  organization in, I: 15
  protective structure of, I: 20
  radionuclide behavior in, I: 129—195
  sensitivity of to radiation, I: 15
  structure of, I: 15, 20
  terrestrial, I: 13, 24, 168
Ecotones, I: 23
Effective atomic number, I: 64
Effective half-time, I: 134; II: 52
Effective loss rate, II: 52
Effective plume height, II: 6
Efficiency
  of fallout interception, I: 170
  trophic level, I: 26
Effluents, I: 90, 181
Eggs
  fish, II: 134, 143—145
  investigations of developing, II: 143
Electromagnetic radiation, I: 11, 41, 42
Electron
  capture of, I: 39
  free, I: 47
Elevation, I: 75
  and cosmic ray intensity, I: 76
Embryos, II: 145
Emitters, alpha, II: 130
Emory University, II: 128, 149, 152
Empirical concentration ratios, II: 61
Emulsions, I: 56
Energy, I: 1, 15, 37
  absorption of, I: 63
  chemical, I: 13, 14, 25
  cosmic ray, I: 76
  flow of, see Energy flow
  kinetic, I: 47
  loss of, I: 13
  nuclear, II: 181, 185
  of photons, I: 41
  solar, I: 13, 25
  transduction of, I: 23
  transfer of, see Energy transfer
  transformations of, I: 26
Energy flow, I: 15, 31
  in ecosystems, I: 25
Energy transfer, I: 26
  linear, I: 64; II: 120
Enrichment
  isotopic, I: 86, 90
  of fissile 235-U, I: 90
  of U, I: 85, 90
Entombment, I: 101
Environment, I: 1
  abiotic features of, I: 23
  unlimited, I: 17
Environmental compartments, I: 133
Environmental contamination levels, I: 3
Environmental gradients, I: 23
Environmental impact
  statements of, I: 11
  uranium mining, I: 87
Environmentalists, I: 12
Environmental pollution, I: 1
Environmental Protection Agency (EPA), I: 11, 82, 91; II: 185, 186
Environmental radioactivity, I: 1, 73—128
Environmental Sciences Division, I: 6
EPA, see Environmental Protection Agency
Epilimnion, I: 142
Equilibrium, II: 12
  concentrations of, I: 131
Erodibility index, II: 28
Erosion, II: 26
  processes of, I: 30
  wind, I: 82
Error
  counting, I: 61
  evaluation of, II: 110
Eskimos, I: 14
Essential nutrients, I: 14
Estimation of dose, II: 108, 131
Estuaries, I: 182
Euler method, II: 109

Evaluation
of dose, I: 67
of error, II: 110
Evolution, II: 184
biological, II: 183
chemical, II: 183
of life, II: 182
process of, I: 11
Excavation, I: 119
Exchange of cations, I: 145
Excretins, I: 133
Experimental data analysis, II: 98
Experimental radiation sources, II: 127
Exploitation and chronic radiation stress, II: 148
Explosions, I: 103
crater-forming, I: 109
neutrons released in, I: 111
Explosives, peaceful applications of, I: 117
Exponential model, II: 102
Exposure, see also Dose; Irradiation
concept of, I: 63
duration of, I: 73
external, II: 126
fraction ofation, II: 120
intensity of, I: 73
internal, II: 126
maximum allowable, II: 186
rate of, I: 66; II: 120
short-term, II: 157, 161
standards for in humans, II: 125
ultraviolet-B, II: 192
External irradiation, I: 67, 167; II: 126
Extinction, II: 165
of uranium, I: 89
Extraterrestrial dust, I: 77

F

Fabrication of fuel element, I: 91
Fallout
airborne, I: 111
behavioral characteristics of, I: 111
delayed, I: 112
dry, I: 115
formation of, I: 105
interception of, I: 170
local, I: 112, 113
particles of, see Fallout particles
patterns of, II: 190
properties of, I: 109
simulated, II: 152
stratospheric, I: 112, 113
transport of, I: 112, 114
tropospheric, I: 112, 113
wind and, I: 115
worldwide, I: 112, 113
Fallout deposition, I: 169
Fallout particles, I: 68
size of, I: 112
water soluble, I: 112
Fe, I: 153
59-Fe, I: 149
Fear
of nuclear accidents, II: 187
of radiation, I: 11
Feedback, I: 133
Feeding patterns, II: 124
Ferns, I: 176
Fertile, I: 86
Fickian diffusion equation, II: 2
Field dosimetry, I: 68
Fireball, I: 107
First-order kinetic process, II: 69
First-order model, II: 52
First-order systems, II: 90
Fish, see also specific types of fish, I: 179; II: 134
mosquito, II: 131, 151
Fish eggs, II: 134, 143—145
Fissile materials, I: 69
Fission, I: 69, 91, 106
products of, see Fission products
Fission fragments, I: 69
Fission products, I: 69, 91, 93, 97
biologically important, I: 109
Fixing capacities, I: 175
Flares, I: 76
Folvic acids, I: 145
Food
caloric value of, II: 43
selection of, II: 124
Food chain, I: 14
detritus, I: 26
Food consumption, I: 138; II: 124
rate of, II: 42
Food habits, I: 140, 141
Food web, I: 14, 26
Forests
oak-pine, II: 155, 156, 158, 160, 162
tropical rain, II: 158
Fossil fuels, I: 29
Foxes, I: 176
Fractionation, II: 120
Fragmentation, I: 75
Fragments of fission, I: 69
Free electrons, I: 47
Free radicals, II: 125
Frequency, population, I: 23
Freshwater, I: 179, 184
Freshwater ecosystems, I: 24, 177
Friction velocity, II: 19
Frogs, II: 164
Frost-heaving, I: 133
Fruit flies, II: 147, 150, 184
Fuel
fabrication of, I: 85, 91
fossil, I: 29
nuclear, I: 84, 85, 101, 121; II: 185
reprocessing of, I: 95
spent, I: 85, 95
Fuel elements, I: 85
fabrication of, I: 91
Functional diversity, I: 20, 26
Functional relationships within ecosystems, I: 25

Functions
of communities, II: 153
of ecosystems, I: 25, 26, 31
of populations, II: 153
Fusion, I: 69, 107

G

Galactic radiation, see Cosmic rays
Gamma-neutron irradiation, II: 157
Gamma point sources, II: 153
Gamma rays, I: 40, 41, 79
acute, II: 137
detection of, I: 52
external, I: 80
sealed point sources of, II: 143
terrestrial, I: 82
X-rays and, I: 48
Gas-bearing formations, I: 120
Gasbuggy, I: 120
Gaseous diffusion, I: 91
of uranium, I: 90
Gases, washout of, II: 25
Gastrointestinal tract, I: 133; II: 125
Gaussian distribution function, II: 2
Gaussian plume models, II: 2
Geiger-Mueller (G-M) detectors, I: 40, 53, 54
Ge(Li) diode, I: 59
Gene mutations, II: 184
General Electric Co., I: 5
Genetic defects, II: 181
Genetic effects, II: 124, 125, 153
Geographic locations of nuclear testing, I: 105
Geologic disposal, I: 101
Glacial cirques, I: 142
Glass dosimeters, II: 152
Global climatic changes, II: 193
G-M detectors, see Geiger-Mueller detectors
Gradients, I: 23
Gravitational settling, I: 130; II: 13, 14
Green plants, I: 13
Grooming, I: 141
Ground squirrels, II: 139
Ground water, I: 88, 101
Growth
impairment of, II: 142
inhibition of, II: 141
phases of, II: 123
Guppies, II: 145

H

H, see also Tritium, I: 136, 147
Habitat, I: 31, 32
destruction of, I: 18—19
selection of, I: 140
Half-lives, I: 43
biological, I: 135—136
physical, I: 135, 136; II: 207—212
Half-times, II: 51
biological, II: 57—60
effective, I: 134; II: 52
retention, II: 57, 61
Halocline, I: 183
Hanford, I: 96, 161; II: 127
Hard component of cosmic radiation, I: 73
Hazards, I: 1
index of, I: 97
He, I: 152
Health, II: 181
risks to, II: 185, 186
standards for, I: 82
Health physics, I: 122
Heat
of Earth, I: 73
solar, II: 182
Heavy metals, see also specific metals, I: 153
Helium, see He
Hematopoetic syndrome, II: 125
Herbivores, I: 14, 26, 81
radioactive material passed to from plants, I: 132
Heterotrophic, I: 14, 19
Hibernation, II: 123
High-level wastes, I: 99
seabed option for, I: 101
High natural background areas, II: 149
Hiroshima, I: 103; II: 189, 190
Historical perspectives, I: 1, 2, 103
Holistic view, I: 12
Horizontal patterns, I: 22
Hot spots, I: 67, 162, 171; II: 121
House mice, II: 138
Humic acid, I: 145, 179
Hydrogen, see H
Hydrospheric dispersion, II: 9
Hypervolume, I: 32

I

I, I: 136, 147, 148
stable, II: 35
129-I, I: 149
131-I, I: 109, 149
uptake of, II: 35
ICRP, see International Commission on Radiological Protection
ICV, see Interphase chromosome volume
Idaho National Engineering Laboratory, I: 6; II: 151
Impaction, I: 130
Implanting of dosimeters, II: 131
Individual value systems, II: 125
Ingestion, I: 180; II: 36
chronic, II: 41
mineral, I: 14
models of, II: 42
rate of, II: 37
soil, I: 141
Inhalation, I: 31, 141
of aerosol particles, I: 132

of airborne particulates, II: 44
of gases, II: 44
Input rates
constant, II: 70
time-variable, II: 74
Insects, II: 133
In situ leaching, : 87
Insoluble radionuclides, I: 133
Instantaneous source, II: 8
Instruments for measurement of radiation, I: 52
Integrals, II: 78, 79
Intensity
cosmic rays, I: 75, 76
elevation and, I: 76
exposure, I: 73
latitude and, I: 76
Interactions, II: 119, 139, 165
Interception of fallout, I: 170
Internal sources of radiation, I: 67, 82, 84; II: 126
International Atomic Energy Symposium on Radioactive Contamination of the Marine Environment, II: 130
International Commission on Radiological Protection (ICRP), I: 152; II: 46, 185
lung model of, II: 46, 48
Interphase chromosome volume (ICV), II: 135, 141
Interrelationships, I: 15
Interspecific competition, II: 123
Interspecific relationships, I: 19
Intraspecific competition, II: 123
Intraspecific relationships, I: 19
Intrinsic rate of increase, I: 16; II: 147, 149
Inventory
of nuclear wastes in U.S., I: 96
of radionuclides, I: 145
Inverse Laplace transformations, II: 83
Inverse square law, I: 68
Inversion, II: 6
Invertebrates, II: 132
Iodine, see I
Ionization, I: 47
Ionization chambers, I: 52
Ionizing radiation, I: 1, 3
biological effects of, II: 120
effects of, II: 119, 120
from natural sources, I: 73
sources of, II: 126
types of, I: 37
Iron, see Fe
Irradiation, see also Dose; Exposure
acute, II: 119
blood changes following, II: 139
chronic, II: 142, 165
external, I: 167
investigations of, II: 131
metabolism of radionuclides and, II: 139
neutron-gamma, II: 157
Island communities, II: 149
Isolation, I: 98
Isomeric transition, I: 42
Isotopic enrichment, I: 86, 90
of fissile 235-U, I: 90

J

Johns Hopkins University, II: 183

K

K, I: 149
exchangeable, I: 175
natural, II: 182
40-K, I: 79, 149
Kangaroo rats, II: 139
Kasli, II: 188
Kidneys, I: 156
Kinematic viscosity, II: 16
Kinetics
compartment systems, II: 1, 64
energy, I: 47
first-order, II: 69
tracer, II: 68
Kr, I: 152
85-Kr, I: 152
Krypton, see Kr

L

La, I: 157
Laboratory of Radiation Ecology, I: 4
Laboratory studies, II: 143
Lakes, I: 12, 179
Lake watersheds, I: 179
Lamprey, I: 4
Land-fills, I: 101
Lanthanide series, I: 157
Laplace transforms, II: 82
inverse, II: 83
Lapps, I: 174
Lapse rate, II: 4
Large gamma point sources, II: 153
Large point radiation sources, II: 128
Larvae, II: 151
Latitude, I: 76
Lawrence Livermore Laboratory, I: 6
Laws on dose limits, II: 186
LD_{50}, see Lethal dose
Leaching, I: 170; II: 50
bedrock, I: 81
in situ, I: 87
soil, I: 81
Leaching pond, II: 104
Leaf-fall, I: 170; II: 50
Lesions, II: 125
pathological, II: 152
LET, see Linear energy transfer
Lethal dose, II: 142
median, II: 131
to 5041 of individuals, II: 123, 131
Lethality, II: 125
Lichens, I: 137, 170, 173; II: 141, 155

Life spans, II: 149
Lightning, II: 183
Limitations, I: 23
Limited Test Ban Treaty, I: 104
Limits
dose, II: 185, 186
release, I: 93
Linear energy transfer (LET), I: 64; II: 120, 121, 125
Linearization, II: 100
Lions, I: 172
Lithium fluoride dosimeters, I: 65
Lithosphere, I: 79
Liver, I: 155
Liverwort, II: 150
Lizards, II: 131, 135, 164
Local fallout, I: 112, 113
Logistic equation, I: 18
Long component of retention curve, II: 101
Longevity, I: 140; I: 138
Los Alamos Scientific Laboratory, I: 6
Loss, II: 49
Loss function, II: 50
Loss rate, I: 134
effective, II: 52
Low-level radiation effects, II: 119
Low-level wastes, I: 98, 101
Lucky Dragon incident, I: 104
Lung model of ICRP, II: 46, 48
Lung tissue of deer, I: 172

M

Ma, I: 153
Magnesium, see Mn
Magnetic field, I: 76
Mammals, II: 136
Management of nuclear waste, I: 101
Manganese, see Ma
Man-generated radioactivity, I: 84
Manhattan Engineer District, I: 4
Marine biota, I: 184
Marine ecosystems, I: 13, 24, 181
Mass, I: 106
Mass action, I: 136
Mass-loading, II: 27
Mathematical models, II: 68
Matrix operations, II: 91
Maximum allowable exposures, II: 186
Mean residence time, II: 49
Mean Value Theorem, II: 103
Measurement of radiation, I: 52
Median lethal dose, II: 131
Meltdown of reactor core, II: 187
Membranes, passage of radionuclides across, II: 34
Metabolic activity, I: 139
Metabolically active elements, I: 138
Metabolic rate, I: 138
basal, II: 42
Metabolism, I: 137
irradiation effects on, II: 139
Metals, see also specific metals
alkali, I: 149
alkaline earth, I: 151
heavy, I: 153
vs. nonmetals, I: 147
Meteoritic dust, I: 77
Mice, II: 136, 149
deer, II: 131, 137, 148, 151
house, II: 138
old-field, II: 138
pocket, II: 138, 164
Microclimate, I: 115, 170
Microflora, II: 155
Microorganisms, I: 15, 177
Microterrain, I: 170
Midge larvae, II: 151
Migration of radionuclides in geological strata, I: 101
Milk, I: 172
Milling of U, I: 85, 88; II: 127
Mill tailings
uranium, II: 188
wind erosion of, I: 82
Mineral baths, I: 82
Mineral cycles, I: 22, 30, 31
of ecosystems, I: 26
Mineral nutrient cycles, I: 15
Minerals, see also specific minerals, I: 30
assimilation of, I: 14
flows of, I: 23
ingestion of, I: 14
Mine water, I: 82
Mining of U, I: 85—87
Minnows, II: 145
Mixing
rate of, I: 183
volume of, II: 11
Mn, I: 153
554-Mn, I: 149
Models
Anspaugh's, II: 29
Brewer-Dobson, I: 114
conceptual, II: 67
deterministic, II: 68
first-order, II: 52
Gaussian plume, II: 2
ICRP lung, II: 46, 48
ingestion, II: 42
mathematical, II: 68
radionuclide transport, II: 106
stochastic, II: 68
transport, II: 1, 107, 108
two-component exponential, II: 102
Molecular attraction, II: 30
Molecular biology, I: 15
Molecular diffusion, I: 179; II: 2, 25
Molecular repair, II: 182
Mollusks, I: 184
Monazite, I: 160
Monazite sand, II: 150
Moose, I: 174
Morphology, I: 137, 171

in *Penstemon*, II: 150
Mortality, I: 17; II: 50
Mosquitoes, II: 164
Mosquito fish, II: 131, 151
Mosses, I: 170; II: 141
Mother substance, II: 66
Mound Laboratory, I: 161
Mountain lion, I: 172
Mountains, I: 168
Mouse, see Mice
Mt. McKinley, I: 76
Mule deer, I: 149
Muons, I: 75
Mushroom cloud, I: 113
Mushrooms, I: 175
Muskrats, I: 156
Mutations, II: 125, 184
Mutualism, I: 34

N

Nagasaki, I: 103; II: 189, 190
NaI detectors, I: 56
NAS, see National Academy of Sciences
Nasopharynx region, II: 45
Natality, I: 17
National Academy of Sciences (NAS), II: 192
National Council on Radiation Protection and Measurements (NCRP), II: 185
National Environmental Policy Act (NEPA), I: 11
National Environmental Protection Act, I: 6
Natural background radiation, II: 143, 182
Natural decay series, I: 156
Natural potassium, II: 182
Natural radiation, I: 73
  dose rates from, I: 82
  geographic variations in, I: 79
Natural radionuclides, I: 79
Natural reactors, I: 79
Natural sources of radiation, I: 73
Natural succession, I: 22
Natural thorium, I: 86; II: 182
Natural uranium, I: 85; II: 182
NCRP, see National Council on Radiation Protection and Measurements
N-dimensional hypervolume, I: 32
Needle ice, I: 133
Negatron, see Beta particles
Neon, I: 152
NEPA, see National Environmental Policy Act
Neptunium, see Np
Net ecosystem production, I: 26
Net primary production, I: 26
Neutralism, I: 33
Neutrinos, I: 39
Neutron flux, I: 93
Neutron-gamma irradiation, II: 157
Neutrons
  activation of, I: 40, 93, 135
  moderation of, I: 40
  proton ratio, I: 39
  release of in nuclear explosions, I: 111
  thermal, I: 69
Nevada Test Site, I: 6, 161; II: 128, 131
Niche, I: 31, 32
Noble gases, I: 152
Nonmetal radionuclide behavior, I: 147
Nonmetals vs. metals, I: 147
Normal death rate, I: 17
Np, I: 158
239-Np, I: 160
NRC, see Nuclear Regulatory Commission
Nuclear accidents, II: 142
  public fear of, II: 187
  Urals, II: 188
Nuclear bombs, I: 106
Nuclear cratering experiments, II: 127
Nuclear detonation, II: 190
  sites of, II: 127
Nuclear energy, II: 181, 185
  history of, I: 4
  peaceful use of, I: 1
Nuclear excavation projects, I: 119
Nuclear explosions, I: 103
  neutrons released in, I: 111
Nuclear explosives, peaceful application of, I: 117
Nuclear fuel cycle, I: 84, 85, 101, 121; II: 185
Nuclear power, I: 104
Nuclear reactions, I: 11
Nuclear reactors, I: 69, 84, 85
Nuclear Regulatory Commission (NRC), I: 11, 82; II: 9, 57, 62, 64, 186
Nuclear stimulation of gas-bearing formations, I: 120
Nuclear testing, II: 127, 128
  amount of, I: 105
  geographic locations of, I: 105
Nuclear transformations, I: 41
Nuclear warfare, I: 1, 105; II: 142, 189
Nuclear wastes
  entombment of in deep geological strata, I: 99
  injection of into outer space, I: 99
  inventory of in U.S., I: 96
  management of, I: 101
  placement of in thick ice sheets, I: 99
Number of samples for radiation counting, I: 58
Numerical approximation, II: 109
Nutrient analogs, I: 109, 136; II: 38
Nutrient cycles, I: 15
Nutrients
  availability of, I: 143
  cycle of, I: 13
  essential, I: 14
  flows of, I: 15
  pool of, I: 14

O

Oak-pine forest, II: 155, 156, 158, 160, 162
Oak Ridge National Laboratory (ORNL), I: 5,6, 121, 161; II: 109, 151

Oaks, II: 160
Obligate symbiosis, I: 34
Observed ratio, defined, II: 35
Ocean, I: 13
depths of, I: 182
dumping in, I: 98
sediments in, I: 77
Oklo quarry, I: 79
Old-fields
diversity in, II: 158
mice in, II: 138
Oligotrophic, I: 179
Omnivores, I: 14
Open pit mines, I: 87, 88
Oppossums, I: 176
Ores
uranium, I: 90
wind erosion of, I: 82
Organic detritus, I: 180
reservoir of, I: 133
ORNL, see Oak Ridge National Laboratory
Orographic barriers, I: 169
Orographic uplifting, I: 169
Osmoregulation, I: 136
Oysters, II: 146
Ozone, I: 169

P

P, I: 147
Pacific Proving Ground, II: 129
Pair production, I: 51
Palanquin, I: 119
Parameters of population, I: 16
Parasitism, I: 31, 33; II: 124
Pathologic lesions, II: 152
Pb, I: 153
Peaceful applications of nuclear explosives, I: 117
Penstemon, II: 150
Percentage similarity (PS), I: 20
Perch, I: 178
Periphyton, I: 180
Permissible radiation dose limits, II: 185
Persistence of radionuclides in ecological systems, I: 135
Pheasants, II: 136
Phosphorus, I: 148
Photoelectric absorption, I: 50
Photoelectrons, I: 51
Photographic emulsions, I: 56
Photon attenuation processes, I: 51
Photon energy degrading process, I: 68
Photons, energy of, I: 41
Photosynthesis, I: 13, 25, 76; II: 141, 182
Phylogeny, II: 133
Physical absorption, II: 30
Physical form of radionuclides, I: 137
Physical half-lives, I: 135, 136; II: 207—212
Physicochemical form, I: 182
Physics, I: 122
Physiognomy, I: 19, 24; II: 153
Physiography, I: 170
Physiological state, II: 123
Physiology, I: 137
Phytoplankton, I: 177
Pikas, II: 139
Pike, I: 178
Pine, II: 161
short-leaf, II: 160
white, II: 152
Pions, I: 75
Plaice, II: 130
Plankton, I: 184
Plant communities, II: 153
Plants, I: 13; II: 140
for fuel fabrication, I: 91
radioactive material passed to herbivores through, I: 132
respiration of, I: 29
Plant/soil concentration ratios, II: 62
Ploidy, II: 141
Plowshare, I: 117
Plume height, II: 6
Plume rise, II: 6
Plutonium, see Pt
Pm, I: 157
Po, I: 153
Pocket dosimeters, I: 52
Pocket mice, II: 138, 164
Poikilothermic animals, II: 44
Poikilotherms, II: 44
Point sources, I: 66
large gamma, II: 153
Poisson distribution, I: 61
Polarity, I: 76
Pollutants, defined, II: 66
Pollution, I: 1
Polonium, see Pu
Polymorphism, II: 151
Polyploids, II: 122, 141
Ponds, I: 12
leaching, II: 104
Population dynamics, I: 16
parameters influencing, II: 147
Populations, I: 2
animal, II: 162
attributes of, I: 16
dynamics of, see Population dynamics
function of, II: 153
investigation of effects on, II: 126
ionizing radiation effects on, II: 119
parameters of, I: 16
patterns of frequency in, I: 23
rate of change of, I: 16
size of, I: 16
stability of, I: 17
structure of, II: 153
Positrons, I: 39, 47
Potassium, see K
Potential birth rate, I: 17
Power function, II: 54, 104
for retention, II: 55
Pr, I: 157

Precipitation, I: 24
Precipitation scavenging, I: 130, 132; II: 24
Predation, I: 31, 33, 133; II: 124, 142
Prediction, I: 129; II: 107
Premature aging, II: 125
Preparation of samples for radiation counting, I: 58
Primary cosmic rays, I: 74, 75
Primary particles, I: 76
Primary production, net, I: 26
Primary succession, I: 22
Primitive proteins, II: 183
Primordial radioactivity, II: 183
Primordial radionuclides, I: 77; II: 182
  distribution of, I: 80
Producers, I: 13, 14
Production, I: 26
Productivity of community, II: 159
Products
  activation, I: 69, 93
  fission, I: 69, 91, 93, 97
  neutron activation, I: 93
Proportional detectors, I: 53
Prostate, I: 155
Protective structure of ecosystem, I: 20
Proteinoids, II: 183
Proteins, II: 183
Protocooperation, I: 34
Protons, I: 40
  recoil, I: 41
PS, see Percentage similarity
239-Pt, I: 111
Pu, I: 158
238-Pu, I: 161
241-Pu, I: 160, 161
Public
  beliefs of about harmfulness of radiation, I: 73
  fear of nuclear accidents, II: 187
  radiation doses to, I: 101
Pulmonary region, II: 45
Purification of U, I: 85, 90
Pycnocline, I: 182

Q

QF, see Quality factor
Quality factor (QF), I: 64
Quantitative aspects of radionuclide transport, II: 1

R

Ra, I: 15, 81, 88, 151; II: 35
226-Ra, recycling of by biota, I: 81
Raccoons, I: 176
Rad, defined, I: 63
Radiation, defined, I: 37
Radiation counting, I: 57, 58
Radiation dose, see Dose
Radiation exposure, see Exposure
Radiation-induced genetic effects, II: 153
Radiation sources, I: 121; II: 127
Radiation stress, I: 22
Radicals, free, II: 125
Radioactive decay, I: 43
Radioactive fallout, see Fallout
Radioactive wastes, see Wastes
Radiobiological effects, II: 124
Radioecologists, I: 3
Radiographs, I: 57
Radiological principles, I: 37—72
Radionuclide burdens, I: 141
Radionuclides, I: 1
  accumulation of in ecological systems, I: 3
  applications of, I: 121, 122
  as model pollutants, I: 2
  behavior of, I: 3, 129—195
  cosmogenic, I: 76, 77
  cycling of, I: 22
  daughter, I: 78
  distribution of, I: 80
  ecological properties of, I: 163—166
  ecosystems and, I: 129—195
  effects of in ecological systems, I: 3
  internal, I: 84
  inventory of, I: 145
  metabolism of, II: 139
  migration of in geological strata, I: 101
  movement of in ecological systems, I: 3
  movement of within ecological systems, I: 2
  primordial, I: 77, 80; II: 182
  reactors and, I: 91
  release limits for, I: 93
  research, I: 122
  solid materials in, I: 131
  solubility of, I: 112
  standards for, I: 61
  uses of, I: 3, 122
  water flow as agent of, I: 11
Radioresistance, II: 136
Radiosensitivity, II: 122, 123, 133, 135
  comparative, II: 162
  of seeds, II: 141
Radium, see Ra
Radon, see Rn
Rainbow trout, I: 4; II: 134
Rain forests, II: 158
Rainout, I: 29, 132; II: 7, 24
  coefficient of, II: 26
Rare earth elements, I: 157—158
Rate constants, I: 134; II: 54, 69
Rate of exposure, I: 66
Rats, I: 176; II: 149
  cotton, II: 131, 137, 164
  kangaroo, II: 139
Rayleigh scattering, I: 49
Rb, I: 149, 150
RBE, see Relative biological effectiveness
Reactions, I: 69
  fusion, I: 107
  nuclear, I: 11

spontaneously triggered fission chain reactions, I: 79
thermonuclear, I: 70
Reactors, I: 69, 84, 85
breeder, I: 86
core meltdown in, II: 187
discharges in, I: 93
effluents of, I: 181
natural, I: 79
radionuclides in, I: 91
Realized niche, I: 32
Reclamation of unused 235-U, I: 86
Recoil protons, I: 41
Recovery, II: 121, 161, 162
Recycling, I: 180
of 226-Ra by biota, I: 81
Redistribution, I: 169
Reindeer, I: 173
Relationships
commensal, I: 33
functional, I: 25
interspecific, I: 19
intraspecific, I: 19
Relative biological effectiveness (RBE), I: 64
Release limits of radionuclides, I: 93
Repair, II: 121
capacity for, II: 123
molecular, II: 182
processes of, II: 120
Replication of cells, II: 182
Reprocessing of fuel, I: 95
Reproduction, II: 136—138
asexual, II: 141
impairment of, II: 142
Reptiles, II: 135
Rescue, II: 152
Residence time, II: 11
Residues of U extraction process, I: 89
Resistance, II: 136
Resources, I: 23
allocation of, I: 21
competition for, I: 17
Respiration, I: 14
of plants, I: 29
Respiratory requirement, I: 26
Respiratory tract, II: 45, 48
Response, II: 121, 123
Resuspension, I: 132, 169; II: 26
Resuspension factors, II: 28, 108
Resuspension flux, II: 28
Retention, I: 137; II: 18, 49, 53
power function for, II: 55
Retention curves, II: 99
long component of, II: 101
short component of, II: 101
Retention half-times, II: 57, 61
Retention ponds, I: 101
Retention rate constants, II: 108
Retrievable surface storage, I: 99
Reynolds number, II: 16
Rio Blanco, I: 121
Risk, II: 181, 187
acceptable, II: 186
biological, II: 121
health, II: 185, 186
Rn, I: 81, 82, 135, 152
Roaches, I: 178
Rock Valley facility, II: 163
Rocky Flats, I: 161
Roentgen, I: 63
Root uptake, I: 155
Roughness height, II: 19
Ru, I: 153, 155
Rubidium, see Rb
Rulison, I: 121
Runge-Kutta algorithms, II: 109
Runoff, I: 170
Ruthenium, see Ru
Rutherford scattering, I: 47

S

Salmon, I: 4, 184; II: 147
Saltation, II: 27
Samples
for radiation counting, I: 58
self-absorption of, I: 59
Savannah River Ecology Laboratory, I: 6, 96
Scattering
Compton, I: 68
of radiation, I: 68
Rayleigh, I: 49
Rutherford, I: 47
Schooner, I: 119
Scintillation detectors, I: 55
Seabed option for high-level wastes, I: 101
Sealed point sources of gamma radiation, II: 143
Seasons, II: 108, 123
Seawater, I: 182, 183
Secondary cosmic rays, I: 74
Secondary effects, II: 119, 165
Secondary succession, I: 22
Secretions, I: 133; II: 50
Sedan, I: 117
Sedges, I: 174
Sedimentation, II: 10
Sediments, I: 180
ocean, I: 77
Seeds, II: 141
Seepage ditches, I: 101
Selection, II: 125, 142
Self-absorption of sample, I: 59
Semiconductor detectors, I: 55
Sensitivity, II: 133, 135, 141
analysis of, II: 107
of ecosystems to radiation, I: 15
Seral stage, I: 22
Sessile algae, I: 178
Seston, I: 177; II: 62
Sex ratios, II: 147
Shielding effects, I: 68
Short component of rententon curve, II: 101

Short-leaf pine, II: 160
Short-term exposures, II: 157, 161
Shrimp, II: 133, 134
Similarity
  index of, II: 152
  percentage, I: 20
Simulated fallout studies, II: 152
Sine function, II: 78
Single compartments, II: 70, 74, 78
Sinks, I: 133, 180; II: 66
Snails, II: 164
Snakes, II: 135
SNAP 9-A, I: 123
Snowfall, I: 169
Snowfields, I: 171
Snowmelt, I: 170
Snowout, II: 24
Social interaction of pikas, II: 139
Soil, I: 19
  characteristics of, I: 175
  depletion of, I: 81
  erodibility index for, II: 28
  ingestion of, I: 141
  leaching of, I: 81
Soil particle classification, II: 33
Soil properties, I: 175
Solar energy, I: 13, 25; II: 182
Solar flares, I: 76
Solar heat, II: 182
Solid materials in radionuclides, I: 131
Solubility of radionuclides, I: 112
Somatic effects, II: 124, 153
Sorption, II: 30
  chemical, I: 130; II: 31
Sources
  experimental radiation, II: 127
  gamma radiation, II: 143
  ionizing radiation, II: 126
  large gamma point, II: 153
  large point radiation, II: 128
  point, I: 66; II: 128, 143, 153
  sealed point, II: 143
Source term, I: 130
Spallation, I: 75
Spatial distribution of dose, II: 120, 121
Spatial patterns, I: 21
Specific activity, I: 44; II: 68
Spent fuel
  disposal of, I: 85
  elements of, I: 95
Sphagnum, I: 170
Spoils, I: 89
Sponge, II: 148
Spontaneously triggered fission chain reactions, I: 79
Springs, I: 82
Spruce, I: 171
Squirrels, I: 176
  ground, II: 139
Sr, I: 151; II: 35
90-Sr, I: 15, 109, 136; II: 146
Sr/Ca ratio, I: 151
Stability, I: 20, 26
  atmospheric, II: 4
  class of, II: 7
  population, I: 17
Stable community, I: 22
Standards
  double, II: 125
  for radiation exposure of humans, II: 125
  for radiation protection for nonhuman species, II: 186
  of health, I: 82
  radionuclide, I: 61
Standing crop, I: 26
Stature, II: 122
Steady-state, I: 134, 135
  defined, II: 66
Sterile-male technique, II: 132
Stewart-Hamilton equation, II: 105
Stickleback, II: 143
Stimulatory effects, II: 158
Stochastic models, II: 68
Strand breakage of DNA, II: 124, 125
Stratospheric fallout, I: 112, 113
Stress, I: 22
  adaption to, II: 123
  agents of, I: 18
  chronic radiation, II: 148
Strontium, see Sr
Structure of populations and communities, II: 153
Subcellular lesion, II: 125
Subcompartments, II: 54
Sublethal doses, II: 142
Succession, II: 162
  natural, I: 22
  patterns of, II: 161
  primary, I: 22
  processes of, I: 22
  secondary, I: 22
Supercritical mass, I: 106
Superposition integral, II: 79
Surface adsorption, I: 180
Surface area, I: 137
Surface creep, II: 27
Surface runoff, I: 170
Surface storage, I: 99
Surface weathering, II: 50
Survival, II: 138
Sverdlovsk, II: 188
Swallows, II: 104
Symbiosis, I: 34
Syndromes, II: 124
  dose-dependent, II: 125
  gastrointestinal, II: 125
  hematopoetic, II: 125
Synecology, I: 19
System
  defined, II: 65
  spatial boundaries of, I: 12
Systems analysis, I: 180

T

Tailings, I: 89
uranium mill, II: 188
Task Group on Lung Dynamics, II: 46
Tc, I: 153, 155
Technicium, see Tc
Temperate mountains, I: 168
Temperature, I: 24, 142
Tennessee Valley Authority (TVA), I: 5
Terminal velocity, II: 14
Termites, II: 164
Terrestrial ecosystems, I: 13, 24, 168
Terrestrial gamma rays, I: 82
Testing, II: 126—128
underground, I: 105
Th, I: 77, 79, 82, 88, 158, 160
natural, I: 86; II: 182
232-Th, I: 152
Thermal neutrons, I: 69
Thermocline, I: 182
Thermoluminescent detectors, I: 57
Thermoluminescent dosimeters (TLD), I: 65; II: 130, 149
Thermonuclear bomb, I: 106
Thermonuclear capability, I: 104
Thermonuclear reaction, I: 70
Thorium, see Th
Three-compartment systems, II: 90
Three Mile Island, II: 187
Thresholds, II: 120
Thyroid gland, I: 149
Tidal pools, I: 182
Tidal volume, II: 45
Time, I: 22
Tissue
distribution of, I: 137
lung, I: 172
TLD, see Thermoluminescent dosimeters
Toads, II: 137
Topography, I: 140, 142
Total attenuation, I: 48
Toxicity, I: 129
Toxins, II: 125
Tracers, I: 121
applications of, I: 122
atmospheric, I: 77
defined, II: 66
kinetics of, II: 68
Tracheobronchial region, II: 45
Tradescantia, II: 150
Training of radioecologists, I: 3
Transduction of energy, I: 23
Transfers of energy, I: 26; II: 120
Transforms, I: 26, 41
Laplace, II: 82, 83
Transition, isomeric, I: 42
Transport
airborne, II: 27
atmospheric fallout, I: 114
factors affecting, I: 133
fallout, I: 112
process of, I: 130; II: 1, 50
quantative aspects of, II: 1
rates of, I: 133
Transport models, II: 1, 10
limitations of, II: 108
utility of, II: 107
Transport pathway, II: 67
Trinity Site, I: 4, 103
Tritium, see also H (hydrogen), I: 15, 28, 76, 107, 136, 147
Trophic levels, I: 172
efficiency of, I: 26
Tropical rain forests, II: 158
Tropopause, I: 114
Troposphere
fallout in, I: 112, 113
Trout, I: 172, 179
rainbow, I: 4; II: 134
Turbulence, II: 2
air, I: 169
Turbulent diffusion theories, II: 2
Turnover time, II: 69
Turtles, II: 135
TVA, see Tennessee Valley Authority
Two-compartment systems, II: 54, 83
Two-component exponential model, II: 102

U

U, I: 77, 79, 82, 160
chemical conversion of, I: 85
enrichment of, I: 85
gaseous diffusion of, I: 90
milling of, I: 85, 88; II: 127
milling wastes and, II: 127
mill tailings dam failure and, II: 188
mining of, I: 85—88
natural, I: 85, 158; II: 182
ores of, I: 90
purification of, I: 85
residues of extraction process and, I: 89
235-U, I: 152
isotopic enrichment of, I: 90
reclamation of unused, I: 86
Ultraviolet-B exposures, II: 192
Underground testing, I: 105
Union Carbide Corp., I: 5
University of North Carolina, II: 128
Unlimited environment, I: 17
Uptake, II: 10
of 131-I, II: 35
root, I: 155
Urals disaster, II: 188
Uranium, see U
Uranium hexafluoride, I: 91
Uranyl fluoride, I: 91
USAEC, see U.S. Atomic Energy Commission
U.S. Atomic Energy Commission (USAEC), I: 4—6; II: 130

V

Validation, II: 110
Value systems of individuals, II: 125
Van der Waals forces, II: 30
Vertebrates, II: 134
Vertical diffusion, II: 6
Vertical mixing rate, I: 183
Vertical zonation, I: 22
Viscosity, II: 16
Voles, II: 139, 150
Von Karman's constant, II: 19

W

Wagon Wheel, I: 121
Warfare, I: 1, 105; II: 142, 189
Washout, I: 29, 132; II: 7, 24, 25
Wastes, I: 96
  disposal of, I: 95; II: 127, 150
  high-level, I: 99
  low-level, I: 98, 101
  seabed option for, I: 101
  uranium milling, II: 127
Water, I: 28
  as agent of radionuclide concentration and redistribution, I: 115
  coastal, I: 182
  ground, I: 88, 101
  mine, I: 82
  ocean, I: 182
Water cycle, I: 28
Waterfowl, II: 136
Watersheds, I: 179
Water soluble fallout particles, I: 112
Weapons testing, II: 126—128
Weathering, I: 170; II: 50
Wet deposition, II: 24
White oak, II: 160
White pine, II: 152
Whitetail deer, I: 176
Willows, I: 174
Wind, I: 141
  and fallout redistribution, I: 115
Wind erosion of ore and tailings, I: 82
Wind rose, II: 7
Windscale Works, II: 127, 130
Wolves, I: 172, 174
Worldwide fallout, I: 112, 113

X

Xe, I: 152
133-Xe, I: 152
Xenon, see Xe
X-rays, I: 40, 41; II: 127
  Bremsstrahlung, I: 42
  characteristic, I: 42
  detection of, I: 52
  gamma rays and, I: 48

Y

Y, I: 157
90-Y, I: 151, 157; II: 146
91-Y, I: 157
Yellowcake, I: 89, 90
Yttrium, see Y

Z

Zinc, see Zn
Zirconium, see Zr
Zn, I: 153
65-Zn, I: 149; II: 146
Zonation, I: 22
Zooplankton, I: 184
Zr, I: 153, 155
95-Zr, I: 171